From the Frontline to the Boardroom, Praise for *Benchmarking for Best Practices*

"Great ideas are everywhere. Understanding best practices can be a cornerstone of how every company competes. If you want to supercharge performance improvement in your organization, read this book! It will show you how."

Sam Foti, President
The Mutual Life Insurance Company of New York

"As a senior Baldrige examiner and a champion for TQM at Marriott, I've seen the power of benchmarking to drive performance improvements. This book is an essential primer on how to use best practice strategies for your organization's success. Keep it close at hand and use it!"

Greg Behm, Vice President of TQM
Marriott Hotels & Resorts

"Choose the masters as your mentors. That's the essence of best practice benchmarking. Bogan and English approach this important business subject with extraordinary grace, insight, and intelligence."

J. Kent Murray, President
Chevron Research & Technology Company

"At Federal Express, the intellectual capital of all our people is our most valued resource. That's why fast learning is so important to Federal Express and other organizations that must manage rapid change. Studying and learning from best practices are of critical importance for every manager who wishes to lead his or her culture to the winner's circle."

Michael E. Reed
Managing Director of Operations Audit & Quality
Federal Express
1990 Malcolm Baldrige National Quality Award Winner

"*Benchmarking for Best Practices* masters an important management subject that should be on the short list of every executive who seeks to lead his organization toward world class performance excellence."

Eric Kennedy
Benchmarking Program Manager
IBM

"Here's how to create an All-Star organization: read *Benchmarking for Best Practices*. This cutting-edge book is indispensable reading for every manager who wants to take his or her team to the championships."

Truman L. Koehler
Chief Executive Officer
Master Builders Inc.

"In a world where increasingly everybody has to compete with everybody else, achieving globally the highest standards in all key company activities is a must. Bogan and English provide clear directions, inspiring stories, and an actionable strategy for seeking out and creatively adapting the best of the best in your company."

Daniel Wagnière, Chief Operating Officer
Sandoz Ltd., Basle, Switzerland

"Learning from benchmarking has been phenomenal at Ameritech. Managers understand how unwise it is to perennially 'reinvent the wheel' when instead we can leverage from others in generating new breakthrough ideas for radical process redesign. In this book, Bogan and English have described the essence of how this convincing approach to innovation pays off. It is recommended reading for all managers."

Orval L. Brown
Process Architecture/Benchmarking Manager
Ameritech

"This book will benefit any organization intent on breaking cycles of mediocre performance whether it is by process reengineering or continuous improvement. Better than any other book I've seen on the subject, useful examples, great stories, and best practice experiences are provided so we gain a much better understanding of how to leverage benchmarking to create learning breakthroughs that pave the road to market leadership."

Dan Hickey, Vice President For Reengineering
Aetna Life and Casualty

"This book is a grand slam for any business reader: important subjects, ideas, and examples fielded from a hall of fame of outstanding companies. It is a must read if your organization cares about change and excellence."

John M. Lavine, Professor and Director
Northwestern University's Newspaper Management Center

"Bogan and English have produced an executive handbook for managing change and continuous improvement. This book provides a handsome return on investment for any manager who wants to accelerate performance improvement in his company."

John Robbins, CEO
American Residential Mortgage Corporation

"Bogan and English provide a blueprint for how to create a we-can-learn-from-anyone culture. Read *Benchmarking for Best Practices* before your competitors do."

Mike Mulligan, Vice President—Sales
SUPERVALU Inc.

"Benchmarking has become one of the key managerial tools for total quality management. The benchmarking process often provides the necessary insights for the radical breakthroughs in performance organizations are so desperately seeking. This book provides a new benchmark for all those interested in benchmarking. It supplies a wealth of examples and specific suggestions for how to select benchmarking targets, how to organize for rapid learning, and most importantly, how to implement these findings."

A. Blanton Godfrey, Chief Executive Officer
Juran Institute, Inc.

"For decades fear of even the possibility of anti-trust prosecution has tended to keep American companies from embracing joint discussions, idea-sharing, and even on-site visits—all foundation activities of benchmarking. Times are changing, and many managers understand the potential benefits of adapting best practices and are gaining approvals from their corporate lawyers to pursue it. Bogan and English's book will enable them with skills, aids, techniques, and directions to harvest its potential."

Arnold J. Lieberman, Ph.D.
Senior Business Development Officer & Vice President
Chase Manhattan Bank

"Like most really useful ideas, benchmarking—studying the best practices of other organizations and adopting them to drive improvement in your own—is not a new concept. However, that doesn't mean that benchmarking is easy to do well. This book is a practical guide to the disciplines any manager needs to follow to speed up how quickly his or her organization learns, changes, and innovates."

Will Miller, Chairman
Irwin Financial Corporation

"To meet the challenges of tomorrow, BC TELECOM knows that it must compete with the best-in-class. Benchmarking pushes thinking beyond traditional solutions and has been essential to our success. *Benchmarking for Best Practices* clearly explains the fundamental elements of benchmarking and will help the reader understand how to use this powerful tool to drive continuous improvement."

Brian A. Canfield
Chairman & Chief Executive Officer
British Columbia Telecommunications Company

"It is rare to discover a book that is both inspirational in how it addresses an interesting business issue, and at the same time is written in clearly understood language and is fun and easily read. This book delivers on all counts. Get it!"

Allen Paison, Chief Executive Officer
Walker: Customer Satisfaction Measurements, L.P.

"Call it wisdom or call it common sense: When you study excellent companies, you come up with excellent ideas for your own organization. This book transforms the common sense of benchmarking into a fine art that's fun to read and filled with practical wisdom."

Gary Mize, Benchmarking Coordinator
EXXON Company, USA

Benchmarking for Best Practices

Winning Through Innovative Adaptation

Christopher E. Bogan

Michael J. English

McGraw-Hill, Inc.

New York   San Francisco   Washington, D.C.   Auckland   Bogotá
Caracas   Lisbon   London   Madrid   Mexico City   Milan
Montreal   New Delhi   San Juan   Singapore
Sydney   Tokyo   Toronto

Library of Congress Cataloging-in-Publication Data

Bogan, Christopher E.
    Benchmarking for best practices : winning through innovative adaptation / Christopher E. Bogan, Michael J. English.
      p.   cm.
    Includes bibliographical references and index.
    ISBN 0-07-006375-3
    1. Benchmarking (Management) 2. Organizational effectiveness.
  I. English, Michael J. II. Title.
  HD62.15.B64   1994
  658.5'62—dc20

94-18076
CIP

2 3 4 5 6 7 8 9 0 DOH DOH 9 0 9 8 7 6 5 4

ISBN 0-07-006375-3

The sponsoring editor for this book was James Bessent, the editing supervisor was Fred Dahl, and the production supervisor was Donald F. Schmidt. It was set in Palatino by Inkwell Publishing Services.

Printed and bound by R. R. Donnelley & Sons Company.

 This book is printed on recycled, acid-free paper containing a minimum of 50 percent recycled de-inked fiber.

Contents

4 The Secrets of Successful Benchmarking

Acknowledgments

Writing a book is much like childbirth—except the labor lasts longer and your extended family contributes much more directly in the delivery. With this in mind, we want to extend heartfelt thanks and praise to all those people who contributed in so many ways to this book.

First and foremost, our thanks should be heaped upon our families, who supported and tolerated long work days, countless stolen evenings, and many emaciated weekends. Thanks then to our spouses, Mary Jo Barnett and Paula English, for their patience, understanding, support, good cheer under sometimes trying circumstances and inspiration. Our children—Evan and Will Bogan, and Wess, Kevin, and Philip Green, and Justin English—all deserve gold medals for being supportive for so many months when you were receiving less of our time and attention than you deserved.

As the comedian once quipped, we owe special thanks to our parents, for without them this book would have been impossible. Thanks to Dorothy Salmon and Stanley Bogan, who have always supported your son's pursuits as a writer and entrepreneur, and thanks to Josephine Vassallo, Steve English, and sister Rose Lotz for being there and believing in your son and brother.

Many friends, colleagues, clients, and "extended" family members shared generously from their experiences, work, and ideas. Leslie Bivens provided valuable editorial assistance, intellectual inspiration, and dedication to this project. Members of the Best Practices Benchmarking and Consulting team contributed valuable research, editing, ideas, and observations; thanks especially to Jimmy Chow, Ruth Tanenbaum, Ken DeGiorgio, and James Kvaal for your support and efforts. Special thanks are also due to Derek Ransley, who contributed generously from his work at Chevron Research and Technology Company and who also proved a thoughtful editor and commentator on the manuscript; Stan Schulz, whose work at Johnson Controls is inspiring and whose attention to manuscript details was always invaluable; Jan Howard, who shared generously with ideas and her experiences at the Mutual Life Insurance Company of New York and who provided a bottomless well of encouragement and good-spirited support. Michelle Yakovac, Susan Prince, Hugh Devine, Allen Paison, and Dave Custable participated in numerous benchmarking collaborations for GTE, especially in customer satisfaction research. Otis

Wolkins provided ongoing encouragement and support for doing this book on the important topic of benchmarking. Elizabeth Murphy believed in the value of benchmarking enough to design IIR conferences at which some of these materials were tested. Cheryl Grohman and The Executive Committee provided the ultimate benchmarking consortium at which to pilot many ideas contained in this book. Jim Staker has contributed time, assistance, encouragement, and valuable contributions to the benchmarking field through his stewardship of the SPI Benchmarking Council. Bob Kirschner provided help and support for many benchmarking projects at GTE. Nancy Burzon and Paul Castorina of GTE's Management Development Center have supported best practice benchmarking as an important course in the 1990s management curriculum. Others who have supported us with their ideas, conversations, and comments include John Robbins at American Residential; Daniel Wagnière, Truman Koehler, Roland Loesser, and Joerg Staeheli at the Sandoz Corporation; Corinne Forti at Forti Communications; Judy Horsfield at AT&T; Eric Kennedy at IBM; Bernice Cramer at Next Frame, Inc., Fred Bowers at Digital Equipment Corporation, Chuck Schallhorn and Jim Madigan at Eastman Kodak, Chris Doerr at Leeson Electric Corporation. Many others—too numerous to name here—shared their thoughts, experiences, and enthusiasm through interviews and collaborations. All of you are the god parents of this book and merit our thanks and gratitude.

Christopher Bogan—Lexington, Massachusetts
Michael English—Irving, Texas

1

Benchmarking for Best Practices: Winning Through Innovative Adaptation

Benchmarking Is an Essential Business Concept

In a world where common sense prevailed, benchmarking would seem prosaic. It is simply the systematic process of searching for best practices, innovative ideas, and highly effective operating procedures that lead to superior performance. What could be more straightforward? No individual, team, or operating unit—no matter how creative or prolific—can possibly parent all innovation. No single department or company can corner the market on *all* good ideas.

In view of this reality recognizing human limitations, it makes eminently *good sense* to consider the experience of others. Those who always go it alone are doomed to perennially reinvent the wheel, for they do not learn and benefit from others' progress. By systematically studying the best business practices, operating tactics, and winning strategies of others, an individual, team, or organization can accelerate its own progress and improvement.

The history of innovative adaptation is arguably as old as humankind. For millennia people have observed good ideas around them and adapted those ideas to meet their needs and situations. Fred D. Bowers, Digital Equipment Corporation's benchmarking program manager, muses that "the second person to light a fire" is humankind's first benchmarker. Bowers' logic: The second fire-starter observed the first fire-starter and then borrowed the practice. Consider a few other noteworthy borrowings from the annals of early benchmarking history:

Lowell, Massachusetts: In the 1800s, British textiles mills were absolutely the best in the world. In contrast, American mills were still in their infancy when it came to producing all types of textiles. Francis Lowell, a New England industrialist, set out to change this situation by upgrading business technology in the United States. Lowell traveled to England where he studied the manufacturing techniques and industrial design of the best British mill factories. He saw that the British plants had much more sophisticated equipment but the British plant layouts did not effectively utilize labor. In short, there was room for improvement.

In 1815, Francis Lowell built a factory that employed much of the technology in the British plants but was designed to be much less labor intensive than the British facilities. It was a splendid example of innovative adaptation. In 1820, this textile mill center became known as Lowell, Massachusetts. By 1840, just two decades later, Lowell had grown to become the second largest city in America and the largest manufacturing complex in the country. This dynamic growth was largely fueled by one man's vision and his ability to creatively adapt practices observed in the world's best mills.[1]

Ford Motor Company: In 1912, a curious Henry Ford watched men cut meat during a tour of a Chicago slaughter house. The carcasses were hanging on hooks mounted on a monorail. After each man performed his job, he would push the carcass to the next station. When the tour was over, the tour guide said, "Well, sir, what do you think?" Mr. Ford turned to the man and said, "Thanks, son, I think you may have given me a real good idea." Less than six months later, the world's first assembly line started producing magnetos in the Ford Highland Park Plant.

Henry Ford articulated his vision in this way: "The man who places the part will not fasten it. The man who puts in the bolt does not put in the nut, and the man who starts the nut will not tighten it." The idea that revolutionized modern manufacturing and automotive history was imported from another industry.[2]

Toyota: In 1950, General Motors was the world leader in the automobile industry, and Toyota was just a small supplier to the Japanese domestic car market. At this time, the founder of Toyota sent his son, Eliji Toyoda, to the United States on a mission to study American manufacturing processes and practices. During his visit, Eliji Toyoda visited General Motors, Chrysler, Ford, and even Studebaker. He took extensive notes describing all that he saw. No detail was too small for his attention. Also during his visit, Toyoda visited American supermarkets, where he was impressed by the speed and precision with which grocers restocked their shelves at night so that supplies were replenished in time for customers to shop during day-time hours. The observations and insights from Toyoda's study visits were transported back to Japan, where they were adopted, adapted, and improved. As history has recorded, these visits planted the seeds for what would develop into Toyota's now famous just-in-time total-quality-control program. Toyota launched its U.S. presence on the west coast and then expanded across the country. During the next three decades, the Japanese car maker flexed its muscles and began challenging the far larger American competitors. By 1983, Toyota had captured 23 percent of the U.S. auto market. In the same year, Eliji Toyoda became chairman of Toyota. In 1984, General Motors signed a joint venture agreement with Toyota to manufacture Toyota products in the United States. "I'm convinced that GM's main reason for this joint venture is to see how Toyota runs a factory," observed a vice president of the Boston Consulting Group at the time. The wheel had turned full circle. Now General Motors was studying Toyota to learn about its winning strategies.[3]

Remington Rifle Company: In the 1980s, the Remington Rifle Company, a division of the giant DuPont corporation, was wrestling with a technical problem. Market research revealed that customers wanted shinier rifle shells. The plant managers didn't pay much attention to what seemed a nit-picking piece of market research. After all, Remington was a company that manufactured some of the oldest and best known rifles in the world. Surely that is what mattered most, they reasoned. But marketing executives persisted in their requests to respond to the customer feedback. Unfortunately, the company's engineering teams made little progress in their efforts to solve the problem.

As luck would have it, a short distance away from the Remington Rifle Company plant in Arkansas was a Mabelline cosmetics plant that produced shiny lipstick cartridges. Remington employees, who drove past the Mabelline plant each day on their way to and from work, surmised that the lipstick maker might have important lessons to impart. After all, the lipstick cases were not much different in size and shape from rifle shells. The site visit to the neighboring plant paid off and enabled the Remington team to solve the production difficulties that previously had proved so nettlesome.

In view of these historic examples of innovative adaptation, the obvious wisdom of studying others' best practices would seem self-evident. Learning by borrowing from the best and adapting their approaches to fit your own needs is the essence of benchmarking. Surely there is nothing new or revolutionary in this prescription for improvement. Or is there?'

For every example of innovative adaptation that graces the halls of history, there are many more examples of organizations, groups, and people that have declined to look outside themselves for solutions. The fact is not remarkable that people have on many noteworthy occasions been inspired through the benchmarking process. By exposing organizations and people to new ideas and approaches, the benchmarking experience often spurs extraordinary insights and breakthrough thinking.

What *is* truly remarkable is that benchmarking has not been embraced as a fundamental business process and skill sooner. Only in the late 1980s and early 1990s has benchmarking become widely regarded as a skill that should be communicated and utilized in day-to-day business operations. Benchmarking has broad applications in problem solving, planning, goal setting, process improvement, innovation, reengineering, strategy setting, and in various other contexts. Quite simply, benchmarking is a fundamental business skill that supports quality excellence.

Benchmarks and Benchmarking

Benchmarking's linguistic and metaphorical roots lie in the land surveyor's term, where a *benchmark* was a distinctive mark made on a rock, a wall, or a building. In this context, a benchmark served as a reference point in determining one's current position or altitude in topographical surveys and tidal observations. In the most general terms, a benchmark was originally a sighting point from which measurements could be made or a standard against which others could be measured.

In the 1970s, the concept of a benchmark evolved beyond a technical term signifying a reference point. The word migrated into the lexicon of business, where it came to signify the measurement process by which to conduct comparisons. In the early 1980s, Xerox Corporation, a leader in the business process of benchmarking, referred to benchmarking in rather narrow terms that focused primarily on comparisons with one's primary competitors. "Benchmarking is the continuous process of measuring products, services, and practices against the toughest competitors or those companies recognized as industry leaders," observed former Xerox CEO David Kearns.

During the 1980s, the definition of benchmarking grew in scope and focus. No longer were the metrical objects or benchmarks of primary interest. Benchmarking came to refer to the outreach activity of comparing yourself against others. Various practitioners offered the following definitions:

> *A process for rigorously measuring your performance versus the best-in-class companies and for using the analysis to meet and surpass the best-in-class.* (Kaiser Associates, a management consulting firm that has actively promoted benchmarking.)
> *A standard of excellence or achievement against which other similar things must be measured or judged.* (Sam Bookhart, former manager of benchmarking at DuPont Fibers.)
> *Benchmarking is the search for industry best practices that lead to superior performance.* (Robert C. Camp, a Xerox Corporation manager, author of *Benchmarking: The Search For Industry Best Practices,* and one of the foremost benchmarking experts at Xerox Corporation.)

The distinction between *benchmarking* and *benchmarks* continues to perplex many managers. In our view, benchmarking is the on-going search for best practices that produce superior performance when adapted and implemented in one's organization. Emphasis should be placed on benchmarking as an *on-going* outreach activity; the goal of the outreach is *identification* of best operating practices that, when *implemented*, produce *superior performance.*

Benchmarks, in contrast to benchmarking, are measurements to gauge the performance of a function, operation, or business relative to others. In the electronics industry, for instance, a benchmark has long referred to an operating statistic that allows you to compare your own performance to that of another or to an industry standard. Operating statistics employed as benchmarks provide incomplete comparisons. In a sense, they are superficial, for they draw attention to performance gaps without offering any evidence or explanation for why those gaps exist. At times, the performance gaps surfaced through benchmark comparisons may reflect significant differences in operating systems and procedures; on other occasions, benchmark variances may reflect differences in the way different organizations track and measure the performance of their systems.

The root causes of operating differences usually cannot be discerned from the benchmarks alone. In this respect, the benchmarks are like divining rods that lead the organization to hidden opportunities to innovate and improve performance. *Benchmarking* is the actual process of investigation and discovery that emphasizes the operating procedures as the things of greatest interest and value.

Consequently, *best practices benchmarking* can be described as the process of seeking out and studying the best internal and external practices that produce superior performance. One measures this performance through various financial and nonfinancial performance indicators. (*Figure 1.1 illustrates the relationship of benchmarks and benchmarking.*)

Best practices benchmarking, which includes but isn't limited to the study of statistical benchmarks, can— and *should*—be applied at many levels of the organization and in many different contexts. The benefits of benchmarking have been well recognized in certain industries and operating areas. For instance, many benchmarking projects have targeted critical technical functions such as distribution and logistics, billing, order entry and fulfillment, and training. However, benchmarking is also an advanced business concept with general management applications for high-level functions such as strategic planning, restructuring, financial management, succession planning, and supplier and joint venture management.

Managing Change

The pace of change is so rapid today that no single organization can ever control or dominate all effective operating practices and good ideas. To be a marketplace leader, one must look outward—as well as inward—for constant improvement and new ideas. Customers everywhere are broadcasting the same message to their suppliers: "Faster, cheaper, better." The old school of thought, which held that "if it ain't invented here, it can't be any good," is a curse in today's high-velocity markets. Don't reinvent what others have learned to do better. Today's rallying cries—"Borrow shamelessly!", "Adopt, adapt, advance!", "Imitate creatively!", and "Adapt innovatively!"—are anthems of business pragmatism.

Benchmarking teams, with a mandate to look far and wide for better operating practices, are arguably one of the best sentinels senior management can post along the watchtowers of the organization. They can sound the alarm when the first signs

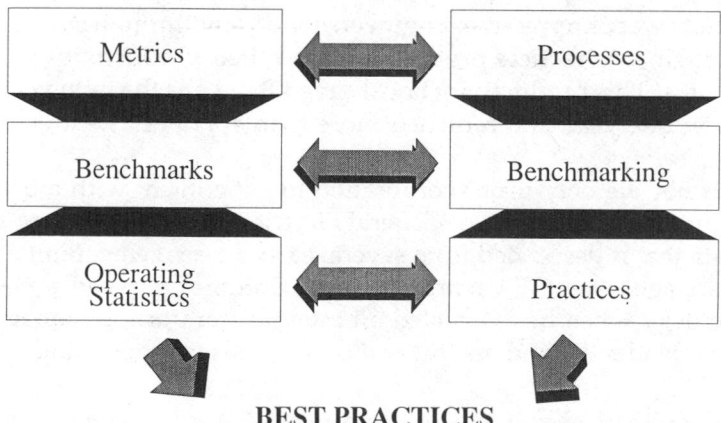

BEST PRACTICES

Figure 1.1. Benchmarking for best practices.

appear on the horizon that the organization has fallen behind the competition or has failed to take advantage of important operating improvements developed elsewhere. Best practices benchmarking provides employees and managers the tool, the rationale, and the process to accept change as constant, inevitable, and good. "Change has no constituency," observes General Electric CEO Jack Welch, who has established benchmarking for best practices as an essential part of GE's on-going management and improvement efforts. "People like the status quo. They like the way it was."[4]

The on-going adaptation of best practices helps an organization avoid being ambushed by unexpected change. A company can accelerate its own rate of improvement by systematically studying others and by comparing its own operations and performance with the best and most effective practices of highly-innovative and successful companies. The search for best practices quickly draws you outside the confines of your own culture and personal habits. Best practices benchmarking is therefore a pragmatic approach to managing change and performance improvement.

Many organizations have demonstrated the power of best practices benchmarking. Bell Laboratories, the research arm of American Telephone & Telegraph Co., has notably showcased the positive behavioral effects of best practices. In a fascinating program focused on AT&T engineers, Bell Labs demonstrated that the company could effectively manage individual behavioral change, which leads directly to performance improvement, through an individual best practices strategy.

During a six-year period, Bell Labs trained 248 engineers in its switching systems business unit to emulate the work and social habits of the unit's best performers. The strategies used by the best practitioners were hardly lofty or the stuff taught at the world's best technical or business schools. These best practices were fundamental skills designed to make people more effective and efficient. The program called for the engineers to learn various important business practices and skills from each other. These practices included how to most effectively manage work "in-baskets" that otherwise filled up with memos, correspondence, mail, and reading, how to accept constructive criticism, and how to seek help instead of wasting time by insisting on solving problems individually.

The results were impressive: Engineers who went through the Bell Labs best individual engineer practices program boosted their productivity by 10 percent in eight months! This productivity boost saved Bell Labs the money spent on the program after one year, and returned more than six times the investment after two years.[5]

AT&T is not the only major corporation to experiment with a best practices strategy in managing its business. General Electric currently embraces a best practices program that is descended from several earlier management initiatives. More than 40 years ago, then GE Chairman Ralph Cordiner pursued a best business practices strategy when he assembled an internal team of top managers and instructed them to identify and institutionalize the era's best operating practices.

In 1951, [Cordiner] assembled a brainy team of GE executives, plus consultants and professors, including Peter Drucker, to recommend ways to improve GE's management. They studied fifty other firms, pored over the personnel records of 2000 GEers, did time-motion studies of executives at work, and interviewed countless GE managers.

Two years later, they emerged with the Blue Book, a five-volume, 3,463-page management bible. Buried in endless pages of stultifyingly elaborate prescriptions are such powerful concepts as management by objective—as well as some of the most revolutionary ideas [current GE CEO Jack] Welch would later espouse. This discussion of decentralization, for instance, sounds a lot like Welch's principle of speed: "A minimum of supervision, a minimum of time delays in decision making, a maximum of competitive agility, and thus maximum service to customers and profits to the company."[6]

GE's Blue Book and every other best practices compendium face the same challenge: How do you avoid bureaucracy when codifying and institutionalizing today's most effective operating practices? The easiest way to fully leverage an identifiable best practice is to declare it a mandatory SOP (standard operating procedure). Paradoxically, an organization risks turning its current best practices into future bureaucratic tendencies as soon as it rigidly mandates and codifies them in hefty operating manuals. The road to competitive ruin is paved with once-effective operating procedures that have outlived their time. Companies implementing best practice strategies must carefully balance the benefits of current SOP compliance with the benefits of future innovation. Tomorrow's best practices will inevitably evolve beyond or diverge from today's best practices. By their nature, best practices are dynamic and progressive. For this reason, best practices benchmarking is often called an "evergreen" process: It renews the organization each time it is repeated. Consequently, best practice champions regard benchmarking as an on-going business process that is fully integrated with continuous improvement in their organizations.

Three Primary Benchmarking Types

Benchmarking has gained tremendous influence and currency in the 1990s. Correspondingly, front-line employees and operating managers have applied basic benchmarking skills in scores of different business situations. Among these applications, three distinct types of benchmarking have proliferated. They include: (1) process benchmarking, (2) performance benchmarking, and (3) strategic benchmarking. (*Figure 1.2 reviews the three types of benchmarking.*)

Process Benchmarking.   Process benchmarking focuses on discrete work processes and operating systems, such as the customer complaint process, the billing process, the order-and-fulfillment process, the recruitment process, or the strategic planning process. This form of benchmarking seeks to identify the most effective operating practices from many companies that perform similar work functions. In recent years, process benchmarking has grown in stature in the United States. Many of the most impressive American benchmarking success stories refer to process benchmarking. Its power lies in its ability to produce bottom-line results. If an

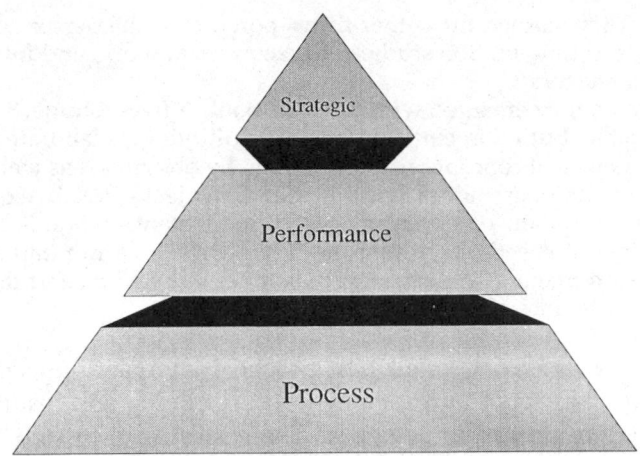

Figure 1.2. There are three primary types of benchmarking.

organization improves a core process, for instance, it can then quickly deliver performance improvements. These performance improvements may be calculated through increased productivity, lower costs, or improved sales, but their net effect frequently translates into improved short-term financial results. For this reason, American managers—seeking performance improvements that will show up on their quarterly score cards—embrace process benchmarking.

Performance Benchmarking.   Performance benchmarking enables managers to assess their competitive positions through product and service comparisons. Performance benchmarking usually focuses on elements of price, technical quality, ancillary product or service features, speed, reliability, and other performance characteristics. Reverse engineering, direct product or service comparisons, and analysis of operating statistics are the primary techniques applied during performance benchmarking. The automotive, computer, financial services, and photo copier industries, among others, regularly employ performance benchmarking as a standard competitive tool.

Strategic Benchmarking.   In general terms, strategic benchmarking examines how companies compete. Strategic benchmarking is seldom industry-focused. It roves across industries seeking to identify the winning strategies that have enabled high-performing companies to be successful in their marketplaces. Numerous Japanese corporations are accomplished strategic benchmarkers. A U.S.-based management consultant who specializes in working with Japanese corporations operating in the United States tells this story: "My clients begin by asking, 'What companies are really good?' Then we set up a trip in which the chairman or CEO of my client will go to visit these really good companies. Unlike American companies that begin a benchmarking project by determining what specific activity or process they want to examine, my Japanese clients are interested in fundamental lessons and winning strategies. They feel as if they already understand their processes." It is not surprising that Japanese corporations, which characteristically focus on long-term time horizons, should be most interested in strategic benchmarking. Strategic benchmarking influences the longer-term competitive patterns of a company. Consequently, the benefits may accrue slowly. Organizations seeking short-term benefits, such as those reflected in quarterly performance reports, usually find that process benchmarking produces results more rapidly.

Applications and Benefits

Benchmarking is a remarkably versatile business tool. Roland Loesser, the chief financial officer of the Sandoz Corporation's American Operations, observes: "Benchmarking is powerful because it can be applied to virtually every function in our companies." Moreover, front-line managers are using it in many new and creative ways. Some of the more frequent applications include:

Setting and Refining Strategy.   Today's markets are in a dynamic state of flux. Consequently, important insights can be gleaned by studying the experiences and competitive strategies of others. Bath Ironworks, for instance, benchmarked the strategies and operations of 10 shipyards in Holland when the Cold War's end rendered the 108-year-old shipyard's business strategy completely out-of-date. Bath, the U.S.'s fourth largest shipyard and Maine's largest private employer, had assumed that the country's need for combat vessels would remain strong for the rest of the century. (Since 1977, 86 percent of the vessels Bath delivered were Naval combat ships.) To quickly rethink its strategy and adjust to the "sea change" in the post-Cold War economy, Bath studied the strategy of the Royal Schelde shipyard in Vlissingen, Holland. Royal Schelde and other Dutch shipyards had already reorganized to accommodate merchant ship building and the manufacturing of other complex structures, such as bridges. "Contingency plans can be developed and implemented much faster and at far less cost [through benchmarking] than if developed from scratch," observes Bath Iron Works' manager of quality improvement William R. "Tip" Koehler.[7] The strategic lessons learned by other organizations and industries can help your own company refine its strategy, project the possible outcomes of changing its present course, and forecast potential cataclysmic shifts brought on by changing market circumstances.

Reengineering Work Processes and Business Systems.   Benchmarking is a necessity for companies engaged in reengineering their processes and systems. Benchmarking gives you the ability to see things differently. It is like setting up a satellite dish outside your offices. Suddenly, signals from throughout the world can penetrate your organization. Benchmarking enables a company to get outside its conditioned responses or customary structures of thinking. When GTE reengineered eight core processes of its telephone operations, it examined the best practices of some 84 companies from diverse industries to help the company rethink the rules of the game for each of its core functions.

Reengineering without benchmarking is likely to produce flat 5 to 10 percent improvements, not the spectacular 50 to 75 percent performance improvements often seen with radical redesign. Benchmarking enables true reengineering. Through the study of outside best practices, a company can identify and import new technology, new skills, new structures, new training, and new capabilities.

Continuous Improvement of Work Processes and Business Systems.
Not every benchmarking project or initiative will yield major-magnitude change and system breakthroughs. Benchmarking also provides a potent source for incremental changes and improvements. Benchmarking exchanges frequently yield "golden nuggets" that are weighed in ounces rather than pounds. KPMG Peat Marwick, the Big Six accounting firm, borrowed the concept of a supermarket's Express Checkout to start an express line in its word processing pools. The change enabled work teams with minor document changes to go through an expedited process. This small change was of great value to word processing departments that handled high-volume work orders. It solved the long-standing and nettlesome problem of work assignments with small changes being stalled in long work queues. Moreover, this innovative adaptation of a supermarket operating practice improved cycle time and boosted internal customer satisfaction. Additionally, it can be applied to many other service functions, such as copying, graphics production, and research.

Strategic Planning and Goal Setting.   The unexpected missteps of blue-chip organizations such as IBM, American Express, Westinghouse, and General Motors have provided the world with an important lesson: In the 1990s, market changes can be swift and powerful, economically hobbling even the most powerful corporations. Consequently, a growing number of companies enable their strategic planning and goal setting process through benchmarking. Benchmarking helps organizations anticipate market changes and validate goals and targets. One can only wonder how much sooner these battered giants might have responded to shifting market realities if they had used benchmarking as an integral part of their strategic planning and goal setting process.

By reviewing the products, prices, practices, strategies, structures, and services of competitors and other industry front-runners, managers can validate the adequacy of their own goals, plans, and strategies. For instance, Mutual Life Insurance Company of New York requires all executives to find benchmarking information on their primary and secondary competitors as part of the company's newly revamped planning process. Says MONY Quality Officer Jan Howard: "Planning without awareness of what your competitors are doing is like flying a plane over the Alps in heavy fog without any instrument controls."

Problem Solving.   Benchmarking frequently demonstrates its value in the problem-solving process. Ironically, most corporate problem-solving processes do not methodically look outside the team or organization for solutions. Standard problem-solving processes provide a structure that make work groups more effective;

they also prompt teams to root their analysis in empirical data, which supports management by fact—rather than by fancy. But most problem-solving processes indirectly encourage teams to reinvent the wheel because they seldom encourage work groups to consider external experience in developing their solutions. As an enabling tool for problem solving, benchmarking frequently produces elegant answers for thorny operating issues. Consider the following case from the Xerox Corporation, where benchmarking has been deeply integrated into the organization's fundamental quality and problem-solving efforts.

Plagued by high associate turnover in its corporate legal department, Xerox looked for solutions both internally and externally. Internal analysis revealed various causes for the problem; external analysis, however, turned up important insights. Xerox's benchmarking partners shared the same recruitment and selection process and all suffered from the same high associate turnover. Once recognizing that its essential recruitment strategy produced suboptimal results, Xerox adjusted its recruitment strategy rather than trying to fine-tune the process. In retrospect, this decision makes excellent sense because the best and brightest law students are usually geared to work for high-paying and high-powered law firms—not for corporate law departments. Many fine lawyers decide to move to corporations after practicing for several years at a firm. The common experience of all the benchmark partners gave Xerox confidence to move away from traditional recruitment on law school campuses to a more radical strategy. This strategy emphasizes recruitment of experienced lawyers, who wish to make lateral career moves away from law firms and into corporate practices. Arguably, Xerox would never have gleaned this insight if not for its benchmarking investigation.

Education and Idea Enrichment.   A Zen-like management riddle asks: "How does a fish know it is wet?" The fish spends all its life in water and knows no other condition. The riddle probes how people, who grow accustomed to operating in certain ways, know there are other approaches—perhaps better ways—of performing the same task. Benchmarking is a tool for achieving idea enrichment and general education. Each benchmarking trip is a learning safari. Successful benchmarkers return to their organizations with valuable trophies—new ideas and approaches for accomplishing old tasks. By regularly benchmarking critical functions, organizations ensure they remain open to new ideas, changing trends, and evolving technology. If seeing is believing, then benchmarking is an effective process to ensure that managers and front-line operators see other approaches to accomplishing the activities over which they preside.

Market Performance Comparisons and Evaluations.   Human nature encourages people to reflect positively on the organizations and colleagues with which they work. Naturally, people want to validate their efforts. Correspondingly, organizations and individuals frequently presume the products and services they provide to customers are also of high quality. Yet without carefully comparing them to competitor's offerings, they cannot fairly evaluate their relative standing. However, fancy gives way to fact when you benchmark your company's products, features, and performance against competitors'. This type of performance benchmarking is common in many industries. Mortgage bankers, for instance, compare their interest rates, service fees, and product types on a weekly basis. *Consumer Reports* has long evaluated the features of various products and J.D. Powers has benchmarked customer satisfaction levels among automobile owners. *Financial World* magazine rates the financial performance and management of America's top cities. All these industry and professional ratings provide a fact-based market performance test that employs the essential skills of benchmarking. By heeding these and other types of performance benchmarks, organizations can assess the adequacy of their products, services, features, and performance. Such information can help them manage by customer-focused facts.

Catalyst for Change.   Al Kuebler of 1992 Malcolm Baldrige National Quality Award winner AT&T Universal Card Services observes an operating truth that every benchmarking manager has observed: "Tell me and I forget, show me and I remember, involve me and I understand."[8] Benchmarking is an effective catalyst for change because it involves employees in the personal discovery of more effective operating practices. Benchmarking exposes people to new approaches, systems, and procedures. It is first-person experience that helps an employee visualize the final goal of prospective change. In this respect, benchmarking demystifies change, making it more tangible and less threatening. Consequently, benchmarking helps manage organizational change.

A Utilitarian Tool

Managers and employees are inundated with a series of highly abstract yet exceedingly important challenges. These include general mandates to oversee such intangible concepts as change management, innovation, creativity, organizational learning, speed or cycle time reduction, process simplification, and reengineering. These concepts seem perplexing for many managers concerned with such matters as serving customers, meeting daily deadlines, reducing costs, and growing revenues. It's difficult to get your thoughts and your hands around these high sounding but abstract concepts.

Consider the company that wanted to establish its position as the leading innovator in its industry. To achieve this goal, it proposed performing an "innovation audit." The concept was powerful: Audit or assess the company's structure, culture, work processes, technology, and managerial systems to determine their positive or negative influences on innovation. The magnitude of this task quickly grew daunting. How would the team define creativity and innovation? What would be the actual goal of such an audit? What systems or processes would be the focus of such an audit? What systems—if any—would be excluded? Arguably since every system and process in the company could influence organizational innovation, the company should consider them all. To avoid attempting to "boil the ocean" on its first foray into this field, the organization refocused on the new product development process. If the company improved this process, it would produce more successful product launches. This success would advance the company in becoming the industry leader in innovation—regardless of how it defined innovation. Benchmarking in the new product development process provided an excellent, results-oriented approach to exploring the larger innovation theme. As the company in the example, any benchmarker needs to focus its study. A focused study turns the abstract concept into the concrete. (*Figure 1.3 reflects the role of benchmarking in managing change and innovation.*)

Benchmarking is an easily grasped, functional tool. As utilitarian as a fireplace poker, it can test and probe the hottest management concepts. Benchmarking doesn't support abstract postulations about arcane management concepts. It promotes the active discovery of systems that embody the concepts in real-world situations. Don't sit in isolation, for instance, while meditating how to compete through cycle-time reduction. Study the best practices of other organizations that have learned to perform critical functions more quickly than your own company. Benchmarking offers a kind of low tech "virtual reality" for organizations eager to simulate operational experiences in their own environments. What better way to project a system's impact in your own company than to examine the performance effects of that system already implemented in another organization? "Ideas are a commodity," observes Dell Computer CEO Michael Dell. "Execution of them is not."[9] As a managerial tool, benchmarking provides a double benefit: It provides a way to access new ideas and to test or evaluate the implementation challenges they may present in your own organization.

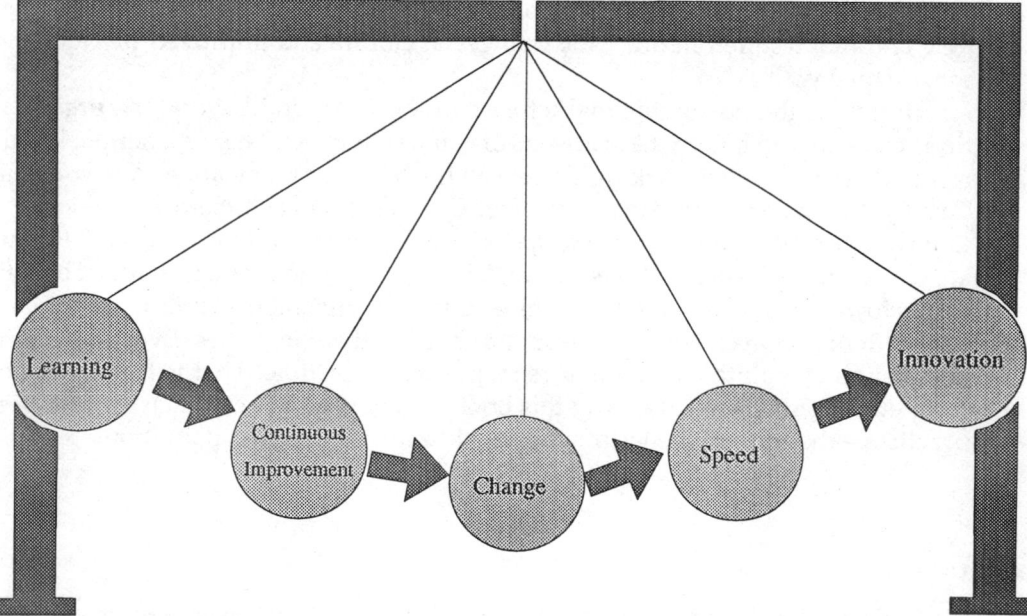

"Borrow Shamelessly"

Figure 1.3. Managing change and innovation.

A full list of benchmarking benefits cited by practitioners would reach as high as Everest. Bottom-line summaries almost always suggest that benchmarking:

- Improves organizational quality.
- Leads to lower cost positions.
- Creates buy-in for change.
- Exposes people to new ideas.
- Broadens the organization's operating perspective.
- Creates a culture open to new ideas.
- Serves as a catalyst for learning.
- Increases front-line employees' satisfaction through involvement, empowerment, and a sense of job ownership.
- Tests the rigor of internal operating targets.
- Overcomes front-line employees' natural disbelief that they can perform better.
- Creates an external business view.
- Raises the organization's level of maximum potential performance.

Finally, benchmarking for best practices generates one more benefit that is arguably the most important of all. It teaches organizations new lessons in competitiveness. "Benchmarking taught the managers how to compete," observed Sam Bookhart, formerly benchmarking manager at E.I. DuPont de Nemours & Company, Inc. "It wasn't just the marketing manager. It was the technical manager and the manufacturing manager and the accounting manager. It taught them how to compete and that resulted in dramatic changes in culture and improved product and service quality."[10]

In the rough-and-tumble marketplace of the 1990s and beyond, no organizations can afford to ignore these lessons of competitiveness. There's a simple litmus test to determine benchmarking's applicability to your organization. Ask yourself: *Can my organization afford to stop improving? Can my organization afford to stop learning? Can my organization afford to stop competing for its position in the marketplace?* If your answer to any of these questions is "yes," then put this book aside; it will not benefit you. However, it's difficult to imagine many organizations—public or private, for-profit or nonprofit—that can respond "yes" to these inquiries. Every organization strives to maintain and enhance its position over time. That is the essence of competitiveness. That's also why this book is dedicated to benchmarking for best practices—the art and science of winning through innovative adaptation!

References

1. Cochoit, Jules, Xerox Corporation, a video broadcast presentation on benchmarking for NTU, August 13, 1991.

2. Munro, A. Sandy, "How Good Is Good? How Bad Is Bad? How Do We Get Better?" A speech presented at the Executive Briefing on Design for Manufacturability: Competitive Benchmarking and Performance Measurements, a conference sponsored by the Management Roundtable, Berkeley, Calif., May 20-21, 1991.

3. Cochoit, Jules, Xerox Corporation, a video broadcast presentation on benchmarking for NTU, August 13, 1991.

4. "Jack Welch's Lessons for Success," *Fortune*, January 25, 1993, pp. 86-93.

5. Rigdon, Joan E., "Using New Kinds of Corporate Alchemy, Some Firms Turn Lesser Lights into Stars," *The Wall Street Journal*, Monday, May 3, 1993, pp. B1, B13.

6. Tichy, Noel M. and Sherman, Stratford, "Control Your Destiny or Someone Else Will," Doubleday, New York, 1993, p. 37.

7. Biesada, Alexandra, "Strategic Benchmarking," *Financial World*, September 29, 1992, p. 30.

8. Kuebler, Al, "The Quest for Excellence 5," annual conference of Baldrige Award winners, AT&T Universal Card Services presentation, February 15-17, 1993.

9. Sherman, Stratford, "The New Computer Revolution," *Fortune*, June 14, 1993, p. 60.

10. Bookhart, Samuel, "Benchmarking, A Powerful Management Tool," a speech made at the "Benchmarking Against The Best" conference sponsored by the International Research Institute, June 13, 1991.

2

Fast Learning Through Innovative Adaptation

The discount idea was the future. We had only two choices: stay in the variety store business and be hit hard by the discounting wave, or open a discount store. So I started running all over the country, studying the concept, from the mill stores in the East to California, where Sol Price had started his Fed-Mart in 1955. I liked Sol's Fed-Mart name, so I latched right on to Wal-Mart. On July 2, 1962, we opened Wal-Mart No. 1 in Rogers, Ark., right down the road from Bentonville. We did a million dollars in a year.[1]
SAM WALTON

Sam Walton understood the power of innovative adaptation. The founder of Wal-Mart always had his eyes open for good ideas that he could borrow, adapt, and refine. The very name of Wal-Mart was a knock-off. But "Mr. Sam" and the Wal-Mart team creatively tailored and improved the ideas they observed elsewhere and admired. Walton, who for a time was the wealthiest person in America, created one of America's greatest retailing success stories. Yet he always remained a very humble man. Walton often drove a pickup truck and until his death from cancer in 1992, he was an indefatigable informal benchmarker. He was always on the prowl for good ideas he could borrow and reuse. The Wal-Mart CEO sought them anywhere and everywhere—among employees, customers, suppliers, competitors, and other companies. "He is notorious for looking at what everybody else does, taking the best of it, and then making it better," observed Sol Price, the founder of Fed-Mart and Price Club, the first wholesale club. More than 20 years after Walton borrowed the

Wal-Mart idea from Price, he followed suit in the late 1970s by creating Sam's Wholesale Clubs based on Price's Price Club.[2]

"You can always learn something from the competition," remarked Walton near the end of his life.[3] In his pragmatic, "I-can-learn-from-everyone" image, Walton created a culture dedicated to fast and continuous learning. Throughout the Wal-Mart empire, borrowing good ideas and applying them to improve performance is reflexive. This creation of a deep-rooted organizational culture that learns and adapts quickly may be the ultimate testimony to Walton's extraordinary leadership. In today's increasingly competitive and fast-moving markets, one of senior management's primary challenges is creating an organization that can react quickly, remain agile and lean throughout sudden market swings, and respond proactively to shifting technology or changing economic conditions. It's the "fast-learning organization" that institutionalizes the capacity for rapid change, constant improvement, and creative evolution. In its many different guises, benchmarking or the search for best practices—both inside and outside one's own company—is a powerful tool for corporate transformation. Benchmarking helps create an organization in which all employees—from board members and senior executives to support personnel and front-line staff—accelerate the constant improvement process by borrowing, creatively emulating, or adapting the best ideas from other successful companies.

Management and the board share a joint responsibility to nurture and develop a corporate culture that supports best practices benchmarking. In organizations that actively encourage the identification and adoption of best practices, a corporate culture springs up that is dedicated to fast learning. The essence of best practices benchmarking is quite simply innovative adaptation. The opportunistic (and legal) borrowing of great ideas, structures, strategies, procedures, and systems is viewed to be healthy and good. Organizations that inculcate this impulse to innovate through adaptation of best management practices become high-charged learning laboratories. The walls erected by factionalism among departments, functions, and divisions tend to crumble when you recognize that it's possible to learn from others outside your own unit, team, company, or industry. In fact, others' successes can fuel your own. "Identifying and sharing your organization's own best practices— that is the essence of synergy in business," observes Jan Howard, quality officer at Mutual Life Insurance Company of New York (MONY), where the organization has studied its own internal best practices in the sales force.

Innovative adaptation is a key strength of many high-performing organizations. In corporations, such as Motorola, "steal shamelessly" is a rallying cry for continuous improvement. This attitude underpins a "we-can-learn-from-anyone" attitude that marks many high-performing organizations. "Best Practices has legitimized plagiarism," is a line gleefully quoted by CEO Jack Welch and other senior executives at General Electric, where the company has institutionalized the study of other high-performing organizations.[4] Welch's observation—not to be taken literally—is meant to pulverize the "if-it-ain't-invented-here-it-can't-be-any-good" syndrome that afflicts many successful organizations. At GE and other modern learning labs, employees are encouraged to borrow and implant excellent ideas that are not trademarked, patented, or proprietary. In Japan, many leading companies possess an instinct for innovative adaptation. However, in Japan, where "study visits" and information sharing among company groups (or *keiretsu*) have been standard operating policy for 40 years, "adopt honorably" is the adage. It reflects a cultural view that embraces the power of learning from others.

Fast Learning and Competitive Advantage

For years, organizations have thought of competitive advantage in terms of superior technology, lower costs, patent protections, regulatory protections, or proprietary products. In the 1990s, fast-learning has been discovered as a unique organizational competency. It may be the ultimate source of long-term, sustainable competitive advantage. Patents expire. Cost positions swing with markets. Technology is in a constant state of flux. The organization that can learn quickly can better manage these changes. "The ability to learn faster than your competitors may be the only sustainable competitive advantage," noted Arie De Geus, the head of planning at Royal Dutch/Shell.[5]

Trace the effects of fast learning. It is like fresh air and water introduced into an arid scab land. It has a revitalizing effect. Fast learning produces a healthy, self-sustaining ecosystem within the organization. Fast learning organizations—call them FLOs—are capable of fast change and creative evolution. With Darwinian vigor, they migrate with their markets and their customers. Consider some outcomes of fast learning:

- *FLOs train employees quickly and effectively.* The result is a work force that quickly implements new and improved practices. Error rates shrink more quickly. They also successfully roll out new programs, products, and services. Consequently, FLOs respond more quickly and easily to market changes.

- *FLOs are peopled by FLEs, that is* fast-learning employees. Fast-learning employees tend to be more open-minded about change, more flexible in responding to market swings, and more open to cross-training. They more readily accept job redefinition, work flow redesign, new skill acquisition, and multiple task responsibilities.

- *FLOs leverage past successes and failures to continuously improve their products, processes, and services.*

- *FLOs adapt quickly.* Because they listen closely to customers, they innovate rapidly. Rubbermaid, for example, launched one new product every day during 1992.[6] To do this, the organization has eliminated complex product trial tests. Instead, the organization has learned to transform customer needs from focus groups and to transform them directly into products.

- *FLOs are fast and time-responsive.* One primary FLO report card is cycle time. Inevitably, organizations that learn to reduce cycle times also learn to simplify processes. Speed and simplification usually connect directly to improved productivity, cost, and customer responsiveness. In this respect, Japanese companies have outpaced organizations elsewhere in focusing on the strategic benefits of speed and process simplification. In a 1991 Best Practices study of 500 businesses in Canada, Germany, Japan, and the United States, the American Quality Foundation and Ernst & Young discovered that 47 percent of Japanese firms reported they "always or almost always" use process simplification. This usage was twice as frequent as in other participating firms from other countries. Only 12 percent of U.S. study participants said they always or almost always use process simplification. In Canada 19 percent and in Germany 6 percent of firms said they use this approach. The gaps were even greater when it came to cycle-time analysis. Fifty-five percent of Japanese firms, 21 percent of U.S. firms, 19 percent of Canadian firms, and 6 percent of German firms said they always or almost always employ cycle-time analysis.[7]

Figure 2.1 depicts the impact of fast learning on a broad range of organizational issues.

Necessity Is Often the Mother of Innovative Adaptation

No natural laws prevent large companies from learning quickly, but size and hierarchy tend to favor control over speed. Consequently, fast learning is a skill that can be difficult for large corporations to cultivate. Borrowing ideas facilitates fast learning. Small companies therefore demonstrate natural tendencies to import good ideas from elsewhere. Why? Small organizations are almost always resource-starved. Survival teaches them to be fast, flexible, opportunistic, quick to borrow

Understanding the Benefits of Fast Learning

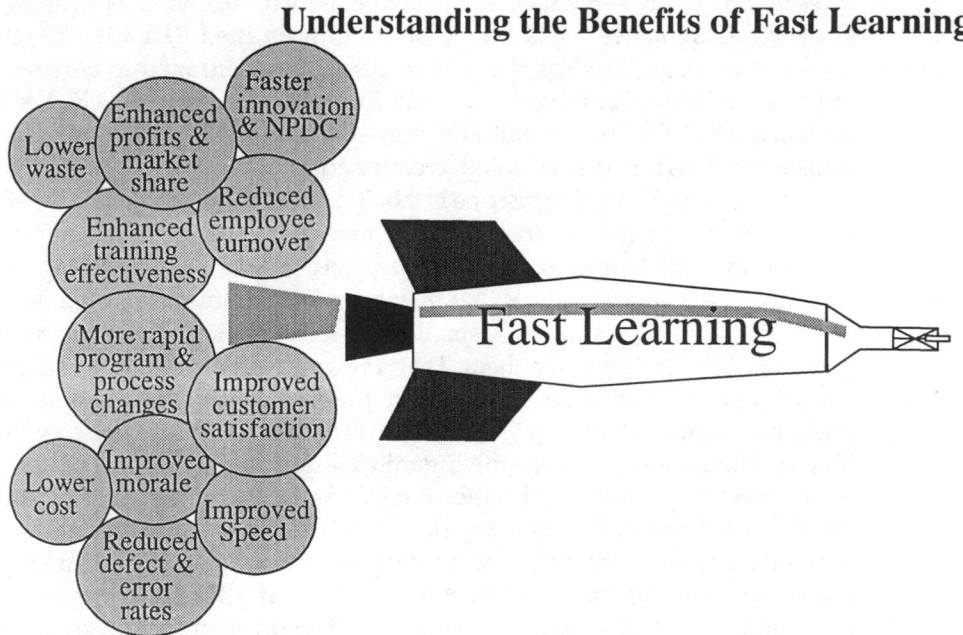

Figure 2.1. Understanding the benefits of fast learning.

and implement. They lack the rich warehouse of resources, staff, and capital to develop many systems in-house. "Some companies are systematically better at borrowing than others are, not least because they approach alliances and joint ventures as students, not teachers," observe Gary Hamel and C.K. Prahalad, two highly-esteemed theoreticians of modern corporate strategy. "Suffice it to say, arrogance and a full stomach are not as conducive to borrowing as humility and hunger. Captives of their own success, some companies are more likely to surrender their skills inadvertently than to internalize their partners' skills. We might call this negative leverage!"[8]

Some larger organizations—Motorola, Milliken, GE and Xerox, to name a few—have also faced prodigious competitive challenges and market pressures that forced them to rediscover the importance of creative adaptation. They use innovative adaptation as an on-going survival strategy. These companies recognize that time, people, capital, and good ideas are always in short supply. Borrowing from the best is therefore a guerrilla tactic for survival. It provides short cuts for continuous improvement.

"We're kind of the beg, borrow and steal gang," observes Robert Marcell, the General Manager of Chrysler's next-generation small-car development team—code named PL. Beaten back by the Japanese auto makers during the 1980s, Chrysler turned the tables by adopting Japanese post-war competitive tactics. The Chrysler team adopted the strategy of borrowing the best ideas from Japanese car makers and implementing them in their new generation of cars. Ford employed the same approach to developing and updating its best-selling Taurus and Sable car lines. The strategy has proven successful for both companies.[9]

Many successful small companies use innovative adaptation to fuel rapid growth and improvements. *Inc.* magazine, which bills itself as "The Magazine for Growing Companies," has a regular column entitled "Hands On" that profiles scores of managerial ideas that *Inc.* readers are encouraged to borrow, adapt, and apply in their own companies. At The Executive Committee (TEC), a nationwide network of CEOs from small and mid-size companies, meeting note stationery reflects this bias for action through creative adaptation. The stationery provides TEC presidents with two columns: one is labeled "notes" and the other is called the "Do List." TEC presidents say their monthly meetings serve as informal benchmarking visits with their fellow executives. They return from TEC meetings and quickly implement effective practices that they have learned from their fellow CEOs.

Ironically, success can stifle small companies too. Rapid growth imposes "large-company" challenges upon them. Their revenues balloon at double-digit rates; staff counts rise; facilities proliferate, and hierarchy—put in place to manage the growth—begins to threaten vicarious learning skills. Chris Doerr, co-CEO of Leeson Electric Corporation, a fast-growing mid-western manufacturer of motors, reflects on the problem: "When my brother and I started this company, concurrent engineering was something we did every day. That's because we all sat in the same room and talked to each other constantly. We grew up being quick to borrow and adapt. Now that we're approaching $100 million in size, it's hard to keep people motivated in the same way. Yet we need those skills and instincts more now than ever. Without them, we won't be able to sustain our growth in today's markets. It's cutthroat."

The Case of Manco, Inc.

Manco, Inc. is a shining example of how a company can use a "we-can-learn-from-everyone" attitude to gird the organization, to forge its strategy, and to develop a strong values-driven culture. This privately owned, Westlake, Ohio manufacturer of adhesive tapes has produced a compounded growth rate of 23 percent during the past 21 years. The company has grown from $1 million in revenues in 1971 to more than $75 million in 1992.[10] That's a sustained growth rate most managers would envy. How has Manco achieved this kind of growth? It has fashioned itself as a fast-learning organization. *Curiosity*, for instance, is a primary value of the company.

"This is how we compete," explains Jack Kahl, president of the producer of duct tape and an avid practitioner of benchmarking. "The fun for us is learning fast."[11] Kahl, the largest shareholder of the employee-owned company, sets the tone for the company and gives vivid testimony to how Manco values learning through borrowing good ideas. "All I know is that if I study excellent companies, I come up with excellent ideas," Kahl opines.[12]

Manco has been a master at learning from larger corporate mentors—frequently multibillion-dollar enterprises with which Manco does business. Manco has formally and informally benchmarked Walt Disney, Wal-Mart, Rubbermaid, and PepsiCo, Inc. These excellent Fortune 500 giants have provided wide-ranging best practices. From Disney it snatched ideas on merchandising and marketing; from Wal-Mart it adapted systems for communication and leadership; from Rubbermaid it borrowed approaches for managing innovation and new-product development, and from PepsiCo it adapted ideas on recruitment and hiring.

Like other smaller companies, Manco is a model of "lean benchmarking." It learns from public-domain sources that do not necessarily require expensive and time-consuming travel. Manco employees identify and adapt highly-effective ideas and operating improvements from books, articles, annual reports, and other public information sources. Some big-company benchmarkers may scoff at such informal "armchair benchmarking." They may correctly observe that such informal benchmarking lacks quantitative rigor and forgoes a highly-structured multistep investigation process. But it works well for Manco. The company has grown rapidly on the success of its partners—customers, suppliers, distributors, and others.

Above his office door, Kahl has placed a sign quoting the immortal Socrates: *"One thing only I know, and that is that I know nothing."* Kahl explains: "This is the message I have chosen to remind myself of every day. It lets all of my Manco partners know of all the learning that is required of both myself and them in order to continue our success. We all make ourselves available to one another and support one another in this constant quest for learning."[13]

To help create an environment that nurtures continuous learning, Manco has developed many managerial systems to support a "we-can-learn-from-anyone" culture. "I wanted to create the idea of a learning environment that never stops," Kahl continues. To that end, Manco reimburses employees for the cost of any educational course they want to take—just "as long as they pass it." At this unusual learning laboratory, "one of the highest values in Manco is to be curious and allow curiosity to take place."[14]

Learning Is the Essence of Continuous Improvement

The basic skills of benchmarking for best practices are the skills of fast-learning organizations. At the most fundamental level, benchmarking means learning from others. The skill is honed into an art by identifying and then comparing your own company to others who are very strong at what they do. The nineteenth-century philosopher Ralph Waldo Emerson put it succinctly: "Hitch your wagon to the stars." When comparing yourself for improvement purposes, compare yourself to the best—or at least to those who are substantially better than you.

In many respects, benchmarking skills are quintessentially networking skills. Many effective managers perform informal benchmarking in meetings, conferences, and discussions with colleagues from other companies. Benchmarking takes informal networking skills and turns them into a science. How? Benchmarking adds process structure, quantitative muscle, research rigor and implementation focus.

"Networking is critical," observes Neil B. Bunis, planning manager at DuPont Co. "There's no point in relearning what someone else in the company already knows."[15] Benchmarking moves this logic even further. There is no point reinventing what others have already perfected. By studying the best practices of other individuals, business units, and organizations, you can accelerate your own learning and pace of change. Benchmarking builds on the foundations that others have successfully tested and constructed.

Mastering How Organizations Learn

Benchmarking's simplicity is beguiling. It leads many organizations to set out on benchmarking projects without much forethought or planning. Such poorly planned investigations usually yield suboptimal results. All effective, structured benchmarking requires good planning. To prepare for your own investigation, it helps to understand how organizations learn.

The power and potential of benchmarking as a learning tool has been borne out by a growing body of academic research. A group of Harvard Business School researchers performed one particularly interesting study concerning how organizations learn.[16] The Harvard group's original work focused on learning in manufacturing, but their observations have general relevance for all organizational learning. They identified and described four primary ways in which organizations learn:

1. *Vicarious learning.* This first learning strategy embraces learning through the study and analysis of others' experience. These "others" may include colleagues within one's organization or people and organizations outside one's own corporation and industry.

2. *Simulation.* This second learning strategy constructs artificial models that help the organization discover what will happen when it implements a new program, introduces a new technology, or rolls out a new service or product.

3. *Prototyping.* This third learning strategy builds and operates the new product, service, or pilot program on a small scale in a controlled environment. In addition to creating sample products, prototyping also includes program piloting and small test market sampling. Similar to a dress rehearsal, it seeks to determine real world reaction prior to full-scale production and rollout.

4. *On-line learning.* This fourth learning strategy examines actual, full-scale implementation while it is operating as part of the normal production or service-delivery process. Similar to opening night, on-line learning raises the curtain before the entire market.

"A clear hierarchy exists among these four methods," observed the Harvard team. "Costs get higher moving down the list, but so does fidelity."[17] Their research reveals that the cost of learning escalates rapidly as organizations advance from vicarious learning to computer simulations to prototyping to learning from the roll-out of a product or service. *(Figure 2.2 illustrates four fundamental approaches to how organizations learn.)*

The Harvard team's research highlights three important benefits of vicarious learning:

1. It requires fewer resources than other forms of knowledge and experience acquisition.

2. It provides the fastest approach to learning.

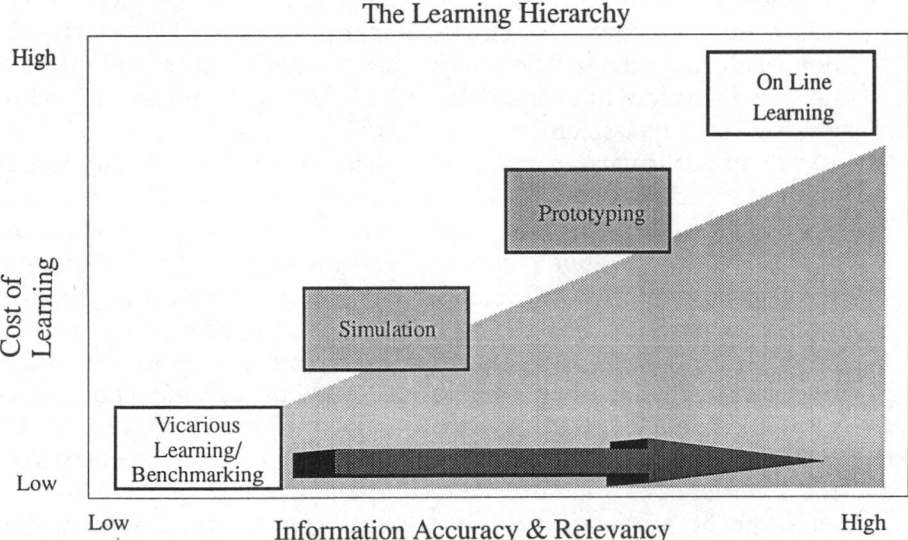

Figure 2.2. How organizations learn. (*Source: Bruce Chew—Harvard Business School; Dorothy Leonard-Barton—Harvard Business School; Roger Bohn—MIT.*)

3. It exposes the organization to outside ideas and experience, often outside the boundaries of current culture and assumptions.

This last observation is especially compelling. We are all prisoners of our own paradigms. Exposure to different approaches and points of view can lead to breakthrough improvements. Unique to organizational learning strategies, vicarious learning necessitates comparisons with others. These comparisons can be of many types, but all involve some form of outreach. Such outreach can lead to rapid, revolutionary progress that far outpaces the incremental or evolutionary progress more typical of other forms of learning. Benchmarking is a principal tool that managers can apply to facilitate vicarious learning. Organizations that harness such learning find it to be a source of great power and resources. "This is our biggest competitive weapon," marvels Manco's Jack Kahl. "We learn from everyone—and we learn faster than anyone else."[18]

Lessons from the National Quality Award

In the late 1980s, the administrators of the Malcolm Baldrige National Quality Award began distilling the lessons learned from the first Baldrige winners. All were high-performing companies; all had developed winning strategies. But one had to wonder: "Were there common characteristics—call them key excellence indicators—that marked these truly excellent companies?" From this general inquiry some important characteristics of organizational excellence emerged. The National Quality Award office identified nearly 40 performance traits that differentiated companies scoring high on Baldrige assessments from companies scoring in the lower quartiles. These traits of high-scoring companies profile the organizational characteristics of corporations that have achieved performance excellence through total quality management.[19] No single company—not even the Baldrige winners—embodies all these characteristics, but, taken together, these indicators of excellence help to identify what constitutes world-class performance in each of the seven Baldrige quality-assessment categories.

Three particular key excellence indicators are rooted deeply and directly in the soils of vicarious learning. They can be described as (1) a "we-can-learn-from-everyone" attitude toward continuous improvement, (2) the active use of "benchmarks and comparative measures," and (3) a "nondenominational approach" to developing performance systems. These three distinguishing traits are observed repeatedly in Baldrige winners and finalists. Moreover, they describe a *radically* pragmatic approach to organizational learning and improvement. This approach regards continuous improvement and vicarious learning as deeply entwined, almost like a double helix. In short, continuous improvement is a central focus of most quality initiatives, and continuous improvement can be accelerated by reviewing the experience of other excellent companies. Simply put, other people's experience can be a powerful catalyst to our own learning and improvement. For instance, at Milliken and Company, a 1989 Baldrige winner in manufacturing, this ethic is captured in the popular corporate slogan: *Steal Shamelessly!* Quoted throughout the company, this credo encourages Milliken employees to borrow and adapt all good ideas and systems that are not copyrighted, patent protected, or otherwise designated as trade secrets or proprietary. At IBM Rochester, a 1990 Baldrige winner, new product design teams went outside the organization for guidance in developing the highly successful AS/400 computer. Executives met with 250 customers to ensure that the product was designed correctly the first time. Then, IBM Rochester adopted Milliken's teamwork approach, Toyota's short-cycle product development system, and Motorola's six-sigma defect prevention program.

This same devotion to vicarious learning also exists at Motorola, a 1988 Baldrige winner. Motorola frequently proselytizes about the benefits of creative borrowing and adaptation. Consider this excerpt from a Motorola public interest advertisement:

OLD TRUTH: **THOU SHALT NOT STEAL.**
NEW TRUTH: **THOU SHALT STEAL NON-PROPRIETARY IDEAS SHAMELESSLY.**

Valuable lessons are not always learned within your own company. It often helps to look to the outside for different perspectives.

That's why Motorola benchmarks the best companies. They can be competitors or they can be in completely different businesses. We also listen to our customers and our suppliers. When your mind is open you can learn something from practically everyone. It's a good way to keep from reinventing the wheel. And, it can also be a reciprocal process, fostering trust and mutual respect.

We use the knowledge we attain to generate ideas and to develop new technologies, products and services, and to improve our cycle time and quality. At Motorola, we recognize that listening to others and learning from them is an integral part of the quality process worldwide.[20]

The Xerox Experience

Other Baldrige winners have also applied vicarious learning principles with great effect. Yet no company has done more to lionize the benefits of benchmarking than Xerox Corporation, a 1989 Baldrige winner. Xerox has long been an energetic believer in the benefits of benchmarking. Benchmarking played a leading role in the revival of Xerox's fortunes in the 1980s. In the late 1970s, Xerox awoke to a real-life market nightmare. The company observed that the Japanese were selling competitive products in the United States at the same prices at which Xerox was manufacturing its own copiers. David Kearns, then CEO of Xerox, has many times recounted the events that followed. Confronted with a formula for extinction (i.e., a competitor selling a quality product at your manufacturing costs) Xerox initially reacted instinctively. Disbelief! Denial! Suspicion that its Japanese competitors were dumping products to unfairly gain market share! Xerox, however, benefited from a relationship that many organizations in similar situations do not enjoy. It had a joint venture in Japan with Fuji-Xerox. Xerox asked its Japanese partners whether it were possible for its Japanese competitors to produce quality products at such low prices. Xerox's partners, who subsequently won the prestigious Deming Prize for quality excellence, confirmed the worst. Not only was it possible, but it was definitely happening. Throughout Japan leading-edge manufacturers were developing production techniques that enabled them to assemble products faster, better, and more cheaply than competitors. This affirmation from Fuji-Xerox came like a giant dose of castor oil. Xerox literally invented the photocopying industry. It thoroughly commanded the industry throughout the 1960s and 1970s and *xerox* crept into the language as a commonly used verb meaning to *photocopy*. For mighty Xerox, it was difficult to accept the news that copycat competitors could actually surpass the market creator.

Xerox is neither the first company nor the last to become blinded by the light of an initial, dazzling market success. "When you're successful," observes the market axiom, "you forget quickly and learn slowly." Xerox's market preeminence discouraged employees from actively reviewing the progress of competitors and learning from their successes. For Xerox, the "if-it-ain't-invented-here-it-can't-be-any-good" syndrome became nearly terminal. During the late 1970s, Kearns recalls, he not only sweated over eroding market share, he feared the company might be driven permanently out of business.

As the pundit long ago observed, "there is nothing like the prospect of death in the morning to focus a man's attention." Such dire situations also capture a chief executive's attention. Certainly the graveness of the company's competitive situation impressed Kearns. He started making study visits to Japan and other countries. With his senior management, Kearns sounded the wake-up call, and slowly Xerox managers began actively probing and learning from the experiences of others.

Xerox was determined to shake off the managerial parochialism that had proven so malignant. Xerox engineers began routinely tearing apart competitors' products and studying them component by component. If they found a design, part, or assembly superior to their own, Xerox teams set out to learn from these superior approaches and to better them. This scrutiny of competitors' products expanded to include the examination of basic business processes. Competitive benchmarking began in 1979 in Xerox Manufacturing Operations, and by 1983 it had become a fundamental business process throughout Xerox.

By the mid-1980s, the formerly entrenched Xerox culture had begun to open. During this period, Xerox discovered repeatedly that others had valuable lessons to impart. In a celebrated collaboration with L.L. Bean, Inc., the Maine outdoor sporting goods retailer and mail order company, Xerox discovered it could even learn valuable lessons outside its own industry. The circumstances at Xerox in 1981, when it began the study of L.L. Bean's warehouse operations, were similar to those at many companies today.

Improving Return On Assets (ROA) was a high-level corporate goal for Xerox, as it is for many companies today. Consequently, any senior manager whose performance evaluation—and perhaps even compensation—are linked to ROA will be keenly interested in distribution-and-logistics functions. They can have a significant impact on the organization's ability to turn inventory, which directly affects sales potential and profits, which both in turn affect the ROA measures.

Understandably, Xerox managers hoped to improve warehousing operations—a lynchpin in its overall distribution process. Xerox understood its warehousing functions were far from world-class, as evaluated by a broad scorecard of performance measures. Indeed, no competitor in its industry was exceptional in this functional area. After studying the issue, a Xerox staff member concluded that warehousing and distribution issues were similar for many companies. Consequently, he believed Xerox might profitably study the practices of companies outside its own industry. Though much smaller, L.L. Bean had developed extraordinary skills at fulfilling customer orders quickly, efficiently, and without many errors or damage to merchandise.

When Xerox study team members set out for L.L. Bean, they might have expected to find highly-automated operations in which advanced technology was the driving force behind L.L. Bean's efficient warehouse operations. What they found was very different. L.L. Bean achieved its superior operating performance through a combination of well-thought-out warehouse layout, efficient work flow design, thoughtful stock placement, effective scheduling, creative use of incentive bonuses, and effective education and training. There was no rocket science and very little high-tech wizardry. The secret lay in thoughtful management of an extremely manual process.[21]

In L.L. Bean's warehouse operations in Freeport, Maine, Xerox fitted itself with many ideas to improve its own operations. The experience further expanded the company's understanding of the power and scope of benchmarking applications. If Xerox, as a large technology-driven manufacturer could learn so much so quickly from a much smaller service company, then from whom else might Xerox learn? Xerox had broken the shackles under which so many companies labor. Those shackles are forged by the common, almost universal, belief that one's own situation is unique. What manager hasn't encountered this nettlesome objection: "We're different from them…. Their approaches just don't apply to us or to our industry…." At times, cross-industry comparisons may not be appropriate. More often than not, however, valuable lessons can be gleaned by studying excellent companies outside your own industry.

In the years that followed its initial benchmarking partnership with L.L. Bean, Xerox expanded its use of best-in-class process benchmarking. By the time it won the Malcolm Baldrige National Quality Award in 1989, Xerox stated it benchmarked more than 230 performance areas and almost always considered excellent companies outside its industry. Shortly afterwards, Xerox spokespeople said the company simply stopped counting the number of functions that they benchmarked. The benchmark projects had become too numerous to track on a scorecard. In short, benchmarking had become a fundamental business process at Xerox. All Xerox employees learn about benchmarking through training and educational materials distributed companywide.

Baldrige and Benchmarking

Benchmarking's rise as a cornerstone quality-improvement concept can be clearly observed in the evolution of the Malcolm Baldrige National Quality Award. The Baldrige criteria are the guidelines by which Baldrige examiners evaluate the quality systems of organizations. Updated and fine-tuned every year, these assessment guidelines embrace new management concepts as they spring into use. The Baldrige criteria represent the national standard for thinking about the concepts, tools, systems, and processes currently driving organizational improvement.

Presently, benchmarking and competitive comparisons are the management concept with the single greatest influence over the 1000 points that can be awarded in a Baldrige assessment. In fact, the full assembly of benchmarking references influenced 550 points in the 1994 criteria. No other business concept, including process management, empowerment, employee involvement, cycle time reduction, strategic quality planning, new product development or innovation, wields such broad-reaching influence in the Baldrige criteria. Even the concept of customer satisfaction, the ultimate goal of the Baldrige system, directly and indirectly influenced only 520 points. Like the rising sun that chases away early morning shadows, benchmarking has steadily extended its reach into all seven Baldrige management areas (Leadership, Information and Analysis, Strategic Quality Planning, Human Resource Development and Management, Management of Process Quality, Quality and Operational Results, and Customer Focus and Satisfaction.) With this growth in scope, the benchmarking concept itself has evolved and matured along with the Baldrige criteria during the past six years.

(Figure 2.3 illustrates the influence of benchmarking on the 1994 Baldrige criteria.)

In 1988, only months after President Ronald Reagan signed the Malcolm Baldrige National Quality Improvement Act into law, the terms benchmark and benchmarking did not appear in the assessment guidelines. The concept's precursor or ancestral twin—in the form of "competitive comparisons"—does appear in the criteria. In this respect, there are seven indirect references to benchmarking in the first Baldrige criteria. In total, 17.5 percent of the criteria's 1000 points were influenced by inquiries about a company's use of competitive comparisons to drive performance improvement.

By 1989, when Xerox received the National Quality Award, seven *direct* references to benchmarks—or comparative operating statistics—along with eight indirect references to competitive comparisons had been seeded into the Baldrige evaluation system. The total influence of benchmarking points crept up to 19 percent. At this time, benchmarking's importance began seeping into the minds of Baldrige examiners—an elite corps of managers, consultants, and academics who were selected to be trained in Baldrige scoring techniques and to evaluate National Quality Award applicants. During site visits to Baldrige Award finalists, examiners began to press companies to determine if they could pass the litmus test of being world-class. This informal test was rooted in the belief that a Baldrige winner must be able to serve as a national role model. National role models, sound logic suggested, ought to be world-class in those areas critical to success in their respective industries. Consequently, examiners increasingly requested Baldrige finalists to demonstrate that they maintained world class operations and quality systems. Site-visit examiners at GTE Telephone operations, a Baldrige finalist in 1989, pressed this line of inquiry.[22] During the visit, examiners explicitly asked the company to demonstrate its world-class standing. GTE responded with evidence of increasing customer satisfaction, improved system speed, enhanced efficiency, and declining error rates—all of which compared favorably to other American telephone companies that GTE actively benchmarked. Examiners pressed for more; they asked the company to provide similar benchmark comparisons with tele-phone companies from throughout the world. In 1989, GTE did not make such comparisons with telephone companies outside the United States. GTE did not win the Baldrige in 1989, but the company's leadership learned an important lesson. GTE began benchmarking itself against telephone companies throughout the world. Demonstrating world-class standing is virtually impossible unless a company customarily compares itself to the best of the best throughout the world.

By 1989, leading companies like GTE increasingly employed benchmarking to improve their quality excellence. Milliken and Co., a 1989 Baldrige winner, trumpeted the fact that CEO Roger Milliken openly embraced the steal shamelessly management ethic; Milliken's chairman and owner endorsed borrowing good ideas that are not patent protected, copyrighted, or otherwise proprietary. Why should Milliken associates reinvent in-house what had already been better perfected elsewhere? To stay at the front of this learning curve, Milliken noted that it benchmarks the products and services of about 400 competitors. These comparisons provided the company with tangible measures for assessing its performance and for identifying improvement areas and marketing opportunities.[23] Vicarious learning was a conscious strategy. Milliken wanted to break open its organizational culture to outside ideas and influences that would supercharge the company's improvement efforts.

550 of 1000 Total Points Are Influenced By Benchmarking

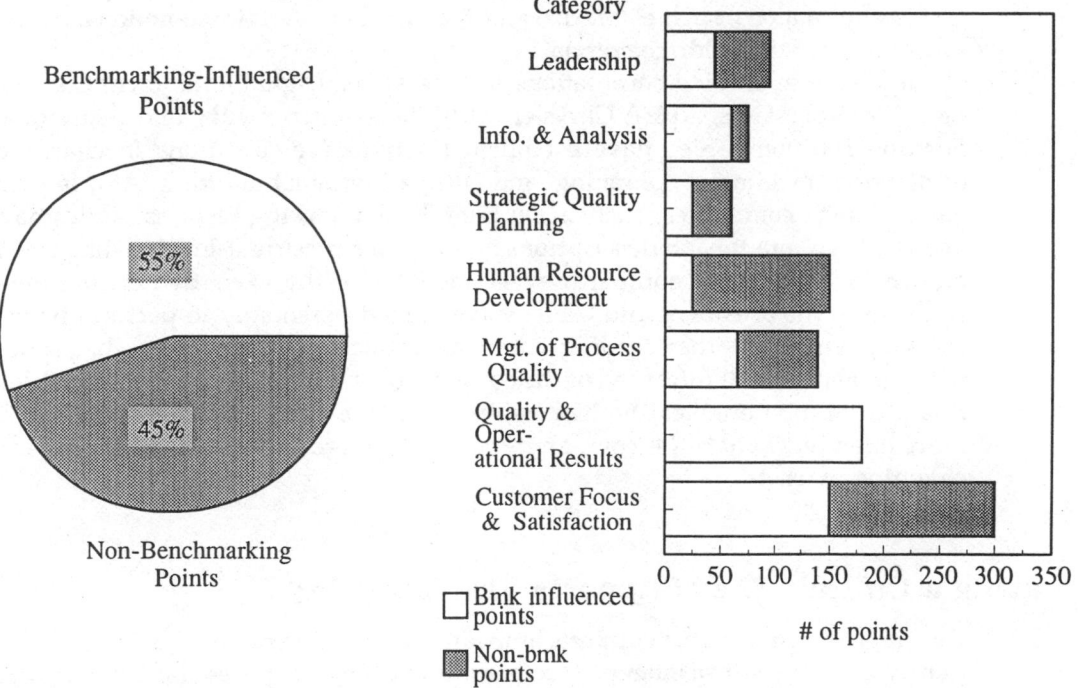

Figure 2.3. Benchmarking and the 1994 Baldrige criteria.

Xerox's 1989 Baldrige victory especially brought benchmarking into the national spotlight. When Xerox began to share its winning strategies with the world, benchmarking was seen to have played a prominent role in the company's revival. Xerox's success story put goose bumps on the hides of the most hard-boiled managers, many of whom worried about slowing productivity or declining market share in their own companies. Not surprisingly, national interest in benchmarking as a management tool gained tremendous momentum in the wake of Xerox's coronation as a Baldrige winner. The updated Baldrige criteria evidence this rise in interest and enthusiasm in benchmarking.

In 1991, the Baldrige administrators added to the assessment guidelines an area devoted specifically to "Competitive Comparisons and Benchmarks." New benchmarking references also proliferated throughout the criteria. By 1994, Baldrige presented 32 inquiries or probes concerning an organization's use of benchmarks, benchmarking, and competitive comparisons. The total influence of these direct and indirect benchmarking references had grown to 55 percent of total eligible points! Not surprisingly, managerial interest in benchmarking has soared as its influence in Baldrige and the national marketplace has expanded. *(See Appendix for a detailed analysis of benchmarking in the Baldrige criteria.)*

The National Quality Award's success has created a "Trojan-horse effect" for benchmarking. In the past six years, nearly one million copies of the Baldrige criteria have been distributed to organizations throughout the world. Hundreds of thousands of American managers and workers have reviewed these assessment guidelines and applied them to their own companies. Consequently, the Baldrige criteria have carried benchmarking, like seeds in the wind, to thousands of companies and to a host of nations, such as Mexico and Brazil, where individual national quality awards mirror the Baldrige criteria.

In response, leading corporations such as Xerox, Digital Equipment Corporation, Motorola, GTE, AT&T, Chrysler, AMP, Texas Instruments, and many other Fortune 500 companies have established senior-level positions in charge of organizing, facilitating, planning, and cultivating benchmarking. At a few advanced-stage companies, such as Du Pont, benchmarking responsibilities have been written into the job descriptions of most vice president-level managers. At one regional trucking company, every member of the executive committee—right up to the president and CEO—have agreed personally to perform benchmarking visits to other leading companies. Such actions ensure that senior management is well informed of the corporation's relative competitive position; this information enables the leadership team to make better decisions and to more effectively chart the company's course through the fast waters of rapidly changing markets.

Creating a Culture That Supports Fast Learning

Developing a culture that supports innovative adaptation or creative imitation is a thorny challenge for managers. Meeting this challenge requires comfort with borrowing, adapting, and importing others' ideas. Yet the barriers to borrowing are subtle and deep-rooted. From childhood, Westerners are discouraged from reviewing others' work and learning from it. "Turn your test over when done! ... No cheating! ... Don't look at anyone else's paper! ... All eyes forward!" Children hear these familiar admonitions in many American and European schools.

The former vice president for quality of a Baldrige winner recognized the potency of this taboo against borrowing others' ideas. He iconoclastically boasts to audiences: "I learned the power of benchmarking early. When I was in third grade, I rapidly improved my performance on spelling tests by benchmarking the work of our class's best speller, a girl with pigtails who—thanks to good fortune—sat next to me." This quip always gets a laugh. But when the laughter subsides, employees and managers at all levels acknowledge they are often hesitant to borrow and apply others' ideas—even when encouraged to do so. Openly borrowing other people's ideas runs against the grain of years of training and education.

The steal shamelessly ethic directly opposes a deep-rooted culture that celebrates invention over creative application of existing ideas. Ironically, as companies grow larger and more successful, the unwritten prohibition against borrowing ideas seems to grow proportionally stronger. This fact forever frustrates successful entrepreneurs. They watch their companies' market successes sedate the steal shamelessly instincts that enabled their companies to move with lightning speed into their original market niches. This instinct is often what allowed them to succeed in the first place against much larger competitors.

Consider this real-life benchmarking pitfall: A vice president-level manager had just completed benchmarking customer-satisfaction measurement systems. By nearly all measures, the project was an extraordinary success. The company previously had no customer-satisfaction measurement process. By benchmarking other corporations' excellent systems, the company accomplished in a fraction of the time and at a fraction of the cost what the others over several years had spent hundreds of thousands of dollars developing and fine-tuning. Yet when it came time to present the soon-to-be-piloted system to the Executive Committee, the manager began with an apology. He nervously explained to his superiors on the executive committee that the team borrowed and tailored a customer satisfaction measurement system from the best ones they had reviewed. He apologized because he was embarrassed that the team members hadn't personally invented the system or its features.

What this very able manager failed to promote and celebrate was the significant time savings, cost savings, and system design elegance that their benchmarking process achieved. Indeed, the teams' efforts immediately saved the organization tens of millions of dollars in on-going revenue. During the system's pilot phase, a major corporate customer blind-sided the previously unsuspecting managers with a scathing response to the satisfaction survey. The customer annually purchased about $15 million in services from the company and it signaled its intention to defect if problems were not remedied. Within a few days of survey receipt from the disgruntled customer, the company had launched a recovery plan that convinced the customer to give the firm another chance. Similar stories, illustrating the barriers to borrowing good ideas, are commonplace across many industries.

For innovative adaptation to become an organizational competency, managerial systems must support and encourage it. Performance evaluations, compensation systems, reward and recognition systems, for instance, will all nurture frequent use of this skill. Currently, though, most companies have not achieved this state of managerial evolution. Consider the results of a 1992 American Productivity and Quality Center survey of 68 leading companies active in benchmarking. Only 2 percent of respondents said that benchmarking was incorporated into most job descriptions, while fully 75 percent observed that benchmarking was incorporated into almost none, few, or some job descriptions. If the maxim "what gets measured is what gets managed" holds true, then many of America's best companies do not manage benchmarking well. The APQC survey revealed that at 84 percent of respondents' companies, benchmarking was incorporated into almost none or few of the organizations' performance appraisals.[24]

Organizations that successfully overcome these cultural barriers to borrowing do so through focus, tenacity, and a systematic approach to shaping and managing corporate culture. Culture is not created through any single act or management system. Aristotle is said to have observed that "Quality is not an act; it is a habit." Correspondingly, corporate culture might be observed to be the result of many things, including managerial actions, policies, statements, stories, and systems. It is the systems that integrate. They cause words, actions, policies, and stories to be frequently repeated. Those actions and stories repeated most often encode the values that guide behavior among employees. These encoded messages in turn create certain types of organizational knowledge and habitual behaviors that are passed on to succeeding generations. No one managerial system produces a we-can-learn-from-everyone culture. It is the result of many smaller systems working together. In such organizations, creative adaptation and innovative imitation are encoded in employee behavior and passed fervently among the generations. Consider a few of the more effective managerial approaches observed in fast-learning companies around the world:

- Invest in employee education that brings outside ideas into your organization.

- Create lending libraries that focus on competitors and other high performers' winning strategies and systems.

- Parade outside speakers and ideas before your employees.

- Review the products and practices of other companies and visit their facilities whenever possible.

- Publicize the benefits of borrowing from the best.

- Sponsor regular meetings and discussions where employees and managers exchange ideas and explore best practices.

- Lay out work areas to encourage impromptu meetings, idea sharing, and information exchange.

- Make best practice information sharing and innovative adaptation a skill evaluated in the performance review and promotion process.

- Make benchmarking competitors and other excellent companies a responsibility clearly described in job descriptions.

- Employ best practice information in the problem-solving and continuous-improvement process.

- Engage high-level executives directly in benchmarking and innovative adaptation.

- Expand reward and recognition systems to include innovative adaptation, creative borrowing, and other types of applied learning through benchmarking for best practices.

- Make competitive information gathering everyone's responsibility, especially including functions such as sales, marketing, and personnel.

- Make someone responsible for being a best practices champion.

- Regularly identify, study, and celebrate internal success stories and best practices with the goal of repeating them.

- Institutionalize learning by making evaluation-and-improvement cycles a required part of performance reviews; focus these reviews on learning from what went right, as well as what went wrong.

When the organizational culture supports borrowing from the best as a worthy operating approach, remarkable things occur. Lots of improvements—small and large, incremental and breakthrough—are quickly imported into the organization. Best practices benchmarking is then a catalyst for fast learning. Organizations that systematically seek out and study best practices experience the beneficial effects of intellectual leverage. They enjoy the compounding effect of good ideas and creativity that spring from a reservoir extending far beyond the boundaries of any one organization.

For example, Chaparral Steel, a company well appreciated for its fast-learning culture, promotes "not *reinvented* here" as a corporate imperative. Yet the company is highly regarded for its innovations. As Harvard Business School Professor Dorothy Leonard-Barton observes about Chaparral:

> Knowledge garnered through such (learning) networks can flourish only in an environment that rejects the "not invented here" mentality. At Chaparral, "not reinvented here" is the operative slogan. There is no value in recreating something—only in building on the best existing knowledge. People in a learning laboratory value the capability to absorb and use knowledge as much as to create it. They understand that all invention is a process of synthesis. As its practices suggest, a key Chaparral value is global outreach—openness to innovation, whatever its origin. Knowledge is valued not so much for the pedigree of its source but for its usefulness.[25]

The Power of Intellectual Leverage

Chaparral and other organizations have created a culture that actively employs intellectual leverage. In these companies, the leverage of ideas, learning, and knowledge enables the organization to achieve lower costs, faster speeds, and higher rates of innovation and improvement. Consequently, management must develop the values and culture that support active learning through borrowing. *(Figure 2.4 depicts the relationship between fast learning and intellectual leverage.)*

"Give me a lever long enough ... and single-handed I can move the world," observed Archimedes.[26] He understood the power of leverage. So have savvy business managers who have applied this concept in many different ways. For better or for worse, the 1980s might be dubbed the "Decade of Financial Leverage." Throughout this period, corporate managers and financiers loaded up on debt as a deliberate financial strategy. As long as they could service the debt from existing cashflow, this strategy drove up the rates of return on their equity investments. This enabled fortunes to be made almost overnight as growth-minded operators often used relatively little equity capital to expand existing franchises and to acquire new ones. The leveraged buyout boom of the 1980s appreciated the power of financial leverage. Financiers used a wide array of debt financing instruments to extend the acquisition power of many corporations.

The 1990s has marked the birth of the generation of intellectual leverage and intellectual capital. Organizations everywhere face the unsentimental ultimatum of global competition: *Learn* to do more with less; *learn* to work smarter, faster, better,

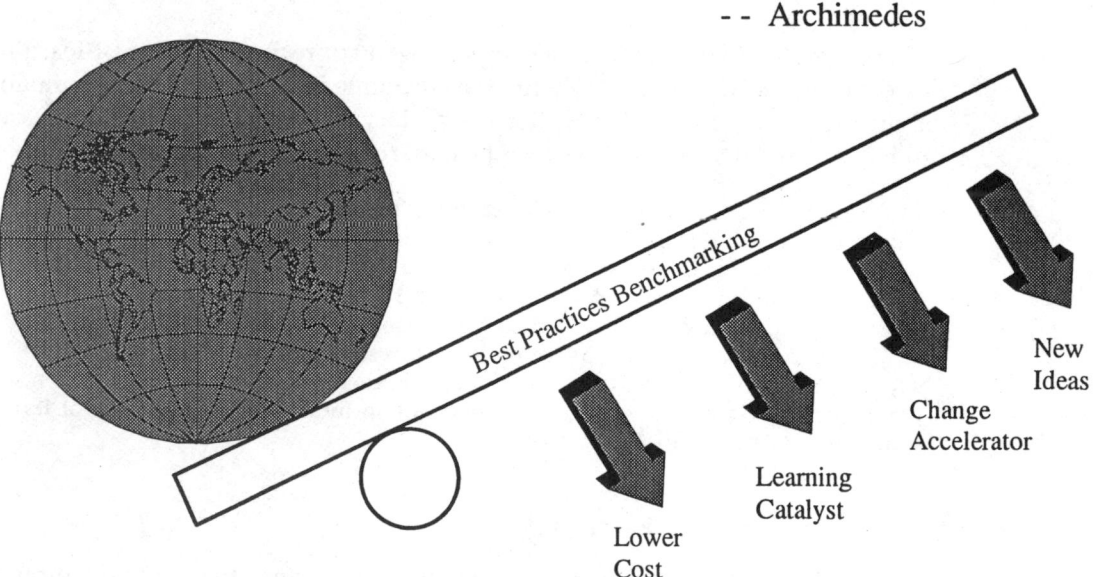

Figure 2.4. Intellectual leverage.

and cheaper, or give up sales and market share to foreign competitors that enjoy cheaper labor supplies, government supports, or other competitive advantages. However, if one gives up too much market share, the game may suddenly end. Corporate downsizings, radical reengineering of work processes, and organizational redesign commonly occur in the American workplace in this era of brutal global competition. Consequently, organizations expect to accomplish more with less, which helps them achieve their ambitious goals for productivity gains. Intellectual capital is enabling this revolution and current productivity boom. Not surprisingly, the 1990s have frequently been proclaimed the era of the "knowledge worker." Learning is the lifeblood of this information age employee. Learning also enables organizations to leverage their intellectual assets. In the end, fast learning may be the ultimate patent protection. It is a renewable source of competitive advantage.

Recent academic discussions have focused on the topic of the learning organization. They portray it as a radically new idea, even though many savvy managers and entrepreneurs already know about the essential concept. Robert Dedman, a philanthropist and entrepreneur who founded the $1-billion-revenue Club Corp. International in Dallas, once observed that the secret of his success rested on two ideas: "OPM and OPB." When pressed for further explanation concerning these two ideas that placed him prominently among the Forbes 400, Dedman replied with complete modesty: "Other people's money and other people's brains." As the developer of country and city clubs throughout the world, Dedman long ago recognized the beneficial effects of financial leverage. He also is quick to say that his success could never have been as great if he had not surrounded himself with smart people and leveraged their wisdom, skills, ideas, creativity, and learning. Successful entrepreneurs and managers have frequently built their companies employing intellectual leverage. A few, such as Dedman and Manco, Inc.'s Kahl, actually develop organizational strategies based on the belief that intellectual capital and learning skills are their organization's greatest assets.

Few companies consider the management of learning as carefully and directly as they address issues such as financial management. Yet the topic of organizational learning, as a vital skill to be cultivated, is gaining currency and reach. The Baldrige criteria, for instance, do not address learning as a specific process or topic. Instead, through many queries about benchmarks, competitive comparisons, and evaluation-and-improvement cycles, Baldrige explores the analytic capabilities, the leadership actions, and the managerial systems that promote organizational learning. The Baldrige criteria's frequent inquiries about "evaluation and improvement" cycles are initially puzzling for many organizations. Many managers have "blown gaskets" trying to figure out what Baldrige means when its criteria repeatedly ask "how the company evaluates and improves" its processes and systems. Evaluation-and-improvement cycles are a litmus test of continuous improvement. Their existence provides evidence that local managers have institutionalized the learning process. Indeed, how can you continuously improve without learning? By Baldrige logic, managers should periodically review *every* organizational process and consider the improvement lessons that past failures and successes can impart. Genuine improvement is arguably synonymous with continuous learning, for improvements without learning cannot be long held.

The old wisdom—"If it ain't broke, then don't fix it!"—is sadly out of date. This view leads to the stifling of creativity. It produces bureaucratic tendencies that are as undesirable in the organization as hardening of the arteries are unhealthy in its employees. Roger von Oech, the noted authority on creativity and innovation, observes how old formulas of success can choke creativity. "There is a saying that Frederick the Great (1712-1786) lost the Battle of Jena (1806), meaning that for twenty years after his death, the army perpetuated his successful organization instead of adapting to meet the changes in the art of war," notes von Oech. "Many rule outlive the purpose for which they were intended. What rule, policy, or way of thinking has been successful for you in the past but may be limiting you now? What sacred cow can you slay?"[27]

The new wisdom—"If it isn't perfect, then improve it!"—views the world from a distinctly different angle. From this view, continuous improvement is the real world embodiment of organizational learning. Fast learning, therefore, is the ultimate competitive weapon. It drives performance improvement.

Consider three incarnations of this modern view:

If you know your enemy and know yourself, you need not fear the result of a thousand battles.
                                SUN TZU
                            The Art of War

Fool you are ... to say you learn by your experience ... I prefer to profit by others' mistakes, and avoid the price of my own.
                        PRINCE OTTO VON BISMARCK

Keep on the lookout for novel and interesting ideas that others have used successfully. Your idea has to be original only in its adaptation to the problem you're currently working on.
                            THOMAS EDISON

None of these statements was originally directed at the topic of benchmarking. Yet each reflects an important dimension of benchmarking and all rest on the foundation concept of vicarious learning. These observations reach across 14 centuries. Nevertheless, each addresses an essential aspect of learning, improvement, and performance management.

Sun Tzu is a great Chinese general and philosopher. His book of meditations, called *The Art of War*, became a best seller on Wall Street and in other business circles in the 1980s. Though his book was written around 500 B.C., his observations on military tactics still seem relevant, especially when extended to the 1990s battlefield of global competition. Organizations that understand their competitors compete wisely. They do not take them on—head to head—in areas where they know the competitor is superior. Instead, they compete on grounds more favorable to themselves. Sun Tzu's ancient lessons are still contemporary among many Japanese companies. Consider the findings of Ernst & Young's Center for Information Technology and Strategy (CITAS). CITAS performed a 1992 benchmarking study that examined the practices of Japanese corporations that were highly successful in information management. Among other things, this study sheds light on Japanese skills in competitive analysis and benchmarking.

"Japanese companies have a 'preoccupation with the actions of their competitors' and literally cannot get enough information on them," the Ernst & Young research team reported.[28] "For example, a recent report by Ernst & Young found that Japanese companies' strategic planning process places more of an emphasis on competitor comparisons than do firms in the United States, Germany, or Canada. The same holds true for information about foreign nations, and global industries." In a top-line finding of this study, researchers Laurence Prusak and James Matarazzo observed: "Japanese management reads. This simple fact is significant. We observed many senior executives actually reading in their corporate information centers and libraries—something one rarely sees elsewhere. At one large library," note Prusak and Matarazzo, "we saw an executive with ten years' worth of annual reports from an American competing firm. He was reading the report's annual letter from the CEO because he 'wanted to get an unfiltered sense of how their CEO thinks.'"[29] No doubt Sun Tzu would have approved of such competitive benchmarking, even if it is conducted from the leather armchair of a corporate library. And even if the Japanese do not call it benchmarking, they are masters of competitive assessment and of the art of innovative adaptation.

The Nineteenth-Century German Statesman, Prince Otto von Bismarck, understood another important aspect of benchmarking. Benchmarking enables the organization to learn quickly by absorbing the experience and progress of others. In this regard, lessons derived from others' mistakes can save critical resources, precious time, and scarce capital. Trial-and-error is not the only school in which to learn. Sadly, this is the primary academy in which many managers enroll.

Thomas Alva Edison, one of America's greatest inventors and the founder of General Electric in 1878, observed that the headwaters of creativity are swarming with clever adaptations. The creative application of proven ideas to solve new problems and to create new opportunities is to be valued equally with pure invention. Creative adaptation is invention with a purpose. Interestingly, more than a century after Edison observed the wisdom of innovative adaptation, the company he founded has come full circle. Under the stewardship of CEO Jack Welch, GE has recommitted itself to the power of learning quickly through creative adaptation. GE has embraced best practices as a central approach to its Work Out program, a key part of the corporate development initiative helping to drive change and improvements at GE.

Though more than two millennia separate Sun Tzu, Bismarck, and Edison, their thoughts share a common lineage: observing competitors and adapting to best them, observing others' experience and adapting to better your own, observing other successful ideas and adapting them to benefit your own situation. All are precursors of the structured business practices embraced through contemporary benchmarking.

In this instance, do not confuse the journey with the destination. Benchmarking is a means to an end. It is a tool or catalyst to facilitate learning. Organizations that learn to learn have an advantage. They harness a renewable resource as powerful as fusion. Benchmarking enables pragmatic organizations to learn quickly and effectively. Sometimes this process of investigation and creative adaptation is informal and sometimes it is highly structured. In all cases, the goal is to improve performance. Benchmarking for best practices is a foundation tool to achieve intellectual leverage. It is the long-handled lever that Archimedes envisioned.

References

1. Walton, Sam, "Sam Walton: Made in America," Doubleday, June 12, 1992, excerpted in *Time*, June 15, 1992, pp. 53-59.

2. Walton, Sam with John Huey, "Sam Walton: Made in America—My Story," Doubleday, 1992, p. 189.

3. Ibid.

4. Tichy, Noel M. and Sherman, Stratford, "Control Your Destiny or Someone Else Will," Doubleday, New York, N.Y. 1993, p. 206.

5. Senge, Peter M., "The Fifth Discipline, The Art & Practice of the Learning Organization," Currency and Doubleday, New York, N.Y., 1991, p. 4.

6. Reese, Jennifer, "America's Most Admired Corporations," *Fortune*, February 8, 1993, p. 53.

7. "International Quality Study^{SM}, The Definitive Study of the Best International Quality Management Practices," a Joint Project of Ernst & Young and the American Quality Foundation, Cleveland, Ohio, 1991, pp. 26-29.

8. Hamel, Gary and Prahalad, C.K., "Strategy as Stretch and Leverage," *Harvard Business Review*, March-April 1993, p. 81.

9. "Detroit's New Strategy to Beat Back Japanese Is to Copy Their Ideas," *The Wall Street Journal*, Thursday, October 1, 1992, pp. A-1, A-12.

10. Kahl, Jack, "Mom, Dad and Mr. Sam," *Cleveland Enterprise*, October/November 1992.

11. Ibid.

12. Ibid.

13. Ibid.

14. Ibid.

15. Verespej, Michael A., "Where Ducks and Fun Mean Success," *Industry Week*, March 18, 1991.

16. Chew, W. Bruce, Leonard-Barton, Dorothy, Bohn, Roger E., "Beating Murphy's Law," *Sloan Management Review*, Volume 32, Number 3, Spring 1991.

17. Ibid.

18. Kahl, Jack, "Mom, Dad and Mr. Sam," *Cleveland Enterprise*, October/November 1992.

19. Bogan, Christopher E. and Hart, Christopher W.L., "The Baldrige: What It Is, How It's Won, How to Use It to Improve Quality in Your Company," McGraw-Hill, New York, N.Y., pp. 225-251.

20. "The New Truths of Quality," a public interest pamphlet published by Motorola, Inc., 1992, p. 13.

21. Camp, Robert C., "Benchmarking: The Search For Industry Best Practices That Lead to Superior Performance," Quality Press, American Society for Quality Control, Milwaukee, Wis., 1989, pp. 285-286.

22. As GTE's Director of Quality, Michael English served as the coordinator of the Baldrige site visit and was asked, along with other executives, to respond to this line of questioning.

23. The 1989 Malcolm Baldrige National Quality Award Winner profiles, published by the National Institute of Standards and Technology, which manages the Baldrige Award for the Department of Commerce.

24. Lambertus, Todd, "Surveying Industry's Benchmarking Practices," a study of the APQC's International Benchmarking Clearing House, Houston, Texas, 1992, p. 23.

25. "The Factory as a Learning Laboratory," Dorothy Leonard-Barton, *Sloan Management Review*, Fall 1992, pp. 23–38.

26. Senge, Peter M., "The Fifth Discipline," Doubleday/Currency, New York, 1990, p. 3.

27. von Oech, Roger, "The Creative Wack Pack," Box 7354 Menlo Park, CA 94026, 1989, Card #40.

28. Prusak, Laurence and Matarazzo, James, "Information Management and Japanese Success," Ernst & Young Center for Information Technology & Strategy, Special Libraries Association, 1700 Eighteenth St. NW, Washington, DC, 1992, p. 1.

29. Prusak and Matarazzo, ibid., p. 4.

3

Benchmarks and Performance Measurement

Revolutions begin long before they are officially declared. For several years, senior executives in a broad range of industries have been rethinking how to measure the performance of their businesses.... At the heart of this revolution lies a radical decision: to shift from treating financial figures as the foundation for performance measurement to treating them as one among a broader set of measures. PROFESSOR ROBERT ECCLES, *Harvard Business School*

In the world of modern management, a dual revolution is underway: the manifesto is to manage processes and to mind performance measurements. In the past, a company only had to "make its numbers" and making the numbers, of course, meant one thing—achieving financial targets. How the organization accomplished its financial goals was unimportant. The bottom line was the primary focus of management attention.

Now managers understand that a myopic focus on short-term financial results—without consideration for the overall health of the operating systems producing those results—can bankrupt the organization over the long term. Consequently, companies are learning to manage systems and processes that reach across traditional departmental or functional boundaries. In this brave new world of process management, many organizations employ broad-reaching performance indicators, including financial and nonfinancial measures, to guide them in managing their businesses. This new generation of process managers creates a balanced scorecard of operating metrics that enable them to carefully monitor, maintain, and improve the health of the systems and work flows. *Benchmarking* represents a versatile process management tool that helps organizations identify and understand what constitutes best operating practices; *benchmarks* are the operating statistics or measures that define the achievement level of any given practice or system. Together these related concepts lie at the heart of the revolution in performance management.

The Whys and Hows of Benchmarks

As in any revolution, confusion often reigns among those engaged in the front-lines of change. Many managers puzzle over the difference between benchmarks and benchmarking. The confusion is revealed, for instance, when a sales management team undertakes a benchmarking project and then articulates its project goal as identifying sale termination rates. The simple metric or benchmark describing terminated sales may provide a useful reference point against which to compare their unit's sales performance, but the metric alone provides no insight into the root causes of performance differences. Without such knowledge of operating practice differences, this sales management team will find it difficult to spearhead performance improvement. Unwittingly, this team continues to worship at the altar of the old bottom-line management philosophy that studies financial results without regard for what differences in organizational sales processes underlie the differences in sale termination rates. Identifying benchmarks takes a team only part-way to its ultimate goal of enhancing performance. In turn, an improvement team will also lose its bearings if it navigates by just studying practice or process differences. Without accompanying benchmarks that contrast the relative performance levels produced by different operating approaches, the team cannot easily evaluate the merits of different practices and systems.

Successful best practice improvement strategies marry the study of metrics and processes; they unite benchmarks and benchmarking. Consequently, a successful benchmarking team will very early determine the benchmarks—the operating statistics or metrics—by which to evaluate two or more systems or operating approaches. One hallmark of a well-designed set of project benchmarks is that they enable measurement and comparisons across systems. Trouble quickly ensues for benchmarking teams that choose esoteric or one-dimensional operating statistics that do not translate well across different organizations and systems. A flexible set of benchmarks reflects full process or system capabilities. Performance indicators may include dimensions such as cost, productivity, cycle time, yields, error rates, waste, and turnover.

Consider the hospital quality team that decided to improve pharmacy operations. They debated which of various quality metrics to employ—error rates, the time it takes to fill a prescription, unit costs, information collection compliance rates, and inventory overages and underages. No single benchmark measure fully reflected the operation's total performance. If the team chose just one metric and then focused efforts only on improving information collection compliance rates or unit costs, the operations might not progress in other critical areas, such as errors or prescription fill cycle time. The team needed to collect and study a group of benchmark measures to help them understand and improve the pharmacy's full operations.

[Table 3.1 graphically illustrates how performance benchmarks may be organized into several different levels of comparison. Each level represents a different type of comparison. The comparisons progress from a first-reference point or baseline (level 1) to best-in-company (level 2), to industry leader, to best-in-country (level 6), to best-in-world (level 7).]

Some important lessons emerge from the experiences of active benchmarkers:

1. *Do not strive to benchmark everything at best-in-country or best-in-world levels.* No company can be best in every function. "If you chase too many rabbits," observes a Japanese adage, "you will catch none." Companies that set out to be best in every function—without regard for the function's strategic importance—dilute their resources and their focus. Like Sisyphus, the mythical Greek figure condemned for eternity to push a large boulder up a mountain only to have it roll back down before he ever nears the summit, companies that develop best-in-class benchmarks for every operation end their efforts in frustration.

2. *Seek best-in-class benchmarks for core processes and functions of the highest strategic importance.* Benchmarks for less strategically important areas may come from com-

Table 3.1. Range of Benchmarks

Focus		Benchmark levels		Type	Improvement benefit
Strategic • Product/services • Business processes • Business function	Best-in-world	7		Generic processes	30%
	Best-in-country	6		Functional areas and processes	30-40%

Performance
• Customer satisfaction
• Output
  • Products
  • Services
	Industry	Leader
		Norm
		Standard
			5
			4
			3
				Direct competitor	15%-20%

Process • Practices • Capability	Best-in-company	2		Internal	15%
Inputs • Material • Supplier	Baseline	1			

parison levels 2 through 5. World and country leadership benchmarks require greater time, resources, and effort to develop. Apply them to strategic or core processes of businesses that compete daily in national or global markets.

3. *Seek internal, regional, or industry benchmarks for secondary and support processes.* For some processes and business activities that are not critical to the organization's strategic advantage, internal, regional, or competitive benchmarks may be most appropriate. Such benchmarks produce incremental improvements that are sub-stantial—even if not radical or "breakthrough" in terms of the size of the expected improvement benefits.

Imprimus Incorporated, an employment service, provides a good example of how a company new to benchmarking sets its benchmark sights. Imprimus includes three core businesses: Wordtemps, which provides personal computer and word processing temporaries; Freeman and Associates, which places admin-istrative support personnel, and Prime Timers, which places temporary and permanent workers aged 40 and over. Imprimus President Valerie Freeman says the corporation has begun by establishing a baseline for all its units. Imprimus gathered its first benchmarks by surveying its customers and analyzing five major competitors. Customers said the single most important feature of Im-primus' business is reliability: When clients call for temporary workers, the workers must show up on time and have the skills to perform the required job. Imprimus therefore has begun gathering internal reliability benchmarks. Next, Freeman says, the company will determine industry leadership benchmarks for reliability.

Imprimus and other organizations employ *benchmarks* as measurements to gauge their functional, operational, or overall business performance relative to others. Positive or negative gaps between benchmark measurements signal poten-tially important operating differences. Yet as mere operating statistics, the bench-marks do not actually explain why those gaps exist. Benchmarks are vital perform-ance management tools to help bring improvement opportunities into focus by identifying relative strengths and weaknesses. The benchmarks still require you to *interpret* their meanings. *Benchmarking* is the systematic process by which an organi-zation searches to identify best practices and to implement those practices to lead your company to superior performance.

Consider the following example of a health insurance provider that undertakes a benchmarking study of billing processes and practices. First, it establishes a measurement or benchmark—*billing inquiries per 100 customers*—by which to evalu-ate billing performance. In this instance, *billing inquiries* are those instances when a customer calls to clarify or dispute a bill; it constitutes a variable numerator and "*100 customers*" is the constant denominator.

The organization next gathers comparative performance statistics from other companies. If necessary, the health insurance provider recalculates the statistics to achieve a common denominator that allows the statistics to be compared across all participating companies. The following comparison matrix illustrates what such a gathering of benchmarks might look like:

Benchmark (inquiries/100 customers)	Company
16.0	1 - Local phone company A
13.5	2 - Long distance telephone company
12.0	3 - Local telephone company B
10.0	4 - Health insurance provider
8.5	5 - Local telephone company C
7.0	6 - Credit card company
6.5	7 - Electric utility company

Next, the company must interpret the benchmarks in order to choose a select few companies for site visits. In this instance, the insurance company eliminates the electric utility company. Even though it has the best performance in this measurement, investigation shows that the utility company bills only for kilowatts used at a flat rate; there is no counterpart to the multiple billing variables used in the health insurers' business. Therefore the utility company's billing practices are not helpful. The other potential best practice companies are comparable in that both bill at least five services and their ratio of billing inquiries per 100 customers is among the lowest. Indeed, they are 15 to 30 percent better than the insurer. Consequently, the insurer chooses the credit card company and local telephone company C for further study and site visits.

The Performance Measurement Revolution

The revolution in performance measurement is spreading rapidly. This revolution is creating a new paradigm for how organizations measure and manage performance. Historically, four important shifts in perspective have helped to elevate the importance of benchmarks in this performance measurement revolution.

First, organizations now more than ever recognize the importance of performance measurements or benchmarks in managing complex systems and processes. Across a multitude of industries managers have observed the same general truth: What gets measured *is* what gets managed and improved. The rising tide of interest in total quality management and the Malcolm Baldrige National Quality Award has highlighted the importance of performance indicators in achieving quality excellence. The Baldrige performance assessment criteria repeatedly quiz companies about the benchmarks, metrics, or performance indicators used to monitor, control, manage, and improve the organizations' processes. As Harvard Business School Professor Robert Eccles observes, "Quality measures represent the most positive step taken to date in broadening the basis of business performance measurement."

Second, customer satisfaction has emerged as a strategic goal for many organizations worldwide. As Eccles observes, "What quality was for the 1980s, customer satisfaction will be for the 1990s."[1] Many organizations are developing a broad menu of measurements to gauge the satisfaction levels among customers. Some leading indicators of satisfaction include customer retention rates, referral rates, repurchase rates, market share trends, complaint rates, satisfaction survey trends, and litigation rates.

Third, leading-edge managers recognize that many other nonfinancial benchmarks are useful in achieving total quality excellence within complex systems and processes. Eccles and his Harvard colleague, Nitin Nohria, argue that financial measures usually "represent outcomes of processes, although they do not always provide the best information about what actually occurs behind the scenes of these processes—or how these processes are related to one another in the big picture."[2] Typical among the growing scorecard of other nonfinancial benchmarks are measures of work process speed, quality, first-pass yields, employee turnover, reliability, productivity, innovation, training, employee involvement, and learning.

Finally, the revolution in information technology places powerful computer hardware and software within the reach of virtually every organization. This technology enables organizations to inexpensively create, distribute, analyze, and store more data about their businesses than ever before. All these data represent potential benchmarks and performance management indicators. Any measurement that is systematically collected can be trended over time. By trending information, managers can transform mute data points into performance indicators that reflect the health and progress of individual processes or systems.

The revolution in performance measurement is spreading rapidly. This revolution is creating a new paradigm for performance measurement. As Eccles suggests, financial measures usually "represent outcomes of processes, although they do not always provide the best information about what actually occurs behind the scenes of these processes—or how these processes are related to one another in the big picture."[3] *(Figures 3.1 and 3.2 illustrate the evolution in performance measurement by*

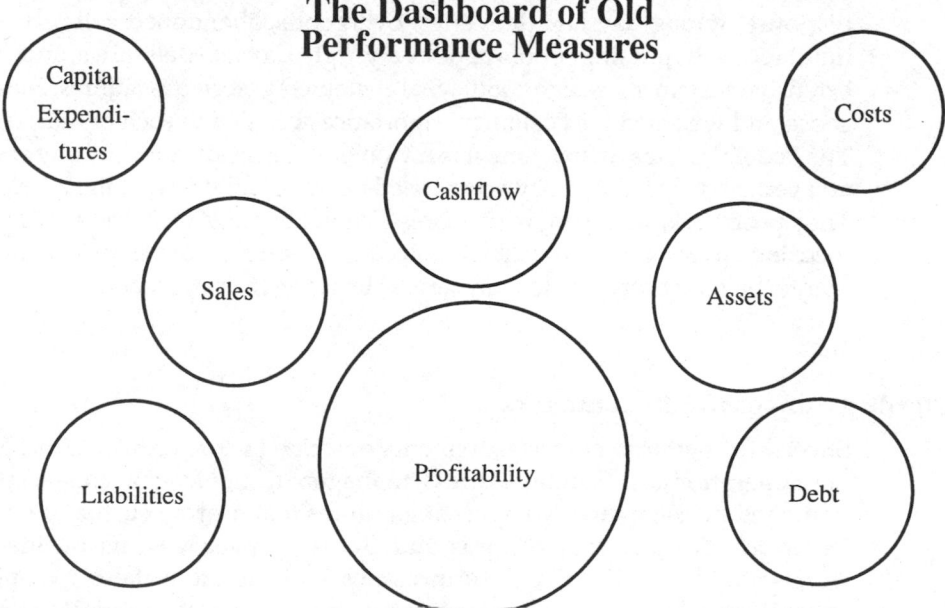

Figure 3.1. The dashboard of old performance measures.

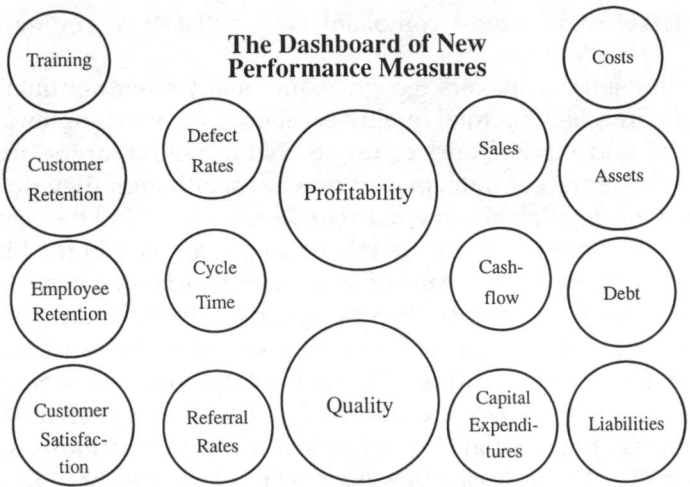

Figure 3.2. The dashboard of new performance measures.

showing both old and new "dashboards" of performance measures by which managers navigate their organizations.)

Federal Express has developed a corporate-wide performance measurement system called the Service Quality Indicator (SQI). FedEx's SQI illustrates the use of nonfinancial, customer-service-oriented indicators to empower an organization's performance improvement efforts. The SQI consists of 12 key service and operations measurements or benchmarks: the number of damaged packages, lost packages, missed pickups, aircraft delays, reopened complaints that were not resolved on first response, wrong day late deliveries, overgoods, abandoned calls, invoice adjustment requests, missing proofs of delivery, right day late deliveries, and traces. After extensive customer research about what customers perceive as failures, Federal Express chose and weighted the customer importance accorded to each of these 12 measures. The Federal Express management team looks at the measures every day and computes an average daily failure point score based on actual failures and their weighted value. They post weekly companywide scores. Federal Express' well deserved reputation for focusing on customer satisfaction is reflected in the nonfinancial performance measurements they use every day to manage and improve their operations.

Designing Successful Benchmarks

Successful performance measurements describe factors critical to successful business operations. At Federal Express, managers recognize that missed pickups and damaged packages are two operating failures that destroy customer satisfaction. If FedEx routinely misses pickups and damages packages, its business will fail. Consequently, FedEx diligently measures and monitors these two performance characteristics. Effective performance benchmarks therefore reflect the most important operating dimensions of a business process, system, or function.

When designing benchmarks, four measurement elements prove especially important; these elements are the measurement focus, the measurement perspective, the degree of measurement control, and the ability to collect data.

Measurement Focus.   Some general principles help sharpen the focus of performance measurements by guiding the organization to develop benchmarks where they are most useful for management purposes. Three general principles are: (1) determine where value for the customer is created in a work area or process; (2) determine where value is detracted through high costs, errors, rework, or accidents; and (3) target benchmarks in areas where performance diverges from designated standards or where variation above and below standards is great.

Measurement Perspective.   Benchmark measures can be broadly classified as reactive and proactive. Performance measures that are proactive or preventive in nature can be called *leading indicators* because they foreshadow or anticipate future system outcomes. Rising employee turnover and rising error rates often precede declining customer satisfaction. Consequently, these two benchmarks can be classified as leading indicators of future system performance.

In contrast, other performance measures are more reactive or descriptive of the actual results of a system or process in a given time period. These measures can also be called *lagging indicators* because they provide perspective on the completed performance of a system. Traditional financial measures such as sales, profits, return on assets, or cost of sales are descriptive of a system's final performance in a given quarter or year. Senior managers, shareholders, and government agencies rely greatly on lagging indicators because these indicators assess the impact of real outcomes versus target outcomes.

Traditional organizations employ lagging indicators, such as quarterly financial results, as the primary tools to manage system performance. High-performing contemporary companies *also* embrace the use of leading indicators to help them better manage performance improvement. Leading indicators are consistent with the preventive focus endorsed by total quality management. They provide valuable information to help an organization intervene upstream, in the early stages of a project or process. Consequently, many nonfinancial performance measures, such as customer retention, cycle time, innovation, and quality, are proving to be more useful than traditional financial indicators in predicting and managing future performance. Comprehensive performance benchmarks include both leading and lagging indicators.

Measurement Control.   Performance benchmarks can be applied at company, unit, and individual levels, but people are always the principal factor affecting the degree of measurement control. Generally, performance reviews evaluate managers and employees based on the performance measurements that they personally control or that they affect as part of a team with a shared objective. Managers fail at performance improvement when they evaluate individual or system performance using benchmark measures that are uncontrollable by the people overseeing the process. Therefore benchmarks that are designed for performance improvement must be crafted to reflect the individual level of authority, responsibility, and skills of those people expected to work with the benchmarks.

Data Collection.   After defining performance measures, managers must be able to readily collect the data from which performance benchmarks are constructed. Many organizations develop interesting performance measures only to discover that they currently do not collect the required information. They are stymied when they discover that the collection costs for some performance benchmarks can exceed the managerial benefit of the information. Consequently, the best performance benchmarks can be collected without excessive investment of time, systems, staff, or capital.

A Benchmark Design Architecture

This first step in designing a performance benchmark system is to create measures that will enable management to achieve the organization's strategic objectives. When creating benchmarks to evaluate your organization's performance, plumb those management categories that are critical to your organization's success. Frequently these categories include customer service and customer satisfaction, product and services distribution, innovation and product development capabilities, quality, people development capabilities, and various aspects of time-based competition or the speed with which core processes are performed.

The second step in designing a benchmark architecture requires managers to create an agreed upon vocabulary describing performance measurement in your organization. Without a common language to communicate about performance improvement, employees will find it impossible to agree on the organization's current status and impossible to build strategies to drive improvements. "Every company has its own language system by which performance is defined," observes Harvard's Eccles and Nohria, "and every manager uses rhetoric to build its meaning and coherence. How *effective* the rhetoric and how *useful* the language system—these are the issues that separate companies like Motorola from the rest."[4] Motorola invented the concept of Six Sigma Quality or the goal of creating 99.99966-percent defect-free products. Every Motorola employee, work team, function and business unit is galvanized by the Six Sigma concept; consequently, across all Motorola units there is clear, consistent and highly-focused communication about improvement efforts, goals, strategies, and progress. Performance measurement vocabularies vary from company to company and from industry to industry. Moreover, corporate values, vision, strategy, and culture greatly influence the individual language of performance measurement. Nevertheless, trying to navigate without any common vocabulary of performance measurement is like trying to govern a nation balkanized by many different languages and cultures.

The third step is to develop plans to collect, process, and analyze the performance measures. Begin by evaluating how your organization currently generates needed performance data. Methods are probably in place to generate a full menu of financial measures. By building on this information management foundation, you

Table 3.2. Designing Your Benchmark Architecture

A benchmark design architecture Ten generic benchmark categories
Customer-service peformance
Product/service performance
Core business process performance
Support processes and services performance
Employee performance
Supplier performance
Technology performance
New product/service development and innovation performance
Cost performance
Financial performance

can expand your system to include a full panoply of nonfinancial performance benchmarks. These additional measures will describe capabilities and performance levels among your organization's core processes—human resources, technology, and physical and intellectual assets. Many managers now place these nonfinancial measurements on equal footing with the financial data. A well-designed performance benchmark architecture provides managers with information that is accurate, easy to obtain, easy to read, statistically valid, and quickly accessible.

A comprehensive benchmark architecture includes many measurement categories, each of which probably includes time and quality dimensions such as error or defect rates and process cycle times.

Some organizations may identify other performance categories for which they wish to develop benchmarks. However, these 10 basic categories can serve as a design framework to help your organization chart the performance measurements that are most meaningful for it. Through discussion, experimentation, and refinement, your organization can and should design its own performance benchmark architecture.

Customer-Service Performance Measures

The best customer-related measures come from objective and valid data collected directly from customers. Consequently, many customer-related benchmarks spring from statistical samples that probe a representative customer group randomly selected from among all customers. Customer-related benchmarks must meet or exceed commonly accepted standards governing statistical factors such as population definition, sample selection rules, sample size, data collection quality, interviewer bias controls, and nonresponse error. Customer-service performance measures include—but are not limited to—satisfaction and dissatisfaction metrics, and customer retention and defection benchmarks.

When boiled down to elemental operating issues, many customer-satisfaction measurements represent statistical probabilities reflecting customers' attitudes and likely behavior. For instance, retention and defection metrics reflect the likelihood that customers will continue to act as they have in the past in terms of repurchasing an organization's services or products. Customer-satisfaction benchmarks frequently employ measurements requiring customers to describe their feelings on a scale which weighs the positive or negative dimensions of their responses. Five-factor scales with responses such as "excellent," "good," "average," "below average," and "poor" usually work well. This type of scale provides survey respondents enough choices to accurately record their opinions and it also differentiates among the range of reactions that is typical of many customer groups.

Customer-service performance measures typically probe organizational performance in the following areas:

- Overall customer satisfaction with products and services.

- Customer evaluations of sales and service representatives.

- Customer assessments of your organization's understanding of customer needs.

- Customer ratings of how clearly your organization communicates cost information and how well the organization suggests customer solutions.

- Customer appraisals of delivery timeliness.

- Customer impressions about the usefulness of your organization's product and service documentation.

- Customer feelings concerning how easy it is to conduct business with your organization.

- The value customers place on your organization's products and services.

Customer-dissatisfaction measures can be as revealing as satisfaction measures. Dissatisfaction benchmarks can be formal or informal; formal dissatisfaction indicators include variables such as written customer complaints, defective-product returns, cancellations, or terminations of service and sales contracts, and payouts sought by customers through guarantees, warranties, refunds, or litigation. Informal dissatisfaction measures include indicators such as telephone calls from customers complaining about product and service features, timeliness or value, disputing billing accuracy, or reporting service and product failures and problems.

Customer retention rates, repurchase rates, and defection rates are critical benchmarks that serve as leading indicators of future customer behavior that will directly influence bottom-line financial results. Customer feelings about services and products are often complex; at times they are even contradictory. Regardless of what customers *say* about a product or service, their *actions* usually speak most directly about their real feelings and intentions. Consequently, retention, repurchase, and defection rates—which reflect customer actions—often represent the most accurate forecasts concerning customers' future market behavior. Accordingly, many companies are expanding their satisfaction measurements to include retention, repurchase, and defection benchmarks.

Product/Service Performance Measures

Product and service performance benchmarks offer a rich palette of operating statistics, including measures of accuracy, reliability, timeliness, order ease, delivery, packaging, ease of assembly and use, documentation, billing, after-sales service, and effective complaint management. These benchmarks may additionally include warranty exchanges and returns, unit productivity and cost, cycle time for key intervals, and market share. Performance measures, not surprisingly, often vary greatly among industries. For example, automobile manufacturers measure time from order to delivery, fuel efficiency, failure times, and rates for key parts, customer complaints, warranty costs, and ease of use for operator manuals. In contrast, overnight package delivery services measures include on-time, error-free deliveries, invoice errors, missed pickups and missing proof of deliveries, deliveries per hour, package damage rates, and cost per delivery.

Differences in benchmark measures across industries are often surprising. These differences underscore the need for each organization to develop a family of benchmark measures that best reflect its processes and operating characteristics. Among many industries, certain types of measures serve as common denominators, such as speed and quality, but the specific applications of these measurement types reflect the organization's specific services, products, and processes.

Business Process Performance Measures

A simple process analysis model can help identify your organization's most important workflows. This model reveals that all work can be viewed in four sequential stages:

1. inputs (including those from both employees and suppliers)
2. processes (including internal operations and support services)
3. outputs (your organization's products, services, and documentation), and
4. customer satisfaction (customers)

(*Figure 3.3 illustrates this input-output process model.*) The model begins with a series of *inputs* that are delivered to the work process. These inputs include *tangibles*, such as supplies, raw materials, and component products, and *intangibles*, such as information. The inputs then enter the work *process*, which transforms them into some final output, which might be a product or service. The goal of the *output* is to create *satisfied customers*. Business process performance measures therefore can be viewed in the context of the process analysis model. Each stage of the input-output model includes performance measures and these measures often include one or all of the following factors:

- Enhanced customer value, often observed through added product features or reduced costs.
- Production costs, frequently described as cost per 100, 1000, or million output units.

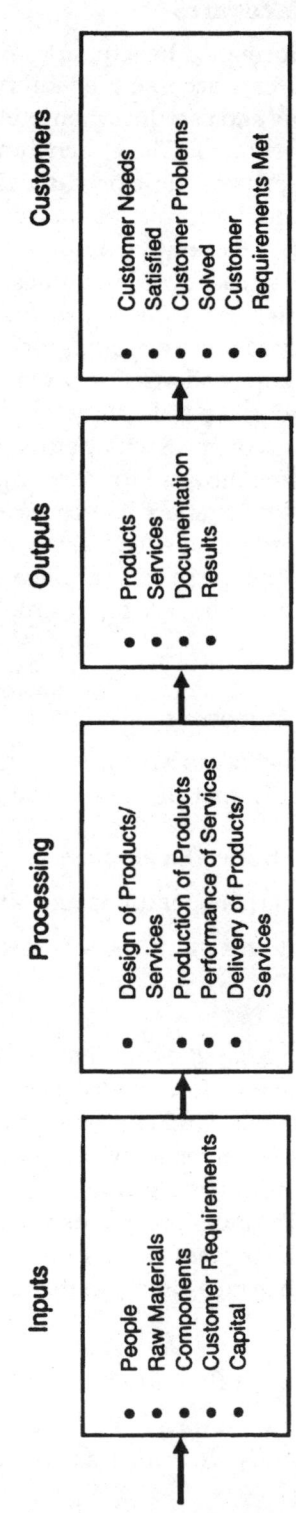

Figure 3.3. Input-output process model.

- Responsiveness and/or process cycle time.

- Defect, error, waste, problem, or failure rates, often formatted as defects per 1000 or million output units.

- Productivity and resource utilization, often reflected in transactions per person, inventory turn rates, or projects operating within budget.

- Public safety and/or legal responsibilities, sometimes observed in accident rates, employee absentee rates, regulatory citations, or litigation rates.

The input-output model, for example, helps clarify what process benchmarks GTE maintains in the dispatch phase of its telephone service repair process. The input in this process is a customer call reporting a service outage. The dispatch clerk receives and records the call, dispatches a technician to correct the outage, and orchestrates the repair process while it occurs. The desired output is quickly repaired telephone service, which results in a satisfied customer. To evaluate how well its repair process is functioning, GTE studies four particular performance measurement areas: customer value perceptions, response times, resource utilization, and safety. To evaluate its current performance and to identify improvement opportunities in the repair process, GTE studies these measurement factors over time and across many units.

The input-output model underscores the difference among critical work processes across industries. Table 3.3 presents key business processes identified by three major international companies: IBM, Xerox, and British Telecom. The performance measures describing the capabilities of these processes vary as greatly as the companies' processes.

Support Processes/Services Performance Measures

Support services are activities and operations that enable your organization's core production and delivery processes. They include functions, such as finance, software services, marketing, public relations, information services, purchasing, legal services, and facilities management.

The table on page 56 identifies some typical providers of support products and services, and also illustrates various performance measures:

Accounting

- Percentage of late payments

- Time to respond to customer requests

- Number of billing errors

- Number of incorrect accounting entries

- Number of payroll errors

Information Services

- Number of errors per line of code

- Percent of reports received on schedule

Table 3.3. Key Business Processes of Three Leading Companies

IBM	Xerox	British Telecom
Marketing information capture	Customer engagement	Direct business
Marketing selection	Inventory management and logistics	Plan business
Requirements		Develop processes
Hardware development	Product design/engineering	Manage process operation
Software development	Product maintenance	Provide personnel support
Development of services	Technology maintenance	Market products and services
Production	Production and operations management	Provide customer service
Customer fulfillment	Marketing management	Manage products and services
Customer relationship	Supplier management	Provide consultancy services
Service customer feedback	Information management	Plan the network
Marketing	Business management	Operate the network
Solution integration	Human resources management	Provide support services
Financial analysis	Leased and capital asset management	Manage information resource
Plan integration		Manage finance
Accounting	Legal	Provide technical R&D
Human resources	Financial management	
IT infrastructure		

- Number of rewrites
- Number of errors found after system accepted by customer
- Number of test-case runs for successful completion

Marketing
- Accuracy of forecast assumptions
- Number of incorrect order entries
- Overstocked field supplies
- Contact errors

Product Engineering
- Project completion cycle times
- Number of engineering changes per document
- Number of errors found during design review
- Number of errors found in design evaluation

Purchasing

- Purchase order errors
- Downtime due to shortages
- Excess inventory
- Cycle time (from start of purchase to receipt in-house)

Quality Control

- Percentage of lots rejected in error
- Number of engineering changes detected after design review
- Errors in reports
- Cycle time for corrective action[5]

Employee Performance Measures

Employee performance benchmarks cover a wide range of employee activities that may include:

- *Employee development* such as percentage of employees in mentor programs or percentage of employees with career development plans.
- *Employee education* such as training hours per employee or percentage of employees who have completed course work in statistical process control.
- *Employee empowerment* such as percentage of customer-contact employees authorized to issue up to $1000 in credit to customers.
- *Employee recognition* such as percentage of employees recognized with 5-, 10-, 15-, 20-, and 25-year service awards.
- *Employee recruitment* such as percentage of employment offers accepted.
- *Employee absenteeism* such as unexcused absences per 100 employees during a given time period.
- *Employee turnover* such as employees terminating employment as a percentage of employment base or voluntary turnover as a percentage of average annualized work force.
- *Employee grievances* such as disputes or complaints about company labor practices per 100 employees.
- *Employee safety/accidents* such as number of workdays lost due to accidents, vehicular accidents per one thousand miles driven, or public liability reports per quarter.
- *Employee involvement* such as number of employees on teams and number of teams.
- *Employee morale* such as employee opinion survey results.
- *Employee performance appraisal* such as percentage of employees receiving an annual performance appraisal or percentage of appraisals completed on time.

- *Employee promotion* such as percentage of positions filled from within the organization.

- *Employee succession planning* such as the increase or decrease in percentage of eligible positions filled through succession planning.

Supplier Performance Measures

Supplier performance measures help an organization qualify or certify the vendors with which it will work. These benchmarks then help the organization monitor and manage on-going supplier performance. Supplier performance metrics often include measures of cost, quality, reliability, speed or responsiveness, agreed-upon service levels, and product specifications. A representative list of supplier certification performance measures might include:

Vendor	Certification indicator	Requirements
Overnight letter and package delivery service	delivery on time	less than 1% late deliveries
	tracing capability	24-hour access to 800 number for status checking
	monthly discount for volume purchases	10,000 or more, 10% discount; 15,000 or more, 12% discount; 20,000 or more, 15% discount
Fiberoptic cable supplier	delivery on time	less than 5% late deliveries
	number of defective or damaged 100-foot spans	less than 0.05 per 10,000
	after-sales service and support	24-hour 800 number access to technical assistance and support
Market research service	adequate data collection	statistical samples at 95% confidence ± error; controls for interviewer bias and nonresponse errors
	valid and reliable analysis	at least two statistical tests
	accurate reports	less than 2% error rate per report
	delivery on time	less than 2% late deliveries

Technology and Innovation-Related Performance Measures

Technology-related measures reflect the productivity, deployment, and effective use of computers and other technology in an organization. Measures range broadly from processing speeds, deployment percentages, network down time and error rates. In turn, innovation-related performance indicators reflect issues such as organizational learning and continuous improvement. Measures may include new product development times, employees' suggestion rates, new product sales as a percent of total sales, and process improvement rates.

Cost Performance Measures

Cost performance measures are broad and flexible. They include balance sheet liability requirements and information drawn from cost centers throughout the organization. Companies can develop useful benchmarks by producing cost ratios for specific products, services, organizational units, processing steps, inputs, and labor. A mortgage company, for instance, might use such measures as cost per loan application, cost per loan processing, human resources cost per loan, data processing costs per 100 bills, and servicing cost per loan.

Financial Performance Measures

Financial measures include performance indicators required by stock exchanges, security analysts, public accounting firms, regulatory agencies, and other organizations that may oversee reporting standards in your organization's industry. Many of these measures make up the items on income statements, balance sheets, and cash flow statements, including measures such as revenue, gross profit, operating income, net income, earnings per share, long-term debt, book value, cash flow, debt/equity ratio, days/receivables ratio, current ratio, and so on.

Benchmarks As Best Practices Geiger Counters

Benchmarks are statistical Geiger counters for performance improvement teams. Benchmarks lead organizations toward improvement opportunities in their processes and systems. However, isolating the *drivers* of superior performance requires a holistic set of measures that describes the entire business operation. A successful benchmark system facilitates critical analysis; it allows managers to evaluate the overall health of a process or system and it also reflects the interrelationships among different parts of the process or system. As training hours go up, for instance, error rates go down. The training-related benchmark is a useful tool for managing process error rates, which strongly influence overall system capability. Moreover, the benchmark trends help managers to more accurately predict future performance of the entire business operation or system.

Inexperienced process managers often focus on improving performance by taking action and tracking the corresponding impact on one benchmarked measure. More experienced process managers, however, construct a broader system of benchmark measures; they manage for continuous improvement by monitoring the effects of improvement actions on the full set of performance measures. Two approaches are especially noteworthy for constructing performance measurement systems that integrate many benchmarks and enable effective system analysis. One approach has been employed effectively at Eastman Kodak; the other is best described by Robert S. Kaplan of the Harvard Business School and David P. Norton of Nolan, Norton and Company.

Eastman Kodak's Measures Matrix Chart

Eastman Kodak is a company with broad interests in consumer and commercial imaging systems, photographic products, chemicals, and diagnostic health equipment. Kodak has developed a comprehensive view of benchmarking and performance measurement. Kodak has refined a performance measurement graphical display tool that enables the company to integrate many related measures into a single viewpoint; in this way the company has created a holistic approach to understanding its own operations and comparing them to others. Kodak calls its approach the Measures Matrix Chart or M^2 ("M-Squared") chart. Other organizations, such as business units in IBM and Chevron, use similar tools but call them spider charts or radar charts. All these charts are circular in shape and consolidate various performance measures by arraying different benchmarks along the radii or spokes of a circle graph. Kodak embraced this approach when the company determined that a single-measure approach often led managers to focus on reaching single number outcomes—sometimes without regard for improving the system or for which best practices would consistently produce superior performance. "Traditional gap analysis allowed for single measure tracking over time but did not anticipate how changing that measure would affect other measures of your operations," says James M. Madigan of Kodak's Corporate Benchmarking Office.[6] *(Figure 3.4 presents an example of the M^2 chart.)* Madigan has refined the M^2 chart so that five features make for easy interpretation:

1. Each radial line or spoke represents a measure.
2. Concentric circles range from 1.0 (at the center) to 0.0 (at the outermost circle.)
3. Data is normalized, so all values fall within the range of 1 to 0.
4. The benchmark for each measure is placed on the center of the arc on its respective line. Consequently, the better an operation's performance, the closer its performance measure moves toward the center of the circular M^2 chart.
5. Greatest opportunities for improvement are the points that fall farthest from the center.

The M^2 chart employs three value types:

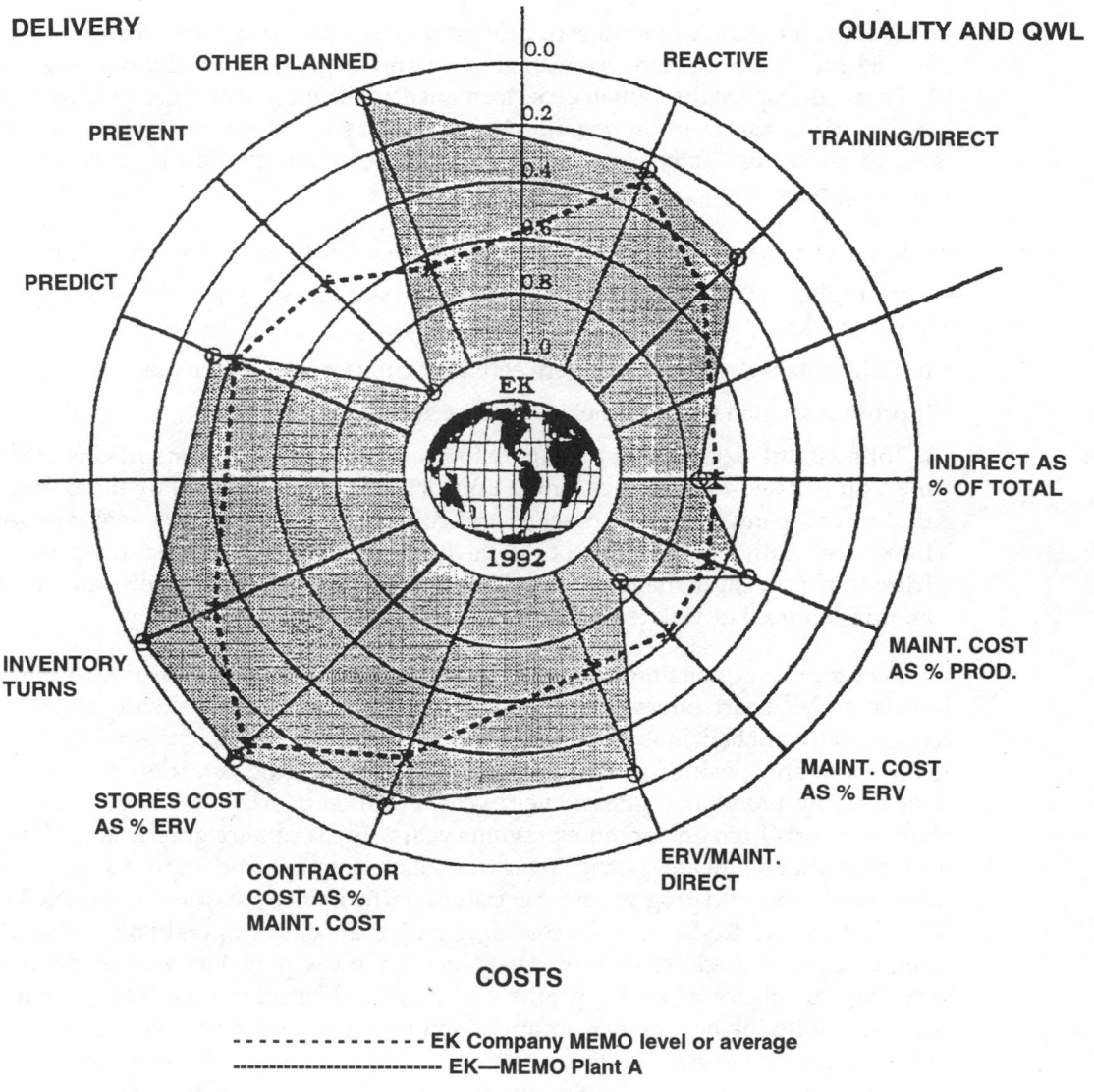

Figure 3.4. Kodak class MEMO benchmarking M² chart.

- Largest value benchmarks where the highest measure is optimal, such as productivity.
- Smallest value benchmarks where the lowest measure is optimal, such as cost.
- Target value benchmarks, where values greater than or less than that of the target value are equally undesirable, such as training.

When constructing performance measures, Madigan observes that it is important to identify what value or data type is appropriate for each measure. Frequently an M^2 chart for a single operation or process will include many various benchmarks and include all three value types—high, low, and target. This flexibility makes the M^2 chart an especially valuable performance management tool; it portrays the benchmark measures of an operation or process as an integrated system—not as isolated pieces or dimensions of the whole. Kodak favors the M^2 chart for the following reasons:

- It illustrates the entire operation or performance system with a single graphic.

- It spotlights performance gaps between one's own measures and the best practice benchmarks.

- It highlights the relationships between the various performance measures.

- It provides a management tool to track performance over time.

- It "blinds" the data so that managers focus on best practice comparisons rather than on isolated measures or single-unit performance produced by a particular unit or company. Such best practice and competitive comparisons evaluate performance relative to the market or to the highest performers among many industries; they prevent managers from viewing their unit's operating performance as an unreferenced or isolated phenomenon.

As a tool for organizing and presenting performance information, the Measures Matrix or M^2 chart offers a single view of Kodak's operations with graphical representation of benchmark ranking, company ranking, and Kodak's relationship to the leadership position in each performance measure. Kodak tracks performance over time by producing a chart every year and then overlaying the current-year chart on charts from one or more previous years. The result is a graphical performance map that delineates change over time, including specific improvements and their correlation with progress in other critical performance measurement areas. The M^2 chart allows Kodak employees to visualize an entire operation's measures integrated into a single, easy-to-read graphic. "People with no knowledge of benchmarking can glance at an M^2 profile and readily determine how that operation compares with the benchmark for any given measure, and how it compares with the average," observes Madigan. "This visualization or profile emphasizes the differences an operation has with respect to the benchmarks and the average … The M^2 chart also preserves the magnitude of the gaps that exist between the measures and their respective benchmarks. This information along with the profile readily shows how the measures interact with each other, providing the user with a very powerful tool describing the operation," says Madigan.

"Experience has shown that the charts are particularly effective in taking presentations with lengthy explanations using many tables and charts and reducing them to a single graphic, saving time and allowing discussions to focus on the entire operation," adds Madigan. "Thus the focus shifts to an operation's strengths and weaknesses, to identifying improvement opportunities, and to track progress toward meeting goals."[7]

Through its refinements of the Measures Matrix or M^2 chart, Kodak has developed an easy-to-use, unintimidating approach to managing performance measurement. The approach integrates benchmark metrics into a system that reflects the complexity of business processes, but still presents the performance data in an easy-to-interpret format that makes them accessible for on-going performance management and improvement purposes.

Kaplan and Norton's Balanced Scorecard

Robert S. Kaplan of the Harvard Business School and David P. Norton of Nolan, Norton and Company have designed another similar approach to managing performance measurements through an integrated system. They conceived of the performance measurement "balanced scorecard," which combines both financial and operational measures into an integrated system of performance indicators. The balanced scorecard approach to performance management assumes no single measure is adequate for managing all companies at all times. *(Figures 3.5 and 3.6 present examples of what Kaplan and Norton's Balanced Scorecard might look like.)*

In a ground-breaking 1992 *Harvard Business Review* article entitled "The Balanced Scorecard—Measures That Drive Performance," Kaplan and Norton described their ideas about creating a comprehensive set of performance indicators that would enable organizations to successfully manage their core capabilities. The balanced scorecard grows out of an organization's central vision of what it must do to be rated number one by customers in terms of total value delivered. In Kaplan and Norton's view, a balanced performance measurement scorecard includes at least four perspectives:[8]

- *The financial perspective* which asks "If we succeed, how will we look to our shareholders?"

- *The customer view* which asks "To achieve our vision, how must we look to our customers?"

- *The internal operating perspective* which asks "To delight our customers, what management processes must we excel at?"

- *The innovation and learning perspective* which asks "To achieve our vision, how must the organization continuously learn, improve, and create value?"

The Financial Perspective

The financial portion of a balance performance measurement scorecard includes three fundamental dimensions: profitability, growth, and shareholder value. Each dimension may include many different types of financial measures. The profitability dimension, for instance, measures both cash flow and real profits versus targeted profits. The growth dimension includes overall sales growth and divisional operating income growth. The shareholder-value dimension employs measures such as market share increases, return on equity, stock value appreciation, price-to-earnings performance, and dividend yield performance.

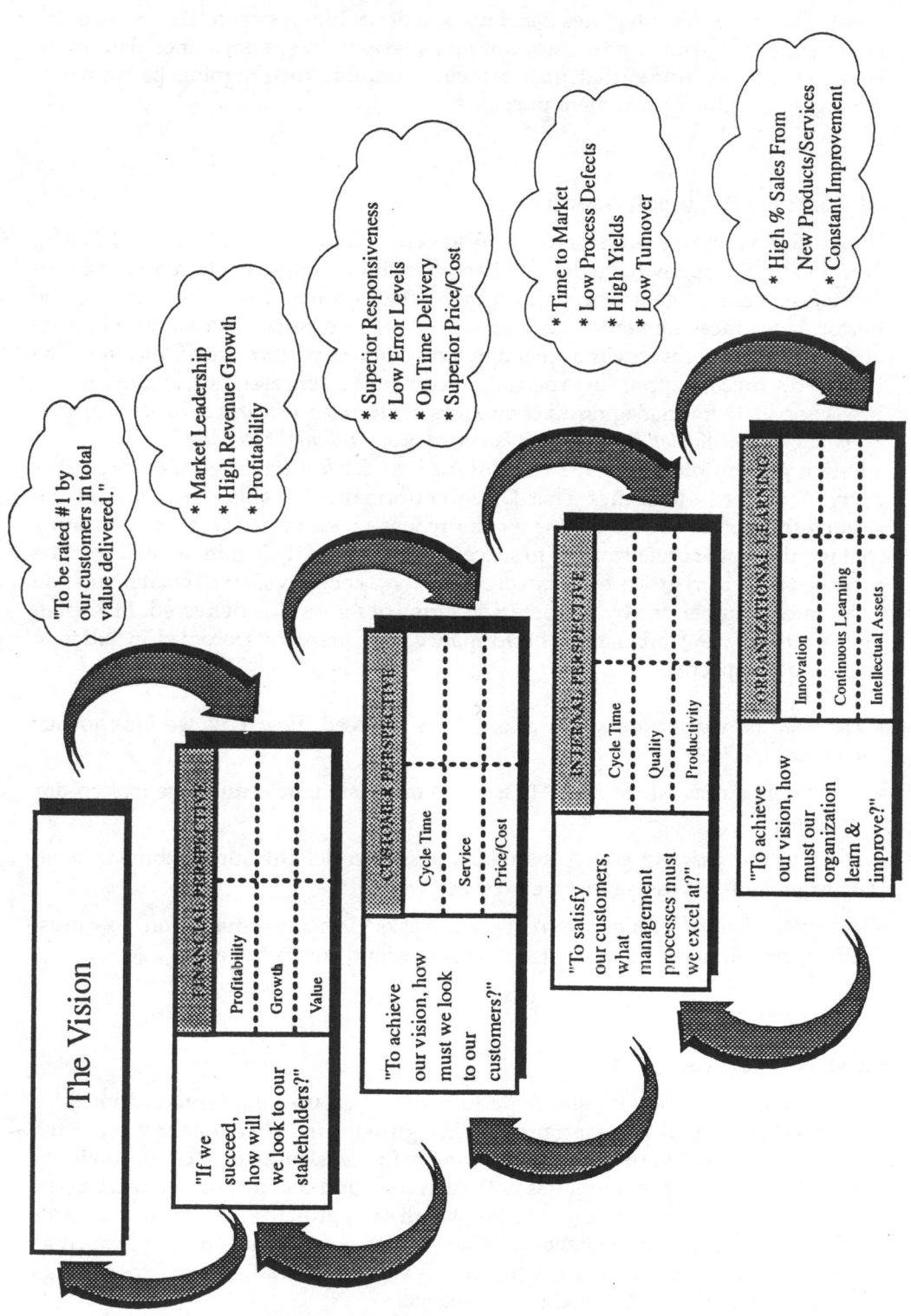

Figure 3.5. Developing a balanced scorecard. (*Source: Prof. Robert S. Kaplan, Harvard Business School and David P. Norton of Nolan, Norton & Co.*)

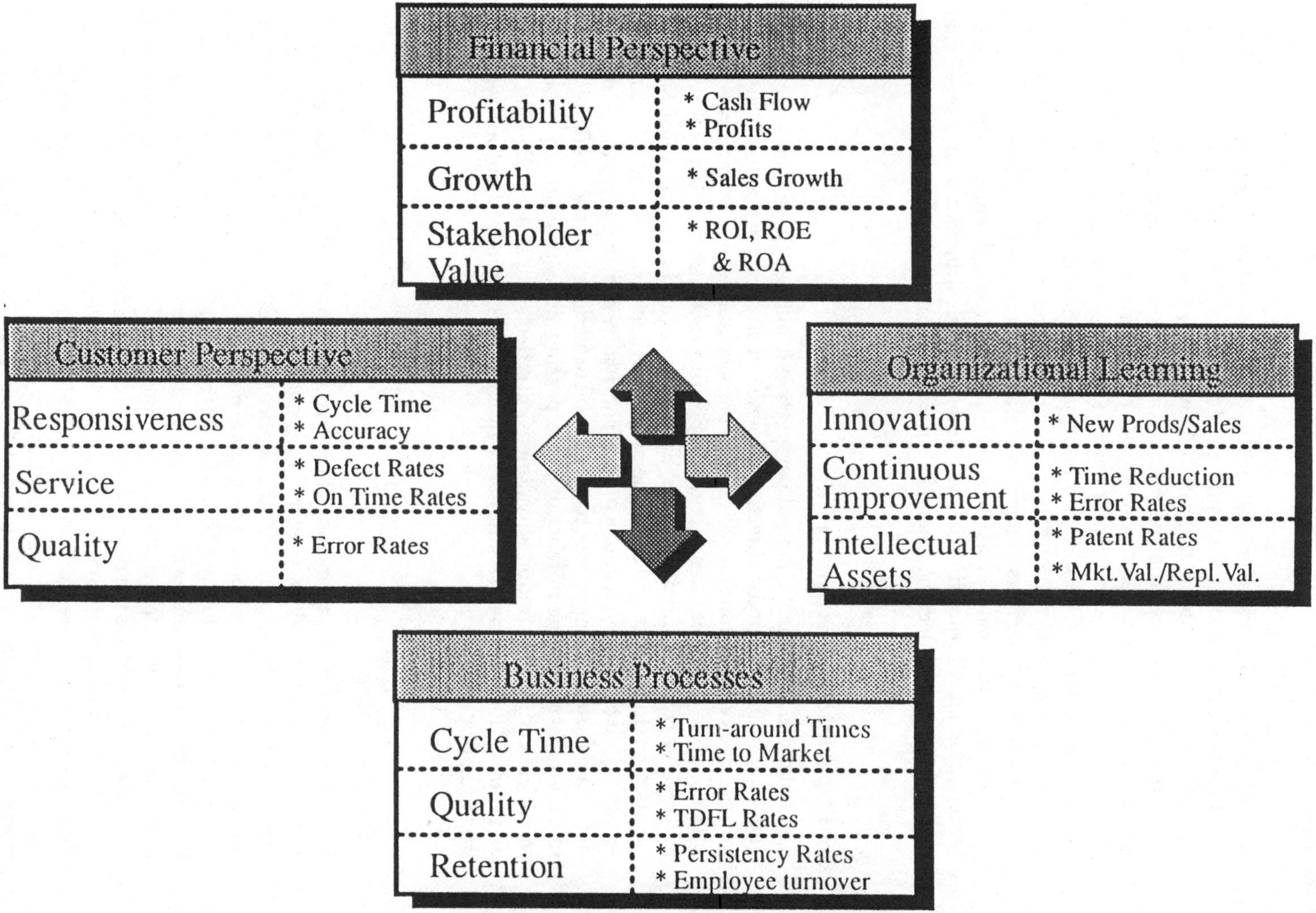

Figure 3.6. Give meaning to the goals with specific measures. *(Source: Prof. Robert S. Kaplan, Harvard Business School and David P. Norton of Nolan, Norton & Co.)*

Financial Perspective

Profitability	* Cash Flow * Profits
Growth	* Sales Growth
Stakeholder Value	* ROI, ROE & ROA

Customer Perspective

Responsiveness	* Cycle Time * Accuracy
Service	* Defect Rates * On Time Rates
Quality	* Error Rates

Organizational Learning

Innovation	* New Prods/Sales
Continuous Improvement	* Time Reduction * Error Rates
Intellectual Assets	* Patent Rates * Mkt. Val./Repl. Val.

Business Processes

Cycle Time	* Turn-around Times * Time to Market
Quality	* Error Rates * TDFL Rates
Retention	* Persistency Rates * Employee turnover

The Customer Perspective

The customer perspective includes four important dimensions: *time, quality, performance and service*, and *cost of ownership*. The time dimension measures cycle time for meeting customer needs. For existing products this dimension expresses the time it takes from receiving an order through delivery. For new products, it measures the time-to-market or time from product definition to first customer shipment. The quality dimension records defects, errors, or problems as perceived by the customer. Defects or errors can be broadly interpreted, ranging from physical product defects, typographical errors, and incorrect information to late deliveries, inaccurate forecasts, or missing information. The performance and service dimension measures how products and services help create value for the customer. The cost-of-ownership includes measures such as invoice costs, repair costs, downtime, and inconvenience. Together, all four dimensions—time, quality, performance and service, and ownership cost—reflect the customer's perception of total value.

The Internal Operations Perspective

The balanced scorecard's internal perspective examines those business processes and operations that most directly influence customer satisfaction. The internal perspective often includes three dimensions: cycle time, quality, and productivity. Internal cycle time measures may track specific process steps, such as time to order and receive materials from suppliers, time to move products and materials between plants, time to produce and assemble products, time to deliver products to customers, and time to process customer orders. The quality dimension may include simple defect measures or it may pick up more sophisticated metrics such as first-pass yield rates, which record the number of items passing through a process without any rework or errors. The productivity dimension reflects employee skills, effectiveness, and motivation, especially as they are evidenced in employees' output per person, per hour, or per day.

Organizational Learning

Innovation and learning include three primary dimensions: market innovation, continuous operational learning and improvement, and intellectual assets. Market innovation records new product and service introduction rates. Product-rich companies, such as 3M, monitor and set goals around the percent of total sales generated by products less than four years old. Other companies track the number of patents they record or the number of major research papers published by employees. In order to meet their goals in these respective areas, these companies manage the processes that produce the new products, services, patents, or research.

Continuous operational learning and improvement measurements record the rates at which individuals and organizations learn. Semiconductor manufacturer Analog Devices, for instance, has developed the concept of the "half-life" curve. It reflects the time it takes to achieve a 50-percent improvement in a specified performance measure, such as error rates, on-time delivery, or new product introduction rates.

Intellectual assets are among the most valuable and most intangible resources of any organization. Companies are beginning to ask how they can better manage and more fully leverage these intangible assets. To this end, managers are developing measures that evaluate skills deployment, training effectiveness, employee involvement levels, employee suggestion rates, cross-functional activity levels, and experience sharing (e.g., frequency with which internal databases are used).

The Best of Both Worlds

Kodak's Measures Matrix or M^2 chart and Kaplan and Norton's balanced scorecard both offer elegant approaches to creating comprehensive performance measurement systems. The balanced scorecard offers a structured framework for thinking about performance measurement and for constructing integrated performance metrics. Managers can easily apply it to an entire organization, a business unit, a function, or a process. The scorecard approach balances financial and nonfinancial perspectives, suggesting four important generic areas for measurement. The balanced scorecard's special elegance lies in its structure and simplicity. However, the balanced scorecard, as envisioned by Kaplan and Norton, does not embody comparative measures. The scorecard elements are single data points that together describe an operating system but do not provide comparative information that establishes the adequacy of individual performance levels. In this respect, Kodak's Measures Matrix much more effectively accommodates benchmark metrics by embodying comparisons in the very structure of the performance reporting system.

Consequently, the best of both systems would combine the balanced scorecard's structure and comprehensive approach to performance measurement with the M^2 chart's use of benchmark comparisons in its normal reporting format. A unified system would combine the scorecard's broad measurement perspective with the M^2 chart's benchmark presentation system. For instance, each scorecard perspective—financial, customer, internal operations, and learning—could be presented in the Measures Matrix format. This combination of comprehensive performance measurements, coupled with comparative benchmark references, yields an effective overview of your company's strengths, improvement areas, and competitive standing.

References

1. Eccles, Robert G., "The Performance Measurement Manifesto," *Harvard Business Review*, January-February, 1991, p. 133.

2. Eccles, Robert G. and Nohria, Nitin, *Beyond the Hype: Rediscovering the Essence of Management*. Boston: Harvard Business School, 1992, p. 156.

3. Ibid., 1992, p. 156.

4. Ibid., p. 154.

5. *Quality Plan Guide*. Dallas: GTE Telephone Operations, 1991, pp. 24-25.

6. Madigan, James M., "Measures Matrix Chart," Eastman Kodak Company, a paper presented to the Strategic Planning Institute, 1992, p. 1.

7. Ibid, pp. 3-4.

8. Kaplan, Robert S., and Norton, David P., "The Balanced Scorecard—Measures That Drive Performance," *Harvard Business Review*, January-February 1992, pp. 71-79.

4

The Secrets of Successful Benchmarking

Lessons Learned

In October 1990, Procter and Gamble conducted a benchmarking project to learn about best practices for product packaging. The P&G team selected eight key measures to study and made four site visits to companies the team had identified as "best-in-class." After nine months, the team realized that "something wasn't working." Their efforts were not yielding operating insights and no significant improvements had been achieved. The team gathered to conduct a post-mortem analysis concerning what was amiss with its project. The team identified the following missteps in its initial benchmarking project work:

- The topic was too broad.

- The project was not clearly focused because the team was studying too many major measures (eight) and their corresponding operations.

- The *best-in-class* companies, chosen because of reputation rather than because of demonstrated performance, provided no true best practices or exemplary operating procedures.

During the second phase of its benchmarking project, the P&G benchmarking team focused only on reliable and dependable operations—that is, operations with the lowest percentage of time to produce a quality product on demand and according to design specifications or standards. The team also narrowed its site visits to six, and two of them were internal—at Procter and Gamble Aircraft Operations. The other site visits took the team outside its industry.

The second-phase results were more promising. The P&G team learned four valuable lessons that apply to all companies conducting benchmarking projects.

First, clearly focus the project. A common pitfall is for teams to set out on projects that are far too large or poorly articulated to allow them to succeed. Teams that set out to benchmark expansive subjects such as empowerment or customer satisfaction might as well try to boil the ocean. Successful benchmarking projects usually start with well-focused project missions that target manageable topic areas. Project teams that cast their nets too broadly bog down; in turn, project teams that focus too narrowly produce trivial results. Table 4.1 provides examples of projects that are focused too broadly, too narrowly, and just right.

The second P&G lesson was to look outside its own industry for important operating lessons. Best business practices don't observe industry, regional, or national borders. In another benchmarking project, for example, P&G studied pit crews at the Indianapolis Speedway and applied their lightning-quick methods of changing parts and tires to reduce some manufacturing changeovers from two days to just two hours.[1] Moreover, most organizations are more inclined to show off what they do well to noncompetitors.

The third important lesson was to focus finally on observable systems, practices, and procedures that proved highly effective in the benchmark organization. Don't obsess over operating statistics and don't become hypnotized by the minutiae of any single subprocess. Superior performance is almost always the cumulative result of many effective actions, procedures, practices, and organizational design factors. Moreover, many best practices require extrapolation from how the benchmark companies perform them and interpretation concerning how they will be translated into a different organization. The road from site visit findings to successful implementation can be steep and winding.

Table 4.1. Focus Your Project on an Appropriate Level of Detail

Process or functional area	Too broad to enable success	Too narrow to enable success	Appropriate level of detail
Customer support services	Best customer satisfaction process	Best phone greeting	Best call center management practices
Human resource management	Best empowerment process	Best refund policy below $20	Best values communication systems
Distribution and logistics	Best distribution process	Best materials receipt stamp	Best warehouse management practices
Training	Best training process	Best classroom configuration	Best needs assessment process
Employee development	Best development process	Best job transfer request form	Best orientation practices
Your function	?	?	?

Benchmarking Critical Success Factors

Successful benchmarking projects bear a Triple-A brand: Adopt, Adapt, and Advance! After searching out and examining highly-effective operating practices, experienced benchmarkers adopt the best, adapt them to their own work environments, and advance performance through careful implementation and continuous refinement of the practices. Several critical success factors enable Triple-A benchmarking processes.

A well-designed performance measurement and benchmark system is essential, of course. Other critical success factors include:

- Senior management support.
- Benchmarking training for the project team.
- Useful information technology systems.
- Cultural practices that encourage learning.
- Resources, especially in the form of time, funding, and useful equipment.

(Figure 4.1, "Benchmarking: Its requisites and benefits," describes factors that are critical to successful benchmarking.)

Senior Management Support

Senior management support is a prerequisite of any benchmarking project that examines a core business process or undertakes major change. Senior management support constitutes more than just cheerleading. Often it requires leadership actions and behavior that signal the importance of the project to the organization. Communication through the leadership's actions is arguably the most effective means to champion a cause. Five types of leadership commitment are especially helpful in championing benchmarking efforts.

1. *Management should visibly promote benchmarking within the organization.* Managers who are effective benchmarking champions both "talk the talk" and "walk the walk." Because of their visibility and positional power, managers influence others and set the stage for employee attitudes. Words and actions are the two networks for broadcasting leadership messages throughout the organization. By consistently advocating the importance of benchmarking for best practices and by supporting this message with deeds, resources, and personal involvement, managers communicate the corporation's need to accelerate performance improvement. Benchmarking for best practices is a principal strategy in this continuous quest to improve. Adopting the most effective business practices of other successful organizations is an all-purpose catalyst for change and improvement.

2. *Management should articulate and reinforce the benefits of benchmarking for best practices.* Continuous performance improvement means an end to "business as usual." Employees therefore need to understand how innovative adaptation strategies benefit them and the organization by migrating more effective business prac-

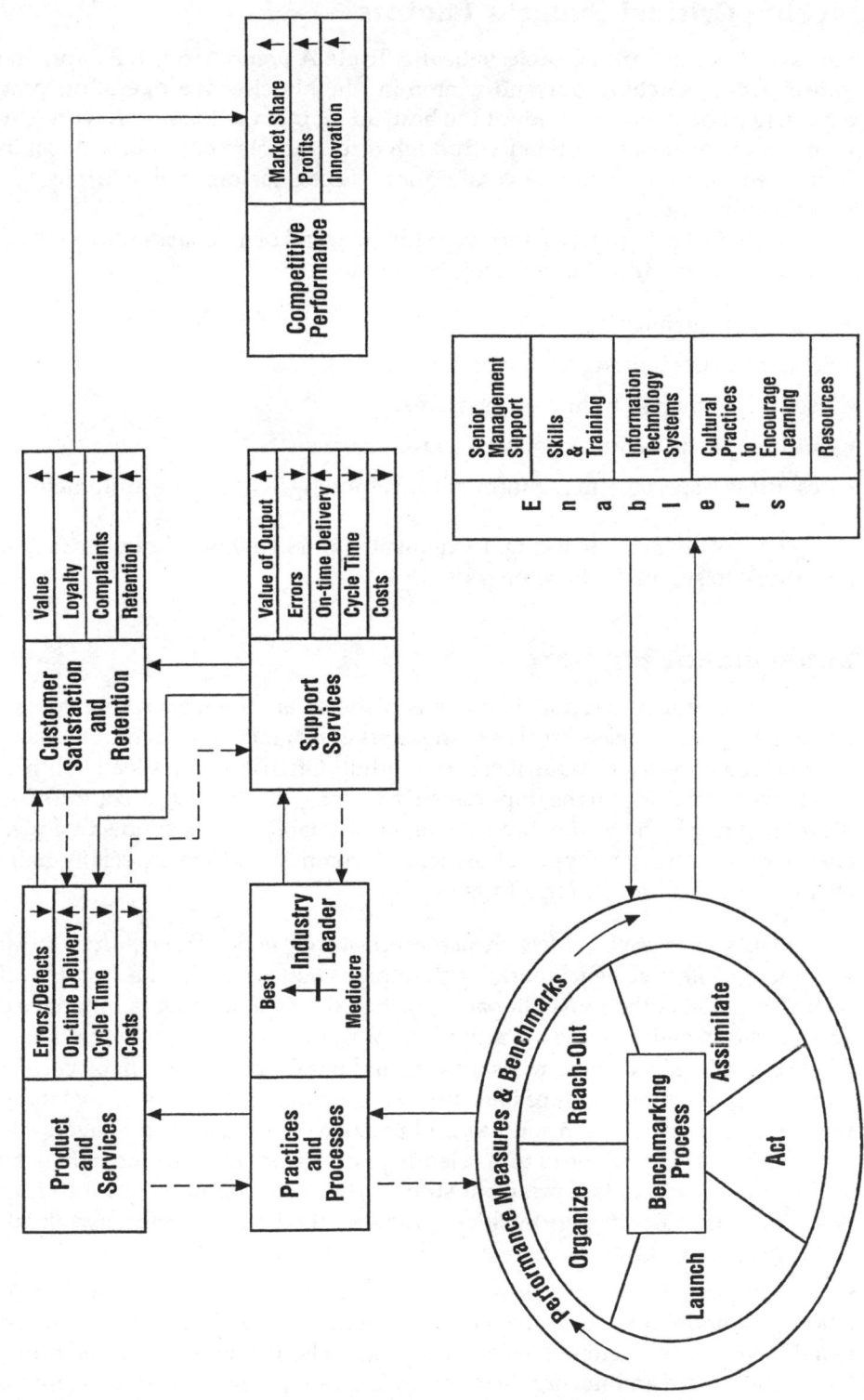

Figure 4.1a. Benchmarking: Its requisites and benefits.

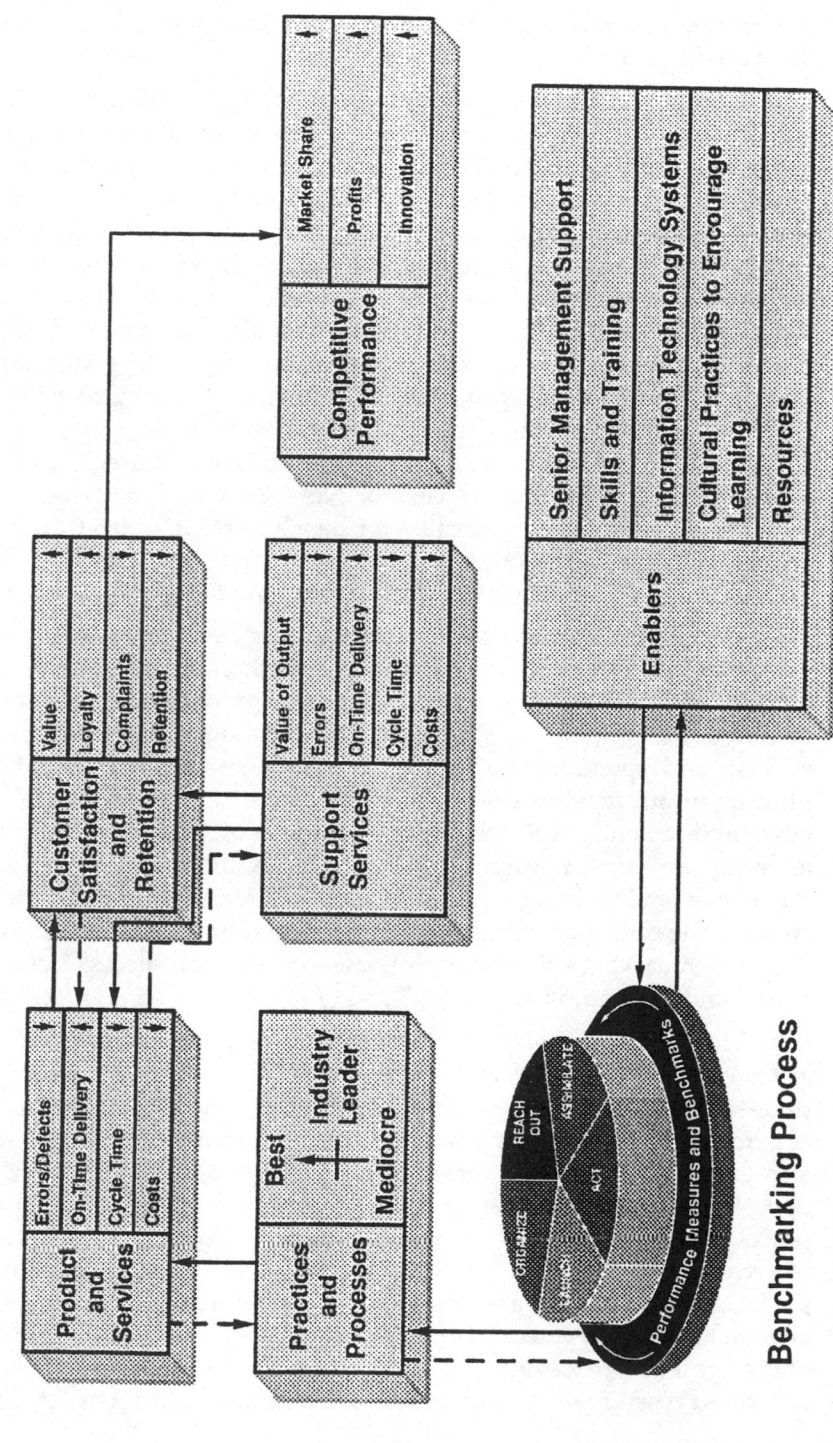

Figure 4.1b. *(Continued)*

tices into their performance systems. Moreover, once benchmarking efforts are seeded, organizations must nurture them. Managers can accomplish this by recognizing the efforts of teams that conduct benchmarking studies and by celebrating and rewarding those who successfully implement best practices in their business units.

3. *Management should translate general support for benchmarking into clear requirements for all managers.* An executive survey of the top 80 managers in the U.S. operations of a multinational organization revealed this fact: Nearly all respondents said they personally supported benchmarking as a business practice to achieve quality excellence; few of these managers, however, *personally* commissioned or participated in benchmarking efforts. The result: Formal benchmarking seldom occurred in this organization. Senior managers must learn about benchmarking techniques so they can actively support and direct others who will conduct best practice studies. One proven strategy to accomplish managerial competency in best practices benchmarking is to establish benchmarking requirements for managers. DuPont, for instance, has written benchmarking into the job descriptions of vice president-level officers. As part of its improved strategic planning process, Mutual Life Insurance Company of New York has made all members of its Executive Committee responsible for benchmarking in his or her responsibility areas. In both of these examples, benchmarking competency rose throughout the organization after executives had clearly defined benchmarking requirements.

4. *Management should ensure that the organizational culture supports and encourages a "we-can-learn-from-anyone" attitude.* The workplace of the 1990s is a brave new world where continuous learning is required for long-term survival. Corporations no longer hire employees who possess the requisite skills to perform complex jobs expertly and repetitively for life. Consequently, the modern corporation is metamorphosing from a repetitive production-line factory into a learning laboratory. In this new environment, employees embrace the idea of lifelong learning, for continuous learning can help employees update the skills, tools, and concepts that will help make them and their organizations successful. Management plays a leading role in creating the corporate culture that views continuous learning—through best practices benchmarking and other techniques—as a critical success factor for the organization and the employee.

5. *Management should empower employees who oversee processes and act as the owners of those performance systems to adapt best practices.* "You get what you expect" is time-tested managerial wisdom. Consequently, managers must trust that employees are capable of improving and managing the systems that they operate. Performance improvement occurs most rapidly when it is driven by those who own and operate the process or system being improved. Tight managerial control, which places all decision rights with managers, deprives front-line employees of any sense of ownership in their jobs. Employees who feel no ownership stake in their jobs and performance functions are less likely to experiment, learn, and initiate positive change and improvements. Good ideas are likely to remain on the drawing board of change if the process owners were excluded from identifying those ideas and best practices. When managers empower line employees to act as "general managers"

of their own jobs, they set the stage for employees to become motivated benchmarkers who actively identify, adopt and adapt best practices to advance their own performance systems.

Benchmarking Training

Benchmarking's greatest danger lies in its deceptive simplicity. What could be more straightforward than finding great companies and adapting their best practices? No doubt this is why so many teams launch benchmarking projects without adequate preparation. Benchmarking is not conceptually difficult, but it is extremely "detail rich." Benchmarking training is the tool to enable a team's success. Without training in the process, tools, techniques, and philosophy of best practices, benchmarking teams are severely handicapped. Different organizations employ different strategies in deploying benchmarking skills to line employees, but successful benchmarking training shares seven common characteristics:

1. *Benchmarking training should familiarize employees with a standard benchmarking process to be used throughout the organization.* Every organization should adopt a common benchmarking process to facilitate training, communication, and collaboration among teams, departments, functions, and business units. Benchmarking processes may vary—usually in superficial details—from company to company. Benchmarking process consistency, however, is critical within the organization if teams hope to learn from and share each other's experiences.

2. *Benchmarking training should familiarize employees with basic tools by which to analyze, understand, and improve work processes.* Benchmarking can be applied to many organizational activities, several of which are discussed in Chap. 5. For many American companies, process improvement is foremost among these uses. Process benchmarking and improvement require team members to be comfortable with basic process management skills and tools. Flowcharting or process mapping is the principal tool used to help teams articulate, analyze, and improve complex work processes. Process mapping—one of the most widely used and powerful of all quality improvement tools—can be applied in simple or sophisticated formats. Simple *relationship maps* illustrate the general sequence of events in a work process; they help to identify the organizational units involved in a process or delivery system. They also define the relationships between these units and describe the inputs and outputs of the system. An example is the creation of a newspaper: Suppliers provide paper, news, and advertisements to the newspaper. The Advertising Department prepares the ads and provides them to the News Department. The News Department prepares the stories and composes the pages, which are sent to the Production Department for printing. After the newspaper is printed, it is sent to the Circulation Department for delivery to readers. *High-level process maps* provide a more detailed—but still overview—picture of a work process. Major subprocesses are noted at each major work sequence. This helps teams identify major process steps, agree on the process scope and begin to define subprocesses. For example, the

subprocesses under News Department in the example might be: report story, write story, edit story, and lay out story. Detailed *flowcharts* show how the precise sequence of work, information, and material flows from department to department in a subprocess. Such finely detailed diagrams help teams identify improvement opportunities. All process mapping formats enable teams to illustrate intangible work flows, material flows, and information flows. These illustrations help teams structure their analysis and focus discussion. Once a process map is constructed, work groups can more easily identify both process strengths and weaknesses and then pinpoint where they occur in the work flow. Process mapping and flowcharting are therefore an essential first step in understanding and improving a work process.

3. *Benchmarking training should prepare the team in performance measurement.* Process management and performance measurement are flip sides of the same subject. Work processes represent a composite performance system that produce a tangible or intangible output or result. Performance measures are the indicators that reflect the health of the performance system. Once a process is described in a flowchart, team members will want to evaluate the relative health or performance standing of their process. Process managers accomplish this "process health check" by monitoring and measuring the most important performance characteristics. These typically are measures of speed or cycle time, quality, customer satisfaction, productivity, and cost. After developing a process map, many process management teams discover that they must develop a more comprehensive set of performance indicators to evaluate their processes' performance over time and against similar processes of other organizations. In this way, benchmark teams must be comfortable with performance measurement if they hope to perform meaningful process comparisons.

4. *Benchmarking training should provide the team with requisite technical skills, techniques, and tools to implement the benchmarking process.* During a benchmarking project, team members need to apply various other technical skills and tools. They may include basic research techniques, survey development, telephone interviews, site visits, team interviews, gap analysis, development of graphical performance charts, action plans, management reports, and implementation plans. The benchmarking training should prepare team members to successfully execute these techniques or it should inform them where they can seek expert help within their organizations. Often case studies, role-playing, scenario-building, and action exercises that require participants to apply benchmarking skills are helpful training strategies. Such hands-on training provides participants with practical experience while they are still in the classroom.

5. *Benchmarking training should prepare team members to be effective problem solvers and solution creators.* Benchmarking teams will be called on to exercise analytic problem-solving skills and synthetic solution-creation skills. Training should therefore prepare the team to be comfortable with both roles. To be effective problem solvers, benchmarking teams must be able to perform process analysis; like other forms of analysis, process analysis requires teams to reduce processes to subprocesses and to break apart the subprocesses into their incremental or component parts. Teams can spot problems and improvement opportunities most easily when per-

forming this kind of process dissection. In turn, solution creation often requires breakthrough thinking or entirely new ways of looking at a problem. For example, a company seeking to improve its logistics and distribution capabilities will probably seek benchmark partners among those companies that are really good at distributing hard goods, companies such as L.L. Bean, Wal-Mart, or SUPERVALU. A creative problem solver might also seek insights from nontraditional logistics champions, such as American Airlines, which is world-class in the way it manages its plane fleet and ticket reservation system, all of which share common characteristics with traditional logistics. Or it might learn from Fidelity Investments, which is world-class at quickly handling tens of thousands of mutual fund transactions and then rapidly dispatching confirmation notices to shareholders. Moreover, benchmarking teams that are comfortable with creative and synthetic thinking are often the best breakthrough thinkers; they bring together insights gleaned at several different benchmark sites to develop new solutions that help their company's performance leapfrog forward rather than making just incremental improvements. A company uncovers leapfrog improvements when another organization's best practices reveal an approach dramatically different from its own. Eric Kennedy, an IBM benchmarking program manager, recalls that when IBM's personal computer business unit wanted to improve its employee-theft-prevention systems, it looked far outside its industry to gambling casinos for process insights.[2] In such instances, benchmark team members must be flexible enough to throw out their own process and replace it with the better approach—or to translate the key parts of the better process into their own system. The benchmarking team's primary role is as a performance improver, and benchmarking training should prepare them to succeed in this role.

6. *Benchmarking training should present the benchmarking process in the context of existing quality improvement initiatives.* Benchmarking is an enabler of problem solving and continuous improvement. In this respect, benchmarking is a means to an end—not an end in itself. Consequently, benchmarking should be presented as an enabling set of skills and tools that support or complement traditional quality improvement techniques already in place.

7. *Benchmarking training should convey the philosophy of best practices as a catalyst for performance improvement.* The best practice approach to performance improvement is rooted in adopting and adapting processes that are uniquely successful. This approach, based on success analysis, is philosophically and behaviorally distinct from traditional quality improvement approaches, which focus on defect analysis and reducing variation in processes. Benchmarking training should communicate the purpose, power, and benefits of best practices as an improvement strategy that complements traditional improvement approaches.

Johnson & Johnson

In January 1991, less than five percent of the operating companies at Johnson and Johnson employed benchmarking—and those that did relied heavily on consultants. Realizing that employees know more about the corporations' businesses than outside consultants, Johnson & Johnson management shifted its focus to create benchmarking teams staffed by J&J employees. The new corporate goal was for J&J teams to perform 70 percent of all benchmarking projects and for consultants to perform the remaining 30 percent. By September 1992, over 60 teams were actively benchmarking. J&J reached its goal by using training and experience sharing to establish benchmarking competency within the corporation. By establishing internal benchmarking capabilities, Johnson & Johnson successfully reduced the cost and cycle time of its benchmarking studies while increasing the quality of project results.[3]

Information Technology Systems

Effective information technology systems are key enablers of successful benchmarking. Current IT systems permit teams to quickly and inexpensively generate, disseminate, analyze, and store vast amounts of information. Current personal computer hardware and software technologies present both large and small companies with extraordinary options at relatively low costs. Some common applications helping teams effectively develop and manage benchmarking information include:

- *Spreadsheet and database software programs* that help teams store, track, monitor, and compare benchmark information and project progress.
- *Process mapping and flowcharting programs* that enable teams to depict detailed work flows.
- *Graphical presentation programs* that help teams organize, display and analyze data.
- *CD-ROM information libraries and on-line databases* that allow teams to perform literature searches in their own offices.
- *Computer networks equipped with E-mail systems and group communications or meeting software* such as Lotus Notes help teams exchange information any time and any place, thus reducing travel costs, meeting times, and project cycle time.

Given the vast number of hardware and software options, companies should also be careful not to permit the technology to rule their work. A computer network, for instance, is not required for team members to share data. A CD-ROM information library is not required to perform library research. First-time benchmarking teams are best advised to use existing technology with which they are familiar to enable their work. Track benchmarking project progress on a simply designed spreadsheet, if team members already know how to use spreadsheet software; don't spend several weeks learning a database program in order to design a customized project tracking template.

The experience of Texas Instruments, whose Defense Systems and Electronics group was a 1992 Baldrige Award winner, illustrates how a company knowledgeable in benchmarking uses information technology to enable its teams. "We utilize an electronic mail distribution system, entitled *TIBench*, to help all our benchmarking teams," notes Laura Longmire, TI's Benchmarking Champion of Worldwide Operations. Team members have access to a wide range of useful information through this

Table 4.2. TIBench

Option	Description
1	Overview of benchmarking process
2	Benchmarking basics
3	Ethical and legal issues in benchmarking
4	Training
5	Listing of TI-wide benchmarking champions
6	Library summaries of benchmarking efforts
7	Benchmarking studies format/forms
8	Sources/industry contacts
9	Benchmarking policy
10	Who's benchmarking Texas Instruments
11	Business process management teams
12	Benchmarking forms
13	Best practices

system. *TIBench's* opening menu, which presents 13 different options and is depicted in Table 4.2, reflects the depth, breadth, and flexibility of the system. It enables Texas Instruments' benchmarking teams by offering on-going benchmarking orientation and training, by keeping teams in touch with the right people, by archiving past projects and by keeping users informed of current TI practices and requirements.

Cultural Encouragement of Learning

Benchmarking proliferates in corporate cultures that encourage learning. Why? Learning organizations evidence some common characteristics that encourage external outreach and innovative adaptation among employees. When General Electric CEO Jack Welch was asked to describe his vision of "the boundaryless corporation," he described his goal of making GE into a company without bureaucracy, where people are curious, open, cooperative, and always breaking down barriers. "It's how open you are about information, how open you are to ideas from other companies. ..." Welch reflected in a *USA Today* interview after being named CEO of the Year in 1993 by *CEO Magazine*. "You'll see charts in GE on the 'Wal-Mart Method' or the 'Lopez Three-Step' process (from former GM purchasing chief J. Ignacio Lopez de Arriortua). You are not a hero at GE for being a Lone Ranger with only your own ideas."[4] The corporate culture in learning organizations focuses more on liberating employees than on controlling them. Commonly these learning organizations encourage prudent risk taking, encourage holistic thinking, expect curiosity and creativity, encourage networking and teamwork, reward meaningful difference, and focus on transformational progress in addition to mere developmental change. The six golden rules of organizations that nurture learning cultures look like the following:

1. *Liberate employee potential rather than control it.* Liberation management encourages prudent risk taking. David Garvin, a Harvard Business School professor who has studied corporate cultures that support learning, observes that "employees must feel that the benefits of experimentation exceed the costs; otherwise, they will

not participate." In addition to developing values that support curiosity, experimentation, and prudent risk taking, these companies develop balanced managerial systems that reward the core values of curiosity but don't penalize failed experimentation. For instance, specialty steelmaker Allegheny Ludlum has devised a system to keep expensive high-impact experiments off the scorecard used to evaluate managers but requires prior approvals for such projects from four senior vice presidents. IBM's legendary founder, Thomas Watson, Sr., also understood the need to balance risk taking and discipline for failed experiments. Company lore has it that a young manager was summoned to Watson's office after losing $10 million in a risky venture. The young man began their meeting by saying, "I guess you want my resignation." Watson replied, "You can't be serious. We just spent $10 million educating you."[5]

Learning organizations turn away from the oligarchic principle of rule by a select few and rule through hierarchies that build rigid controls into the organizational structure. Such rigid corporate structures do not support change, experimentation, and rapid progress to the degree they are required for survival in the permanent white water business environment of the 1990s. Liberation management seeks to free employees to learn, create, improve, and innovate to best fulfill customer needs. Taco Bell represents another company that has gained by liberating its management. In the past, Taco Bell's focus was on manufacturing a product (i.e., shredding lettuce, grating cheese, and assembling tacos); in this environment one regional supervisor directed five stores. Taco Bell corporate managers, however, shifted their focus to training store managers in teamwork, customer service, and sharing. Now one regional supervisor coaches and supports 20 stores. By reengineering its production process and by liberating the people who are closest to the customer, Taco Bell has grown sales by 60 percent, has increased profits by more than 25 percent, and has lowered prices by more than 25 percent.

2. *Encourage holistic, systemic thinking and learning in addition to narrow problem solving.* Traditional problem solving is inwardly-focused: employees analyze problems and then brainstorm their own solutions to those problems. This process usually neglects to build upon the proven ideas, experiences, innovations and improvements of other organizations that have already addressed similar situations. Moreover, standard problem-solving techniques tend to target narrowly defined operating difficulties. Rearranging the deck chairs on the Titanic would not have created a better overall result. Continuous improvement of rotary telephone dials would not have staved off the advent of easier-to-use touch-tone systems. Breakthrough improvements, major innovations, and comprehensive solutions to complex operating problems are often achieved through process simplification, redesign, or reinvention. Just as most core work processes flow across multiple functions and departments, process solutions usually require a wide-angle perspective of the full performance system. Consequently, a holistic or systemic approach to learning and problem solving is highly beneficial.

3. *Expect creativity, innovation, and continuous learning—rather than conformity—from managers and employees.* By expecting creativity, innovation, and continuous learning, the organization's leadership reinforces its commitment to liberating the

true potential of its employees. Growing numbers of organizations support pay-for-knowledge systems, job rotations, apprenticeships, on-going training, career pathing, innovation awards, and other managerial systems that institutionalize knowledge as a valuable asset. These systems help establish senior management's expectation that every employee—no matter what his or her role and job function—has the privilege, right, and responsibility to continuously learn, innovate, and improve. All these managerial systems can be further expanded and integrated to establish innovation, creativity, and continuous learning as requisites of personal and corporate success.

4. *The corporate culture should encourage networking and information sharing.* Best practice companies have learned that traditional bureaucracies encourage isolationism. In a world where speed, agility, and close customer contacts are critical for marketplace success, pyramid-shaped hierarchies breed behavior that is now regarded by the marketplace as dysfunctional. Strict adherence to the reporting and information-sharing channels of the bureaucratic organizational structure produces unresponsiveness. By breaking down the walls inherent in bureaucratic organizations, corporations create a freer, faster flow of information. At a minimum, networking across bureaucratic lines discourages duplications of effort. At the very best, networking encourages sharing, cooperative learning, team-focus, and creativity. In turn, these elements support the evolution to flatter and faster organizational structures.

5. *The corporate culture should nurture integrated, cross-functional teamwork.* Hierarchical structures try to perform project improvements through the vertically-oriented departmental functions that support the hierarchy. Work, however, is performed through processes that flow horizontally across functions. A customer order, for instance, may be booked by sales, recorded by order fulfillment, administratively processed in accounting, handled in the warehouse, and delivered through transportation. Most work is cross-functional. Consequently, nearly all major improvement efforts are cross-functional too. Observes William G. Stoddard, director of Andersen Consulting's process reengineering practice: "Departments are stovepipes. We work in sewer pipes."[6] Best practice companies, therefore, labor hard to minimize bureaucratic tendencies. They routinely unite the best talent from different divisions, business units, departments, or work groups in cross-functional teams that are empowered to create the changes required to improve performance. Sometimes these cross-functional teams work together for short periods to complete special one-time projects. At other times, these process-focused teams have long-term charters and responsibilities. Some pioneering companies regularly employ job rotations, unit-to-unit site visits, multiunit sharing rallies and line-to-staff temporary assignments as tools to transfer knowledge and nurture an integrated view of the organization. On-going cross-functional teams offer several special benefits: They provide the continuity needed to successfully implement multiyear projects; they help institutionalize an integrated, process-oriented view of the organization, and they reinforce the leadership's commitment to networking information, human and intellectual resources. So committed is General Electric to cross-functional behavior that CEO Welch says: "We take people who aren't boundaryless out of jobs."

Integrity is clearly the most important value, but boundarylessness comes next. If you're turf-oriented, self-centered, don't share with people and not searching for ideas, you don't belong here."[7]

6. *The corporate culture should reward meaningful differences or departures from the status quo.* Continuous improvement is a cornerstone of best practice companies—but fast companies also embrace the need for periodic radical, large-scale improvements. These break-through improvements frequently occur under one of several circumstances:

- The competitive marketplace changes the rules of the game and organizations that hope to compete must change or die; the dramatic shift from mainframe computer systems to client server computer networks is a good example of one such shift.

- The leadership mandates major magnitude change, often articulated through "stretch goals" or "leapfrog goals." Organizations such as Motorola, Texas Instruments, and selected other companies have mandated breakthrough improvements through Six Sigma defect reduction programs, which aim to reduce defect rates to 3.4 errors per million opportunities, and through cycle time reduction programs that cut elapsed processing times in 50-percent increments. To achieve such large-scale improvements in relatively short time periods, these organizations have challenged themselves to radically rethink what is possible and to redefine the status quo.

- The organization undertakes a major effort in core process redesign or reengineering. Organizations that reengineer their most important work processes begin by asking themselves a few simple questions: *If we were to reinvent the way we work today, how would we design our work systems? If we were to start with a clean sheet of paper, what would our process map look like?* Such ventures in redesign exploit powerful new technology and a thorough understanding of customer preferences. Reengineering teams are formally licensed to diverge from established paradigms and work traditions and to reinvent the organization's work systems.

Whatever the catalyst driving major-magnitude changes, best practice organizations reward individuals who can visualize how to completely change a process to improve it. They reward those who can see the future in different terms than those employed by the status quo. Without such visionaries in the organization, change comes much more slowly and incrementally.

Resources

Without adequate resources, even the most enthusiastic teams may find themselves handcuffed. For benchmarking teams, resources are of three primary types: time, funding, and equipment.

Time.   Team members must be allowed the time for training, research, and team meetings. Teams that never have the time to meet are almost always ineffective. On some occasions, fulfilling team duties may require team members to be released from other work commitments. Corporations should view the benchmarking team's work as just as vital as the team members' full-time jobs and support them accordingly. The benchmarking process will succeed only if people are given time to do the job well. Expecting team members to learn, plan, analyze, report, and act on their benchmarking findings while they are 100 percent occupied with their regular jobs is a sure formula for failure.

Funding. Resource support must translate into adequate funds to provide for project expenses such as research, training, site visits, data acquisition, report production, and interviews. Like any division or business unit, benchmarking teams must operate under budgetary restraints, but without adequate funding, benchmarking teams cannot fulfill their charge.

Equipment. Equipment—often in the form of technology support—is critical to the benchmarking team's success. For instance, teams often must perform research through public databases that typically have limited points of access in an organization. Teams must also be able to gather, store, retrieve, and share best practice information. Effective benchmarking teams are therefore supported with the computer time and applications that will enable their success.

Designing Your Benchmarking Process

Newcomers to the field of benchmarking often puzzle over the great variation among benchmarking processes used at different companies. Some organizations use a four-step benchmarking process while others use a six-step process, seven-step process, eight-step process, or some other variation. The differences are cosmetic. Most companies employ a common approach that helps them plan the project, collect and analyze data, develop insights, and implement improvement actions. However, each company breaks this process into different numbers of steps depending on how much detail they wish to describe at each step of their template. The best approach reflects common sense: Adopt a benchmarking process that suits an organization's culture and existing quality-improvement initiatives! Successful benchmarkers have found it's far more important to build on the managerial foundation and culture in place than to blindly adopt another organization's specific process. One company's effective benchmarking process design may even fail at another organization with different operating concerns. Indeed, organizations that excel in benchmarking almost always customize their own benchmarking process to reflect the organization's culture, infrastructure, and leadership philosophy. The following four benchmarking process designs, in place at some prominent U.S. companies, dramatize this point:

Motorola, a Baldrige Award winner and frequent benchmarker, uses a straightforward five-step process model which is easy to use and can be tailored to specific benchmarking projects.

The Motorola Five-Step Benchmarking Process

Step	Description
1	Decide what to benchmark
2	Find companies to benchmark
3	Gather data
4	Analyze data and integrate results into action plans
5	Recalibrate and recycle the process

Bristol-Myers, Baxter International, and several other corporations use a seven-step benchmarking model.

The Seven-Step Benchmarking Process

Step	Description
1	Determine which function(s) to benchmark
2	Identify key performance variables to measure
3	Identify the best-in-class companies
4	Measure performance of best-in-class companies
5	Measure your own performance
6	Specify programs and actions to meet and surpass
7	Implement and monitor results

Robert C. Camp, Manager of Benchmarking Competency Quality and Customer Satisfaction at Xerox, advocates a 12-step organization-wide benchmarking process that is presented in five phases.

The Xerox 12-Step Benchmarking Process

Step	Description
Phase 1—Planning	
1	Identify what to benchmark
2	Identify comparative companies
3	Determine data collection method and collect data
Phase 2—Analysis	
4	Determine current performance gap
5	Project future performance levels
Phase 3—Integration	
6	Communicate findings and gain acceptance
7	Establish functional goals
Phase 4—Action	
8	Develop action plans
9	Implement specific actions and monitor progress
10	Recalibrate benchmarks
Phase 5—Maturity	
11	Attain leadership position
12	Fully integrate practices into processes

AT&T, which also has two Baldrige winners among its operating units and which is an active benchmarker, uses a nine-step organization-wide benchmarking process.

The AT&T Nine-Step Benchmarking Process

1	Identify what to benchmark
2	Develop a benchmarking plan
3	Choose a data collection method
4	Collect data
5	Choose best-in-class companies
6	Collect data during a site visit
7	Compare processes, identify gaps, and develop recommendations
8	Implement recommendations
9	Recalibrate benchmarks

The Customized Benchmarking Process

Some managers may feel frustrated by the lack of standardization among benchmarking processes. Benchmarking experts, however, agree that no common benchmarking process standard is likely to establish itself soon.[8] Corporate benchmarking managers contend that the benefits of a benchmarking process design that suits one's individual culture are far greater than the benefits of establishing a national process standard. They identify three motivating factors that throw the weight of benefits onto the side of each organization designing its own benchmarking process:

1. *Customized benchmarking processes complement the individual corporate philosophy toward performance improvement.* The leadership of every organization must establish benchmarking's role within the organization's continuous-improvement efforts. In some companies, this role is tactical: benchmarking enables problem-solving and process-improvement efforts. In other organizations, benchmarking plays a more strategic role: Benchmarks are viewed as critical performance measures required for all core processes; benchmarking is viewed as a fundamental business process employed by all effective managers and line employees; benchmarking drives continuous improvements and enables managers to evaluate their units' progress against competitors; the leadership embraces best practices as primary motivators of organizational change, learning, and improvement. Whether benchmarking assumes a tactical or strategic role within the organization, a customized benchmarking design ensures that the process complements the organization's existing TQM system.

2. *Customized benchmarking processes support cultural differences among organizations.* Organizations, like people, have different personalities. Some organizations

are dominated by people from engineering and technical backgrounds who appreciate structure and detailed procedures; such organizations may successfully promote a benchmarking process that has many steps when fully and meticulously articulated. Organizations that are dominated by people from sales, marketing, and service backgrounds may prefer less structured management procedures. These organizations will probably promote a simpler benchmarking process that reflects their preference for flexibility and individual improvisation.

3. *Customized benchmarking processes accommodate the need for organizations to feel unique.* Many organizations believe their operating circumstances are unique. Therefore they distrust all "off-the-shelf" solutions because they were "not invented here." These organizations favor customized approaches because they help managers and employees "buy into" and accept the benchmarking process more quickly.

The Simple, Consensus Model

Successful implementation frequently favors simplicity. Consequently, this book employs for discussion purposes a very simple benchmarking process. One such model has been articulated by the members of the Strategic Planning Institute's (SPI) Council on Benchmarking. The SPI Council is a loosely-knit confederation of corporate benchmarking managers who meet quarterly to share benchmarking experiences. To facilitate discussion and learning among its members, more than 50 SPI member organizations crafted a simply-worded five-step process that summarizes the essentials of the most successful corporations' individual benchmarking processes.

The SPI Council on Benchmarking Model

Step	Description
1	Launch
2	Organize
3	Reach out
4	Assimilate
5	Act

The SPI model represents a user-friendly template for designing your own benchmarking process. It is deliberately articulated in generic terms so that virtually any benchmarking process can be mapped into its five phases. Xerox and Motorola's benchmarking processes are mapped onto the SPI generic model as illustrations.

Mapping Xerox's Benchmarking Process onto the Generic Process

	Description	Launch	Organize	Reach out	Assimilate	Act
1.	Identify what to benchmark.	✓				
2.	Identify comparative companies.		✓			
3.	Determine data collection method.		✓			
3b.	Collect data.			✓		
4.	Determine current performance "gap."				✓	
5.	Project future performance levels.				✓	
6.	Communicate findings and gain acceptance.				✓	
7.	Establish functional goals.					✓
8.	Develop action plans.					✓
9.	Implement actions.					✓
10.	Recalibrate benchmarks.					
11.	Attain leadership position.					✓
12.	Fully integrate practices into processes.					✓

Mapping Motorola's Benchmarking Process onto the Generic Process

	Description	Launch	Organize	Reach out	Assimilate	Act
1.	Decide what to benchmark.	✓				
2.	Find companies to benchmark.		✓			
3.	Gather the data.			3		
4.	Analyze data and integrate.				✓	
5.	Recalibrate and recycle the process.					3

The GTE Case Study

This process model provides a useful framework for planning and managing benchmarking projects. The experience of GTE Telephone Operations reveals how the model can be used by a benchmarking team. In this example, the benchmarking team sets its sights on improving major-account sales.

The Launch Phase

At the outset, management must decide what improvement opportunity areas have the greatest impact or potential for the organization. This decision where the benchmarking team should set its sights is addressed immediately in the *launch phase.* *(Figure 4.2a illustrates the process, purpose, and individual steps that compose the benchmarking launch phase.)* Frequently benchmarking projects arise when senior executives identify high-level needs during senior management activities. Consider a few examples:

- *During the strategic planning process, management must evaluate business factors critical for the organization to sustain its competitive advantage.* Issues surrounding technological competency, core process capabilities, cost competitiveness, and fast response capabilities—all issues that could prompt the redesign of core business processes—often command immediate benchmarking. The findings help the leadership evaluate the adequacy of the organization's current position and ensure the company's future plans will enable it to remain competitive in the marketplace.

- *During monthly and yearly operating assessments of core business processes and functions, management identifies opportunities for continuous improvement.* Every organization has core processes and functions that are comparable to the engine of a car: They make everything go. During quarterly operations performance reviews, management may identify some element of the core business system that appears overloaded, obsolete, or inadequate. Benchmarking becomes a tool for accomplishing three fundamental objectives: adapt best practices, get the process back under control, identify and adapt current best practices, and accelerate the journey to improved performance.

- *Current benchmarking projects trigger awareness of additional opportunities.* In many instances, one benchmarking project simply gives rise to another as new improvement areas are surfaced.

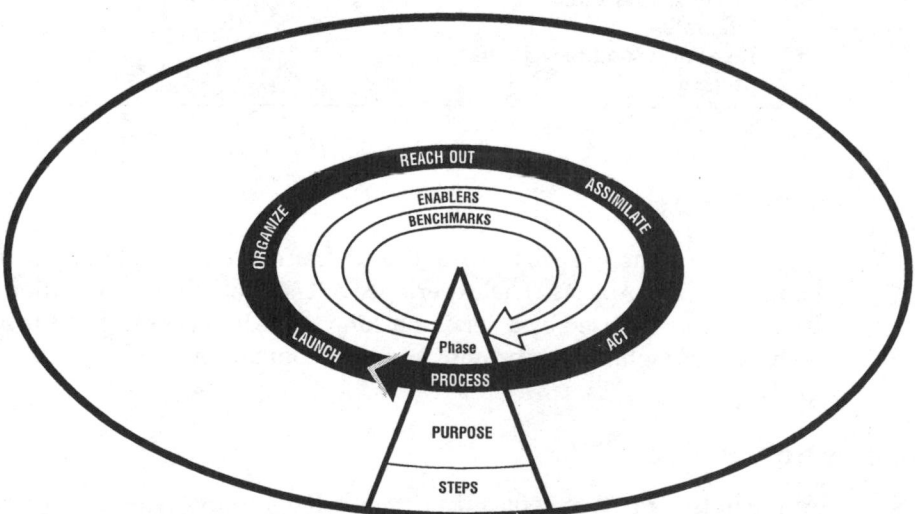

Figure 4.2a. Benchmarking process.

At GTE's telephone operations, the strategic planning process and the company's operations performance reviews were both catalysts that prompted the corporation to benchmark its major-accounts sales. The strategic plan, for instance, articulated goals for improving capabilities among several professional and technical groups, including the sales force. For sales, one question emerged above all others: *What would it take to be the major-accounts sales leader in the telecommunications industry?* Operating reviews created another kind of improvement pressure: They judged the sales force's performance to be suboptimal. "Due to the ever changing marketplace and the environment in which GTE operates," the company observed, "there have been many needs identified regarding its sales force which are imperative for GTE to address." GTE decided to employ benchmarking to identify the most important gaps between its major-accounts sales practices and those of sales leaders. By understanding its practice deficiencies, GTE planned to mount a campaign to close the gaps and thereby strengthen its market share and improve its profitability.

The Organize Phase

During the second benchmarking phase, the *organize phase*, management organizes the project to ensure a clear project focus. At this time, senior-level managers identify important issues, weaknesses, and improvement opportunities; they prioritize the specific functions or processes to be studied; they obtain approval and support from process owners and stakeholders; and they assign the project to a benchmarking team that is usually comprised of front-line employees from all functions that help manage the process under study. The team then prepares a benchmarking project plan. *(Figure 4.2b illustrates the launch and organize phase of the benchmarking process.)*

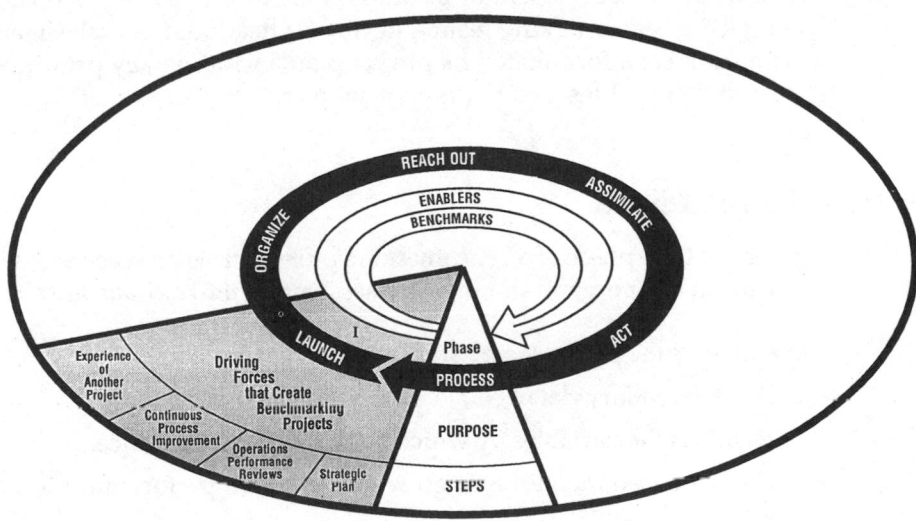

Figure 4.2b. Benchmarking process.

Team Size

An important proviso for organizations forming benchmarking teams is to keep membership size within manageable limits. Three to eight team members are optimal. If it is the team's first benchmarking effort, an experienced benchmarking facilitator should ideally be included on the team to answer questions and guide the group in its efforts.

The team's first duty usually is to produce a benchmarking project plan that serves as the team's road map. A benchmarking plan often includes the following components:

- Statement of purpose.
- Justification for the study.
- Statement of problem(s) or opportunities to be addressed.
- Baseline survey or poll data from those involved in the process or practice.
- Relationship of the project to company business priorities and goals.
- Proposed benchmark measures.
- Statement about benefits of the benchmarking study.
- Methodology and scope statement, sketching what level of benchmarking is optimal for the circumstances (internal organizations, competitors, regional companies, industry leaders, functional, or best-in-class leaders without regard for industry).
- A resources planning statement, identifying the time, people, and equipment resources needed, as well as estimates of study and implementation costs.

GTE's operating reviews led its senior management to target four aspects of major-account sales management: sales-force structure, sales-force costs, sales-force performance improvement, and sales-force awards programs. The process owner, the Assistant Vice President of Business Sales, approved the project. Management assembled a benchmarking team and chose a headquarters sales manager to head it. The team soon formulated its project plan, including key priorities, a schedule, expected deliverables, and required resources.

The Reach Out Phase

During the third phase of a benchmarking project, the team reaches out to understand its own and other organizations' processes. During the *reach out phase*, the team:

- Documents the process to be studied (based on customer needs).
- Collects secondary data.
- Determines the variables by which to evaluate performance.
- Designs a questionnaire through which to solicit performance information from other organizations.
- Conducts telephone interviews and general information gathering.

- Selects benchmarking partners.
- Conducts on-site visits among the best performing partners.

(Figure 4.2c illustrates the launch, organize, and reach out phases of the benchmarking process.)

Understanding the Process

After project formation, team members usually immerse themselves in an intensive effort to operationally understand the process or practice they will benchmark. This immersion process begins with the team's reviewing—or even learning for the first time—of the process or practice area's mission, goals, and objectives and how they fit within the context of the corporate strategic plan. The team then identifies the customers who receive the process's product or output and the team further defines the customer's needs and requirements for the process. Next, the team gathers critical performance measures that enable benchmarking team members to evaluate the relative perform-ance excellence of similar processes among different organizations. Finally, by observing, discussing, and chronicling the practice area with the front-line operators or process owners, the team creates a functional flowchart, often including finely detailed subprocesses, operational boundaries, and hand-off points.

Performing Research

Once the team thoroughly understands the target process, it collects information about prospective benchmark partners and their best practices in the study area.

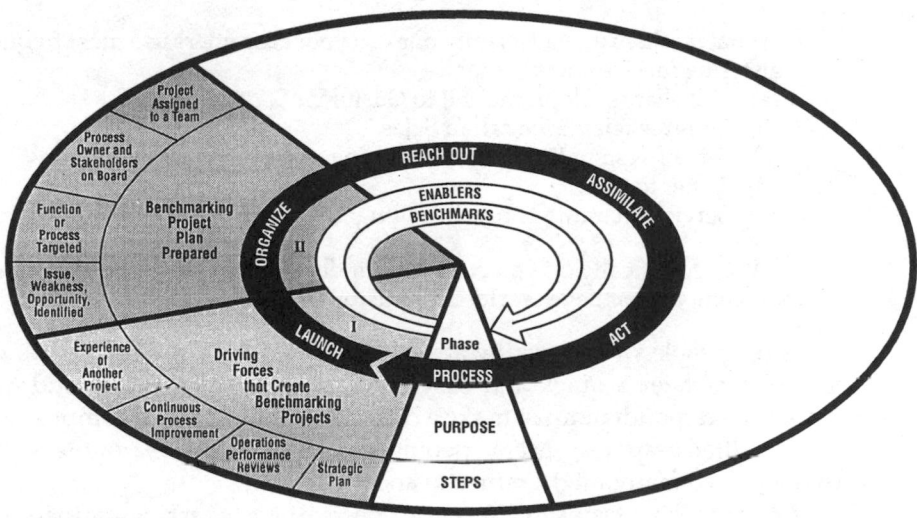

Figure 4.2c. Benchmarking process.

The first line of research on best practice partners is usually mined from among four primary sources: internal corporate information, public domain information, information from outside experts, and original research.

Don't underestimate the wealth of information that resides within your own organization. Corporate libraries, data bases, competitive reports and published papers, current studies, and market research analysis are often gold mines of information. The Human Resource department's list of internal experts, including long-range planners, technical experts, and fellow employees who have worked for other leading organizations, represent potentially rich information sources.

Public domain information (usually available through local public or university libraries) includes articles published in various journals and periodicals, position papers, business directories, conference proceedings, and company publications such as prospectuses and annual reports. Other external best practice information sources include trade and professional societies, consultants, industry experts, academics, trade journalists, Wall Street analysts, vendors and service bureaus, and public seminars. Many benchmark teams also conduct original research, including telephone interviews with prospective benchmarking partners, surveys mailed to experts, customers, and suppliers, personal interviews with subject matter experts, third-party studies, reverse engineering, and, of course, on-site observation.

The first round of research enables the benchmarking team to narrow its sights on a few really good benchmark partners and to create a list of site-visit questions for each performance area under study. A qualifying survey or questionnaire often proves invaluable as a screening device and time-saver when evaluating whether a prospective benchmarking partner warrants a site visit.

Conducting a Benchmarking Survey

Please answer one of the following two questions:

1. What mechanism (select only one) do your customers use most frequently for registering service complaints?
    (a) Toll-charge telephone call to Customer Service.
    (b) Toll-free telephone call to Sales.
    (c) E-Mail system linked to Purchasing.
    (d) Letter to CEO.
    (e) Service guarantee on your billing invoice.

2. What specifically do you believe your department's role should be in helping your company achieve world-class customer service?

Most people would readily answer the first question. The inclusion of specific choices facilitates a quick and easy response. However, the second question is a terror. The respondent must make a complex evaluation and compose a potentially long, written response. Many people would likely have trouble with Question Two—indeed many might refuse to answer it at all.

Benchmarking teams that want prospective benchmarking partners to respond to their surveys must consider the style, substance, and organization of their questionnaires. When should a team employ a benchmarking survey? Written-response surveys are especially useful for gathering general information, for identifying potential benchmarking candidates from among a larger class of excellent companies, for developing baseline performance comparisons, and for gathering specific information before conducting a site visit. Many experienced benchmarkers also use surveys when seeking information from internal operating units. Inexpensive to administer and summarize, the survey presents a fast, easy method of reaching a large audience.

Surveys also possess some drawbacks as information-gathering tools. Surveys are best suited for gathering quantitative information; they are often weak for capturing qualitative information or longer narrative responses. When survey respondents answer open-ended qualitative questions, their responses are often vague and fragmentary. Compared to interviews conducted by telephone or in-person, surveys prove to be clumsy tools for developing detailed process information that may require in-depth discussion and clarifying questions. Moreover, surveys often produce low response rates—30- to 40-percent survey response is considered excellent.

Despite these drawbacks, benchmarking surveys are essential benchmarking tools when appropriately employed during the research and information-gathering process. Experienced benchmarkers offer the following eleven tips for developing effective benchmarking surveys:

1. *Develop a target population—by company type and known contact individuals—for your survey.* Many organizations can sell or provide mailing lists of prospective respondents. Some of the most useful mailing list sources include trade associations, trade publications, mailing list vendors, conference planning companies, and internal corporate departments such as Marketing, Customer Service, and Research.

2. *Carefully select the type(s) of questions to be used in the survey.* Most questions fall into one of four categories:

1. Multiple choice: *What type of billing does your law firm most often use?*
   a) fixed fee, b) hourly rate, c) contingency fee.

2. Forced choice: *Are you satisfied with your law firm's current billing practices?*
   Yes ___ No ___

3. Scaled: *Rate the following on a scale of 1 to 5 with 1 being "I strongly disagree" and 5 being "I strongly agree." I can easily understand my bill.* 1 2 3 4 5

4. Open-ended: *What are suggestions to improve the billing practices of your law firm?*

Respondents more readily reply to multiple choice, forced choice, and scaled questions because answer options are provided for them. Open-ended questions pose more difficulty because the respondent must develop and write down original ideas. A good survey often employs more than one question type, using different types to develop different kinds of information.

3. *Organize the survey into a logical sequence with an introduction, a body, and a conclusion.* A standard introduction includes the survey's instructions, scope, and purpose. The body encompasses general, objective, and easy questions followed by difficult, subjective, and open-ended questions. Thus, open-ended questions follow multiple choice,

forced choice, and scaled questions. Once respondents have begun thinking about a subject in the preliminary questions, they will more likely respond to open-ended questions. Since they have already expressed their beliefs in the earlier questions, the open-ended ones allow them to expand on their answers. The open-ended questions lead to the conclusion of the survey. This section contains demographic questions covering respondents' types of business, geographic location, and organization.

4. *Keep it short and simple.* The ideal length of a benchmarking survey is hotly debated. Unfortunately, there are no answers cut in stone concerning survey length. A two-page survey filled with all open-ended questions can prove very difficult to answer, while some four-to-six-page surveys composed of all multiple-choice questions are easy to complete. Here are two useful rules of thumb: *First, sparingly employ open-ended questions. Second, the shorter the survey, the higher the response rate.* Business people often start a survey when they first receive it. If they can complete it quickly, without great difficulty or thought, they are more likely to answer all the survey's questions and return it to the sender. When questions are complex or difficult, busy managers are apt to throw the survey away or put it on the back burner. Many surveys initially placed on the back burner eventually end up in the garbage: The momentum is lost—or respondents return to the survey after the requested response deadline.

5. *Design the survey for easy transfer of data.* Poorly designed surveys may generate responses but fail to capture or format the information in a way that enables easy transfer to computer for comparison. The more difficult the data transfer, the more time it takes and the higher the likelihood of distorting or invalidating information during transfer. Some benchmarking teams, for instance, have succeeded at gathering lots of qualitative survey responses, only then to find the narrative responses difficult to summarize and compare. The more summarizing that occurs, the greater the likelihood that errors or misinterpretations will be introduced into the narrative answers. Survey experts design questions that are easy to answer and they also employ survey formats that enable fast, easy data transfer from the response sheets.

6. *Involve internal process owners and beneficiaries during survey design.* Internal process experts can provide valuable feedback and reality checks during the survey design process. Moreover, these process owners and project beneficiaries are more likely to support project findings if they've played a role in designing, reviewing, and validating benchmarking surveys and other information-gathering instruments. Pretesting a benchmarking survey among your own field units is a good tactic to ensure the success of both the survey and the project.

7. *Alert recipients before sending them a benchmarking survey.* An advance call to the respondent personalizes the survey and informs the recipient of the survey's purpose. During such initial contact calls, benchmarking team members should identify themselves and explain how the respondent can reach them to ask follow-up questions.

8. *Provide incentives to motivate survey recipients to reply.* Increasingly, benchmarking teams must provide incentives, such as survey summary results, small prizes, mementos, or even cash premiums, to induce some other organizations' managers

to participate in the host company's benchmarking survey. A respondent who might normally put aside a survey is more likely to reply if he or she believes participation will benefit the individual or the company.

9. *Enclose a self-addressed, stamped return envelope to facilitate the survey return.* A return postcard—asking for the respondent's name, company, response date, and permission to discuss the survey in greater detail—helps overcome a primary weakness of surveys: the lack of follow-up questions.

10. *Place follow-up calls to ensure high response rates and to seek clarification or further detail concerning responses.* Even the best surveys often require personal follow-up calls to ensure that respondents complete the task requested.

11. *Acknowledge survey receipt, provide promised incentives, and say thank you.* A simple way to say thank you to respondents is to acknowledge receipt of their surveys and provide any promised incentives.

Conducting Benchmarking Telephone Interviews

The benchmarking survey is just one of several data-gathering tools that benchmarking teams often employ. Many times a telephone interview is especially valuable for developing and supplementing best practice information. Telephone interviews can be both time savers and time wasters. In the hands of well-prepared benchmarkers, the telephone is an effective tool for developing, clarifying, expanding, and validating information. When ill-prepared benchmarking team members "let their fingers do the walking," they are apt to waste prospective benchmarking partners' time, embarrass themselves, and sully their own company's reputation. Consider the following real-life scenario:

"Hi, my name is Mary and I'm conducting benchmarking interviews for Company X. Will you spend a few minutes answering questions about your purchasing department?"

The busy manager agreed to spend a few minutes with the caller. Mary then read questions and multiple choice answer options from a long list. After about 20 questions, the busy manager asked how much longer the interview would take. Mary responded, "Only a few more minutes," as she began a series of open-ended questions. When asked to clarify a question, Mary responded, "I'm sorry. I don't know anything about this subject. I'm just doing what my boss asked me to do."

Five minutes later, the busy manager said he had to conclude the interview. Mary pleaded … if he could just answer a couple more questions. She then explained that if she didn't complete the full interview, none of the previous responses could be used and she'd get into trouble with her boss. The busy manager agreed. Four questions later, when asked for additional comments on an especially important open-ended question, the manager tersely said he had no additional thoughts. By the interview's end, the busy manager was annoyed. He hung up the phone and promptly directed his secretary never again to put through such calls and never to purchase any products from the company initiating the interview.

Not all phone interviews cause as much frustration as Mary's. Easy to plan and conduct, phone interviews provide a simple method of accessing a wide geographical area. When conducted properly, phone interviews provide a fast, efficient, and cost-effective way of collecting information.

Like other information-gathering tools, the phone interview has its shortcomings. Individuals may be hard to contact and often do not return phone calls. Callers must be persistent; they will likely need to call the interviewee more than once. When the caller reaches the desired person, frequent interruptions may occur. Other calls or distractions may disrupt the respondent's train of thought. In addition, many respondents have limited time and consequently may only respond briefly to questions.

Despite these weaknesses, telephone interviews present another fast, flexible, and cost-effective approach to information gathering and benchmark partner screening. Since benchmarking team members do not have to hop a plane to fact find, they can traverse the country many times in a single day, making contacts coast-to-coast or even outside the country.

In certain respects, the telephone interview is similar to the survey. Each one is tailored to the project and specific benchmarking subject area. Interviews can also employ several question types, including multiple choice, forced choice, and scaled questions for producing quantitative results. Interviewees respond most easily to these question types. Yet open-ended questions also play an important role: They provide the benchmarking team with the opportunity to develop qualitative responses complete with explanations and opinions.

Effective telephone interviews have their own set of critical success factors (CSFs). Among the most important CSFs are the following:

- *Prepare the interview questions in advance and pretest them on the phone in one's own company.* Just as with surveys, the benchmarking team must develop telephone contact numbers for its target population.

- *Prioritize the list of interview candidates and contact the most important candidates first.* The "most important" candidates often provide the most relevant and useful information.

- *Coordinate calls so that two interviewers do not call the same person.*

- *The interviewer should always try to ask for a specific individual.* If a caller does not know a contact's name, he or she should call Public Relations or Personnel for the name of a subject expert in the field. A caller is unlikely to unearth desired information through random telephone calls into any large organization.

- *Callers should introduce themselves and explain the purpose of the interview.* Respondents generally are more eager to talk to a friendly and informative caller.

- *Use referrals whenever possible.* Mentioning the person's name who suggested you call usually makes the respondent feel more at ease and open.

- *Provide an estimate of the time needed for the interview.* Busy managers often prefer to schedule a convenient interview time that matches the requested time, rather than pinch a few minutes here and there spread over several phone conversations.

- *The interviewer may need to assist the respondent by sharing information and explaining what information the company needs.* If a respondent does not understand a question, the caller may need to help clarify it; at times this is best done by sharing early research findings and operating experience that provide examples of the type of information your company may be seeking. The caller can also provide answer options or ranges to the respondent.

- *Offer to send the study findings to the respondent as an incentive to participate in the benchmarking interview.* Some respondents may decline to share information if they doubt they'll gain anything in return. Many managers, though, are eager to learn what other companies are doing in their function. Consequently, they often participate solely so that they can receive a summary of the project findings.

- *The caller should express sincere thanks at the interview's conclusion.* The benchmarking team member wants the respondent to hang up the phone with a positive feeling. More than just reflecting good etiquette, a thank-you note and summary of the interviews usually persuade subjects to answer follow-up questions should they be necessary.

Recognizing Benchmarking Partners

Answering a few fundamental questions can help the benchmarking team identify those partners whose experiences are likely to be most relevant; consider the following generic questions that benchmarking teams will want to address before selecting benchmarking partners:

- Is the benchmark partner comparable financially (similar revenues, sales, profits)?
- Is the benchmark partner comparably sized (similar number of employees and market share)?
- Does the partner engage in comparable functions (similar work process, methods, practices)?
- Does the partner have comparable outputs (similar products and services)?
- Does the partner have comparable requirements (similar customer expectations)?
- Does the partner have comparable logistics (similar set-up and work flows)?
- Does the benchmark partner have comparable inputs (suppliers and ingredients)?
- Is the benchmark partner part of a comparable industry (similar products and markets)?
- Does the partner have comparable organizational and divisional structures?
- Is the partner part of a comparable market sector (public, private, governmental)?

Benchmarking Question Sets

In addition to these generic questions, the benchmarking team will also develop a more technically focused set of questions. This benchmarking question set is a valuable tool to help the benchmark team focus on the most critical considerations, such as how the site-visit partner performs its process, work practice differences, cost differences, performance advantages, technology differences, structural differences, training approaches, and other important operating issues. A good benchmark question set can make or break the site visit. Gregory Watson, a benchmarking expert at Xerox Corporation, shares two "war stories" that are especially instructive. In the 1980s, when Watson worked for Hewlett-Packard, a company arranged to benchmark HP's successful new product-development process. One small problem: The company presented 57 questions to be covered in a four-hour meeting, allocating roughly four minutes per question. When the company arrived at the site visit meeting with HP, they presented Watson with a pared-down list of 21 questions. He subsequently put away information for the 36 other questions he'd already prepared. "That company didn't get very much," Watson recalls.

In contrast, Watson also remembers "very productive" benchmarking visits at HP from a Ford Motor team that focused intently in a four-hour period on four key questions: (1) How did Hewlett-Packard measure performance in the benchmark area? (2) What were the performance results? (3) What was the HP process? and (4) How did HP handle a host of areas that were troublesome for Ford?[9]

Accessibility of Benchmarking Partners

Not all site visits require extensive travel. Often, by examining excellent companies outside one's own indigenous industry, companies can find benchmarking partners from which they can learn a great deal and which also lie close to home. Geographic proximity, of course, should not be an overwhelming consideration when seeking benchmarking partners. Performance excellence counts most. But in times of shrinking budgets, accessibility is a worthwhile consideration. Benchmarking teams are encouraged to consider a partner's total accessibility, which includes factors such as the cost of collecting data, available time, willingness, or ability to participate in the full scope of the project, distance to the partner's site, likelihood of receiving needed data, and confidentiality.[10]

This last consideration—confidentiality—is critical for initiating and maintaining openness and trust between benchmarking partners. Several dozen corporations that belong to The Council on Benchmarking of the Strategic Planning Institute and The International Benchmarking Clearinghouse of the American Productivity and Quality Center have together adopted The Benchmarking Code of Conduct, which codifies some general principles governing ethical behavior when benchmarking.

The Benchmarking Code of Conduct

To contribute to efficient and ethical benchmarking, individuals agree for themselves and their organizations to abide by the following seven principles for benchmarking with other organizations:

1. *Principle of legality.* Avoid discussions or actions that might lead to or imply an interest in restraint of trade: market or customer allocation schemes, price fixing,

dealing arrangements, bid rigging, bribery, or misappropriation. Do not discuss costs with competitors if costs are an element of pricing.

2. *Principle of exchange.* Be willing to provide the same level of information that you request in any benchmarking exchange.

3. *Principle of confidentiality.* Treat benchmarking interchange as something confidential to the individuals and organizations involved. Information obtained must not be communicated outside the partnering organizations without prior consent of participating benchmarking partners. An organization's participation in a study should not be communicated externally without their permission.

4. *Principle of use.* Use information obtained through benchmarking partnering only for the purposes of improvement of operations within the partnering companies themselves. External use of communication of a benchmarking partner's name with their data or observed practices requires permission of that partner. Do not, as a consultant or client, extend one company's benchmarking study findings to another without the first company's permission.

5. *Principle of first party contact.* Initiate contacts, whenever possible, through a benchmarking contact designated by the partner company. Obtain mutual agreement with the contact on any hand-off of communication or responsibility to other parties.

6. *Principle of third party contact.* Obtain an individual's permission before providing their name in response to a contact request.

7. *Principle of preparation.* Demonstrate commitment to the efficiency and effectiveness of the benchmarking process with adequate preparation at each process step, particularly at initial partnering contact.

Reach Out

After the benchmarking team organizes completely, a smaller subteam often visits a few select benchmarking partners to collect on-site information and view operations. Frequently this detailed, on-site review helps the benchmarking team learn what specific practices and operating principles enable the partner to achieve its superior performance. *(Figure 4.2d.)*

Conducting Site Visits

To gather more detailed information than that developed in telephone interviews or surveys, many companies conduct site visits. The most interesting and credible method of gathering information, the site visit affords firsthand feedback. Site visits also often result in long-term relationships that foster an on-going exchange of information between benchmarking partners. Site visits are the most time-consuming and expensive method of gathering information due to the travel required. However, they produce the highest quality of information because companies view

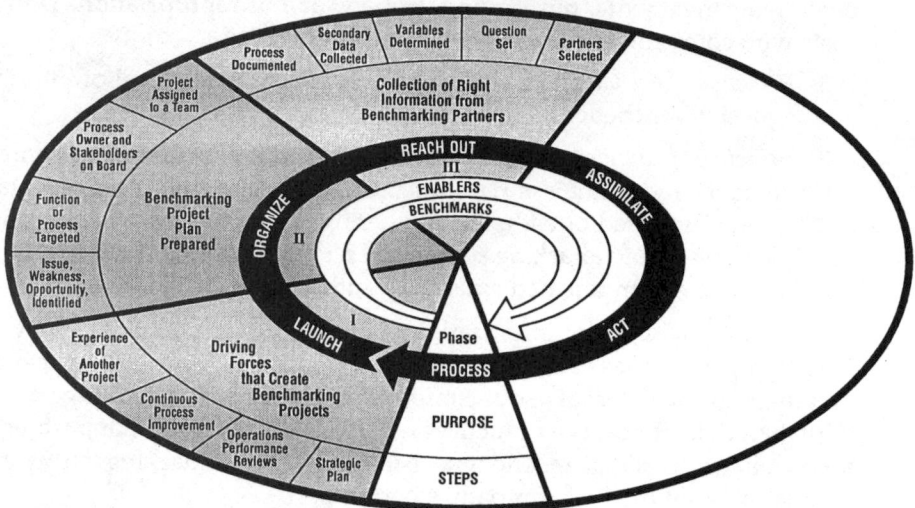

Figure 4.2d. Benchmarking process.

work processes, methods, and practices in action. The following nine guidelines can help your site-visit teams be more effective:

1. *Collect and absorb information about the partner before the visit.* Advance research helps the benchmarking team more quickly understand its partner's process.

2. *Prepare a question set in advance to guide and structure your team's visit.* These questions will help the team stay focused and develop best practice information in the areas of greatest interest. In addition to benefiting itself, a prepared benchmarking team demonstrates respect and courtesy for the partner's time. Partners can easily detect unprepared visitors. Though they may not show their disappointment during the visit, many host organizations will feel insulted that the visitors wasted everyone's time by not investing the time to learn rudimentary facts about the host company. Some companies will decline to participate in future exchanges with a company if a benchmarking team seems ill-prepared and unprofessional.

3. *Prepare the benchmarking partner for your team's visit.* To ensure that the partner will be prepared, the visiting company should send its question set, agenda, and a project information package before the visit.

4. *Use the questionnaire to structure your site visit and discussions.* Before any tours and in-depth discussions begin, ask for flowcharts, diagrams, or other published materials that explain the process. A few flowcharts or diagrams can save hours of discussion about how an intangible work system is structured.

5. *Travel in pairs or small groups during the site tour.* Traveling in small groups affords the team several advantages: One person can take notes while another converses; the group can split up to view different operations; group members can validate discussion points and observations.

6. *Arrange for follow-up conversations should there be additional questions.* This provides a way to quickly clarify any unclear points or observations that may surface in debriefings following the site visit. Experienced teams almost always send a summary of site visit observations to the partner to ensure the team has correctly interpreted the work system's key practices.

7. *Conduct a post-session debriefing to discuss the team's observations and ideas.* Site visit participants should debrief immediately following the visit while their thoughts and observations remain fresh and clear.

8. *Prepare a trip report summarizing site visit findings and conclusions.* This report will provide the means to transfer the site visit team's learning to other team members and colleagues.

9. *Send a thank-you note and confirm the accuracy of site visit notes describing the benchmark partner's operations.* The thank-you note shows the visitors' appreciation for their partner. Validating site visit notes ensures the benchmarking teams' observations and conclusions are 100-percent accurate.

The GTE Case Study—The Reach Out Phase

During the *reach out phase*, GTE's major-accounts sales benchmarking team set its critical performance variables, collected data, selected four benchmarking partners and, using a formal questionnaire to guide and focus in-person interviews, completed site visits to the four companies. Among the benchmarking partners, one company was a leader in information technology major-account sales; one company was based in the aluminum industry; another was an electronics corporation, and the last was a computer hardware company. "This was the most enthusiastic part of the study," recalls GTE project manager, Bob Kirschner. "It gave me an opportunity to meet with and develop relationships with companies that had been identified as the best in the class in my particular areas of interest. These companies' enthusiasm was amazing. It allowed them to brag about their company and how well they perform the tasks I wanted to talk with them about."

Assimilate

During the fourth phase, the *assimilate phase*, the benchmarking team assimilates the best practice information it has developed and prepares this information and corresponding improvement recommendations for senior management's review. During the assimilation phase, the team normalizes any measures that may still be in different reporting formats, studies and highlights performance gaps produced by different operating approaches, targets future performance goals, and develops change recommendations. The outcome of the assimilation phase is a best practices report. *Figure 4.2e illustrates the launch, organize, reach out, and assimilate phase of the benchmarking process.)*

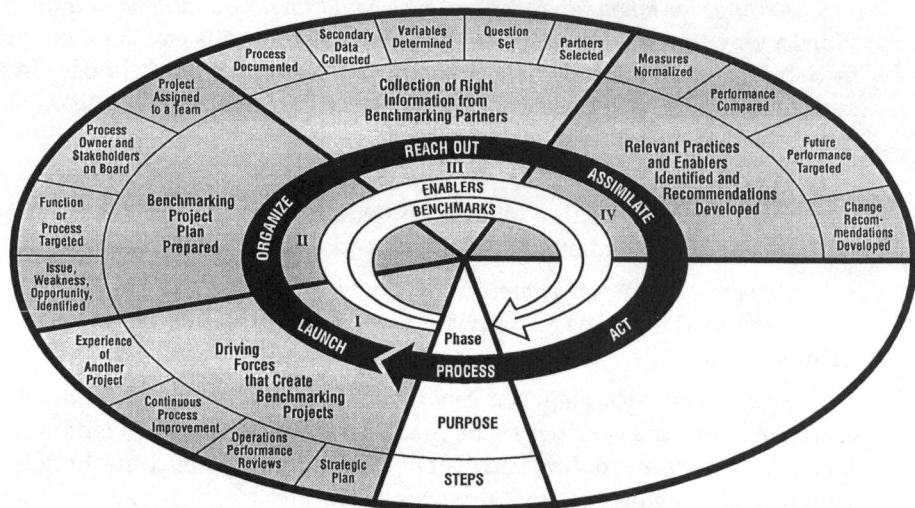

Figure 4.2e. Benchmarking process.

Normalizing Data

To ensure management can easily review benchmark information, the benchmarking team normalizes all measures used to compare performance across companies. This is accomplished by mathematically deriving common denominators for all performance indicators. For instance, if the partner measures billing productivity per week and the benchmarking team's company measures productivity per month, the measures are normalized to daily, weekly, monthly, or some other common denominator. Once the measures are in a form that can be easily compared, the team charts its company's measures against the benchmarking partners' measures. Graphical displays like those discussed in Chap. 3 often best highlight performance gaps between the host company and the benchmark partners. Based on the performance gap analysis, the benchmarking team develops a recommendation for change. Typically, the team adopts some version of four basic process improvement strategies. These strategies, which are discussed more fully in Chap. 9, include:

- *Immediate tactical change.* The organization refines the existing process through simple process improvement.

- *Immediate strategic change* or *organizational restructuring.* The organization seeks to quickly leapfrog forward by adapting the best practices of the best companies and by reforming itself in a new, more competitive image.

- *Long-term tactical change* or *managed reform.* The organization emulates the benchmark companies and gradually adapts their more effective practices in a way that suits its existing culture.

- *Long-term strategic change* or *process reengineering.* The organization seeks to rewrite the rules of the game in its industry by redesigning its core processes based on a brand new operating model that frequently is inspired by other excellent companies operating outside the organization's indigenous industry.

Preparing the Benchmarking Report

After concluding research, surveys, telephone interviews, and site visits, the benchmarking team completes its duties by preparing a best practices report that summarizes its findings and recommendations. The benchmarking report consists of several key components. It includes a mission statement that describes the purpose of the benchmarking study and a statement of the problem or opportunity under review. A description of the study process then follows and includes the people involved in the project, the departments and operating units represented, and the methods of gathering information. The sources of information can include: companies, industry professionals, academic experts, government experts, books, magazines, reports, consultants, industry associations, and primary and secondary research.

A key to the best practices summary report lies in the performance indicators and practice findings. Performance indicators may include measures of productivity, quality, speed, yield, staffing levels, customer satisfaction, utilization rates, error rates, and cost. Key findings include best practice descriptions, process diagrams, critical success factors and general observations.

To display performance measurement results, benchmarking teams can use several different graphical tools. Each tool has a unique function. The *matrix* effectively summarizes information and presents comparisons. The comparison matrix relates performance criteria (listed in left-hand column of Fig. 4.3) to a group of benchmark partner companies (listed in the top row). For example, it might present performance of companies A, B, C, and D for profitability, cycle time, defects per million, industry type, and process management techniques. The matrix allows for the inclusion of descriptive text and metrics. (*Figure 4.3 presents an example of a comparison matrix.*)

The *bar graph* usually illustrates performance differences in a single area in contrast with the comparison matrix that may review many different performance dimensions. Using a bar graph, the benchmarking team can compare itself to companies A, B, and C for training hour per employee per year. A pareto format, where the bars are prioritized into ascending or descending order, makes the graph especially easy to read.

If not presented in pareto format, the graph is often plotted chronologically. (*Figure 4.4 presents an example of a bar graph.*)

The *line graph* also usually depicts comparisons in a single performance dimension. For example, the host company might compare its performance for on-time delivery to that of a world-class company from another industry with similar activities, or to its toughest competitor, or to its industry average. (*Figure 4.5 presents an example of a line graph.*)

Some benchmarking teams favor a version of the line chart called the *Z-chart*. The Z-chart relates a company's performance over time to that of a benchmark partner. The

Criteria	Your company	Company A	Company B
Profitability	Excellent	Excellent	Excellent
Cycle Time	reduced 10%	reduced 40-50%	reduced 25%
Defects per million	900	500	200
Industry Type	Banking	Pharmaceutical	Electronics
Process management techniques	Flow charting & process mapping	Value analysis & cycle time focus	FMEA & systematic process simplification

Effective for summarizing and presenting comparisons.

Figure 4.3. The comparison matrix.

Z-chart therefore depicts historical performance and it also projects future improvement rates for all comparison companies; consequently, the Z-chart is a helpful tool to relate four critical change variables: (1) How much improvement will be required to match or surpass the benchmark partners? (2) What rate of improvement will be required to catch or surpass the benchmark partners? (3) How long will it take to match or surpass the benchmark? (4) Where will the company stand in relation to its partners if it successfully implements gap-closing improvements? *(Figure 4.6 presents an example of a Z-chart and Fig. 4.7 an example of the spider chart.)*

The *spider chart* (also called a radar chart or an M^2 chart, which was discussed in Chap. 3) displays performance positions and gaps between the host company and its benchmark partners on one or many performance measures. The spider chart is especially effective because—in a single view—it illustrates performance gaps along many different dimensions, highlighting the best practice company along each measure.

These chart formats are the essential tools of a benchmarking team seeking to summarize in quantitative terms the performance in the benchmark area. In addition to the analysis of performance gaps, the benchmarking report presents descriptions of the best practices and the critical success factors that enable the superior performance in the benchmark partners' processes and work systems. Following the performance summary come various support materials developed in the study. Such materials might include questionnaires, interview summaries, site visit reports, key research articles, study cost figures, and comparison flowcharts. Some benchmarking teams also provide a summary of best practice benefits, scenario analysis, improvement recommendations, an action plan and cost estimates for implementation, an implementation schedule, and performance goals and milestones. The level of final detail, of course, depends on the expectations of the senior managers who have requested and championed the benchmarking project.

The GTE Case Study—The Assimilate Phase

After completing its fact-finding and site visits, GTE assimilated a large body of collected information and focused on the best practices of the site visit partners. "This phase required the most time due to inconsistent measurements among our benchmarking partners," recalls GTE project manager, Bob Kirschner. "I had to take painstaking steps to analyze, digest, and understand the meaning of our findings." The success of this phase is highly dependent on how well focused and detailed the question set is that's developed during the reach out phase.

In GTE's case, the project results supported a host of proposed changes, including:

- Establishing higher ratios of sales people to sales managers
- Setting turnover targets for sales
- Increasing investment in specific sales training for employees
- Setting higher requirements for the amount of time spent by each sales employee with customers
- Reducing the amount of time sales people spent on administrative functions.

Act

During the fifth and final benchmarking phase, the *act phase*, the team works with management and the process owners to prioritize recommendations and agree on an implementation strategy. Agreement leads to formalization of action plans, implementation schedules, measurement and tracking mechanisms, and recalibration plans. When the action plans are completed, on-going responsibility for managing the improvement efforts often shifts to an implementation team. The benchmarking team still supports the implementation team, helping place tracking mechanisms, develop process standards, cycle-time measures, quality measures, and statistical control charting. Over time, the process or system is continuously fine-tuned and improved. Meanwhile, the organization must eventually recalibrate its benchmarks based on the newly established best practices. *(Figure 4.2 illustrates the launch, organize, reach out, assimilate, and act phases of the benchmarking process.)*

The GTE Case Study—The Act Phase

At GTE, the change recommendations prepared by the major-accounts sales benchmarking team were fully compatible with the parallel findings of a corporate process reengineering team that was evaluating GTE's customer contact processes. Consequently, the benchmarking team's improvement recommendations were incorporated into the action plans of the process reengineering implementation team. The implementation plan included monitoring and tracking the impact of the changes; consequently, improved performance benchmarks were put in place. "It is critical during this phase," warns GTE project manager Kirschner, "to link and

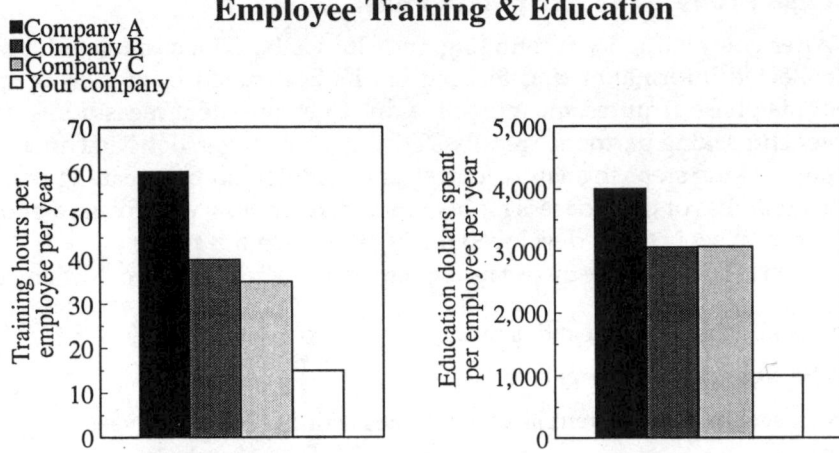

Graphically illustrates performance differences.

Especially easy to read when presented in pareto format.

Figure 4.4. The bar graph.

integrate action plans coming out of the benchmarking project with action plans coming out of other company initiatives—which in our case involved business process reengineering."

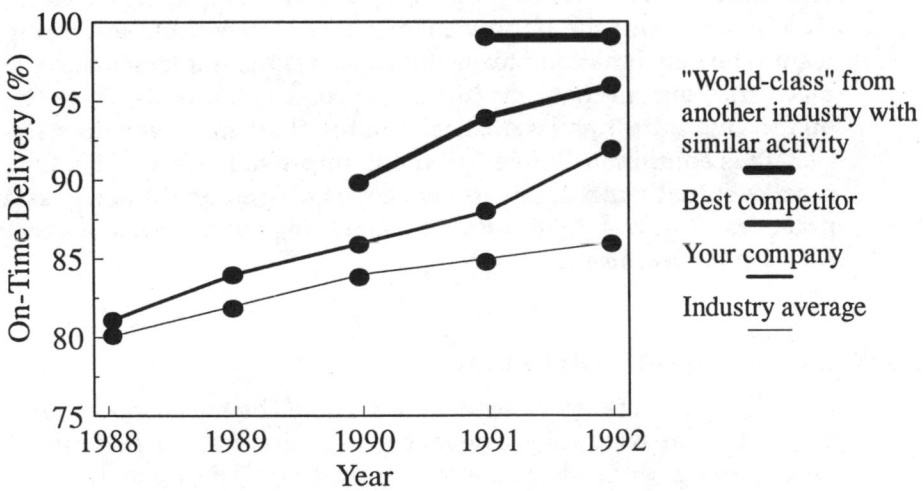

Particularly useful for comparing different companies' performances over time and relating your company to best practice companies.

Figure 4.5. The line graph.

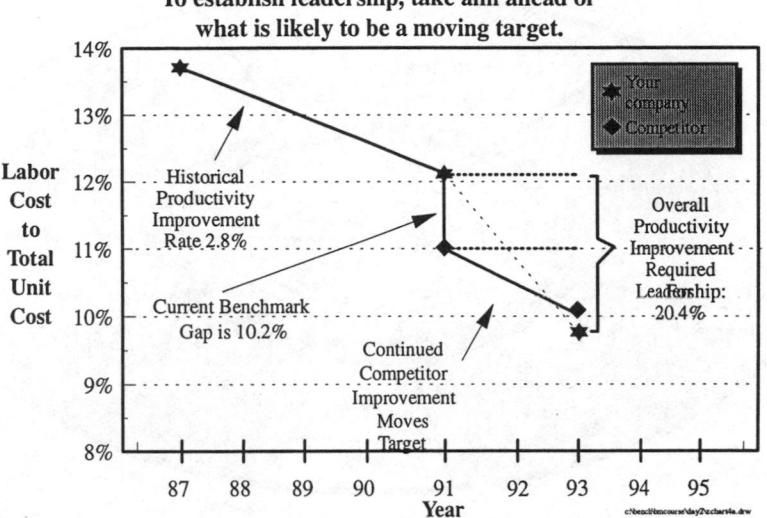

Figure 4.6. The Z chart.

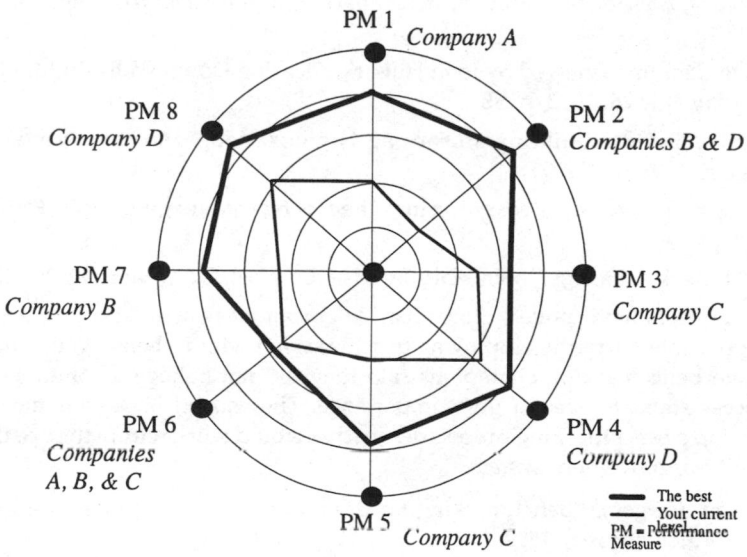

Figure 4.7. The spider chart.

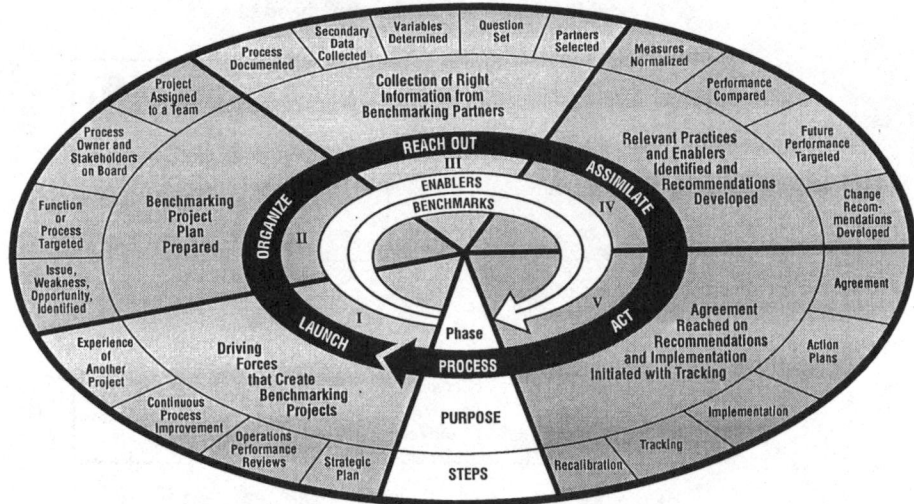

Figure 4.8. Benchmarking process.

References

1. Schiller, Zachary, "Procter & Gamble Hits Back," *Business Week*, July 19, 1993, pp. 20-22.

2. Kennedy, Eric, "Developing a Benchmarking Program: IBM's Journey," a presentation made at the "Making Benchmarking Work" conference sponsored by *Industry Week* magazine and the International Quality and Productivity Center, July 28, 1993, San Francisco, California.

3. Biesadra, Alexandra, "Strategic Benchmarking," *Financial Week*, September 29, 1992, pp. 34-35.

4. Welch, Jack, interviewed by John Hillkirk, "Tearing Down Walls Builds GE," *USA Today*, Monday, July 26, 1993, p. 5B.

5. Garvin, David, "Building a Learning Organization," *Harvard Business Review*, July-August, 1993, pp. 78-91.

6. Stewart, Thomas A., "Reengineering: The Hot New Managing Tool," *Fortune*, August 23, 1993, pp. 41-48.

7. Welch, Jack, "Tearing Down Walls Builds GE," *USA Today*, Monday, July 26, 1993, p. 5B.

8. The authors have polled more than 20 corporate benchmarking managers at major corporations currently employing benchmarking within their organizations. These corporate benchmarking champions said they did not believe a common benchmarking process standard was of great importance. They stated it was far more important to develop a benchmarking process design that would work within the existing culture and quality-improvement system.

9. Watson, Gregory, "Benchmarking for Competitiveness," a conference speech recorded by Productivity, Inc., 1992.

10. Zions, Bernie, *Benchmarking: Staying Competitive in the 1990's*, The Quality Network, 1991, 2:(14–16).

BENCHMARKING PROCESS

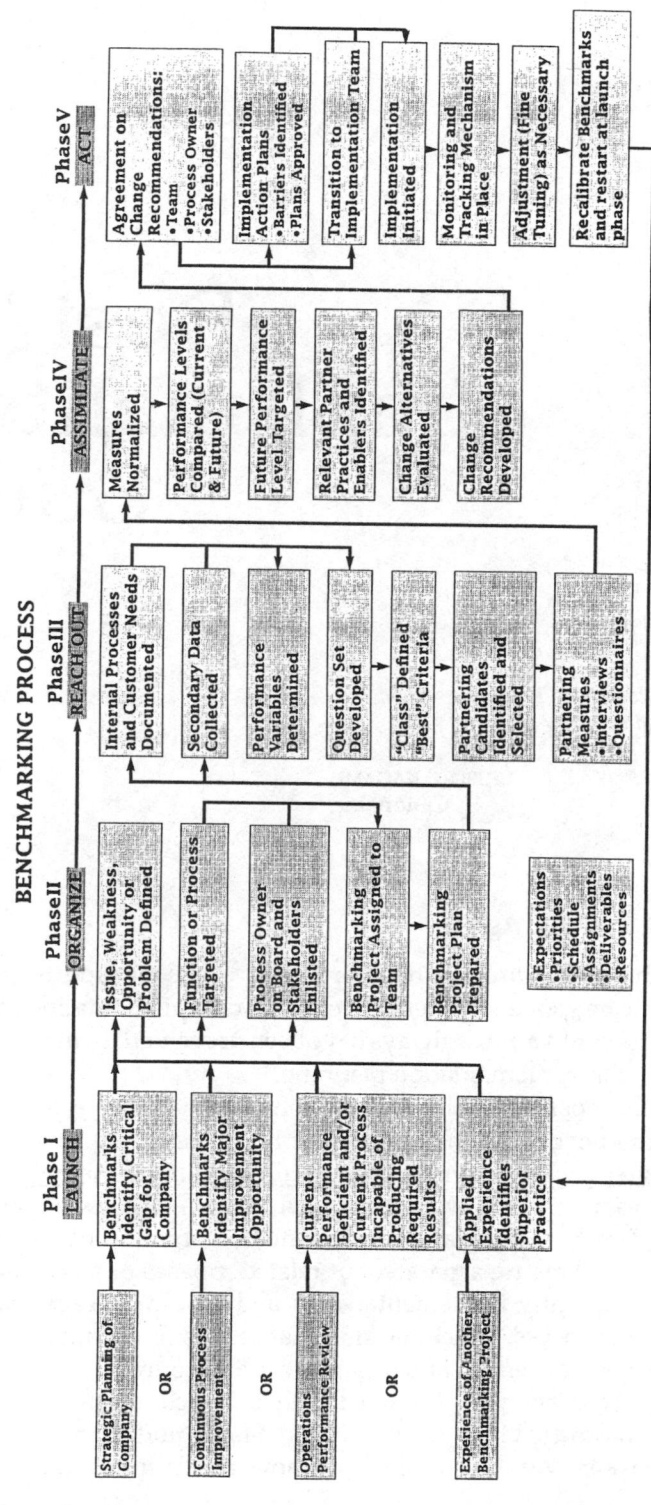

Figure 4.9. Benchmarking process.

5
Design for Implementation Success

*If clear water flows from the upper stream,
there is no need to purify it farther
downstream.*

HIROTO KAGAMI,
Canon, Inc.

Implementation Planning

Many benchmarking projects have produced excellent reports and recommendations—only to be spoiled by ineffective or failed implementation efforts. The causes for the failures could almost always have been prevented. In many cases, the failures result from poor implementation planning.

One of the most common pitfalls is failure to plan the research and fact-finding phases of the benchmarking process early so that they reflect implementation realities. Segregating benchmarking planning and implementation planning is like separating Siamese twins who share vital organs—the results are almost always calamitous. Benchmarking and implementation should not be considered as two distinct events. They are separate, but related, phases of a common improvement process. Consequently, implementation is, and should be, regarded as one with the benchmarking process. Benchmarking that results in no improvements is of little value; it wastes time, effort, and organizational resources.

The Japanese are well known for their benchmarking skills, practiced for decades from study visits to the United States and Europe. Consequently, an American visitor was taken aback by how much information a Japanese host company shared during the American's site visit. At the visit's end, the American guest expressed his surprise and complimented the Japanese on their openness to sharing. His surprise quickly turned to astonishment when the Japanese confided that they were not worried about sharing, because they did not believe the American company could implement what they saw.

This incident reflects how the Japanese view implementation as a critical phase of meaningful benchmarking. For many, benchmarking may imply only the acquisition of knowledge. But from a pragmatic managerial view, learning that does not translate into improvement actions should not be classified as learning at all. That is why every benchmarking project should be designed with an eye toward successful implementation.

DFM Corollaries

In manufacturing, the concepts of *design for manufacturability* or *concurrent engineering* have proven a powerful approach to producing better products. In simplest terms, DFM and concurrent engineering recognize the benefits of working in cross-functional teams. Design engineers, manufacturing engineers, equipment operators, marketers, and financial staff work together to design products. By bringing together all these points of view concurrently in the design process, companies develop better products more quickly. In this fashion, design engineers don't produce products that are overly difficult or unnecessarily costly to manufacture. Manufacturing staff help the design experts understand what they can most effectively manufacture in the factory, and early designs realistically reflect these capabilities. In turn, marketing staff communicate what customers require and what they believe will most successfully sell. Working together, these teams dramatically reduce the time it takes to develop products, with corresponding decreases in cost and improvements of product quality.

The principles of integrated communication and prevention-based planning that inspire concurrent engineering and IFM can be extended to benchmarking project management. This process of cross-functional teams concurrently planning all aspects of the benchmarking project might be called *design for implementation*—or DFI for short. DFI recognizes that successful research investigations identifying best practices are of little value if no improvements result through successful implementation. Consequently, benchmark teams must address critical implementation issues even as the benchmarking research and information gathering are being planned. DFI produces better project results by involving representatives from all operating areas that will ultimately implement any improvements. At times, more people may be beneficiaries of a benchmarking project's findings than can effectively work on a single team. In these situations beneficiaries or those who will be influenced directly or indirectly by the project's findings, should still be involved regularly—even if only through communications—in the project's planning and progress. In this way, they can still feel ownership for the project's directions and findings, even if they personally did not execute the work.

Common Pitfalls

To experienced benchmarkers, DFI is a compelling concept. Experience and research demonstrate that poor planning is a primary cause for benchmarking failure. In a survey of 88 companies conducted by the American Productivity and Quality Center's International Benchmarking Clearinghouse, poor planning was the most frequently cited reason for unsuccessful benchmarking efforts, followed by no top-management support, and no process ownership.[1] Other common obstacles to successful benchmarking include:

- Process owners who are not involved in the benchmarking process.
- Staff members who try to run a line project.
- Team members who don't represent all relevant functions and beneficiaries.
- A project mission that is poorly communicated to the team.
- Beneficiaries who are not adequately informed of the project, its purpose, benefits, and progress.
- An effort that is too broad or has too many parameters.

DuPont Case Study

The importance of early planning is highlighted in the experience of many companies. Consider DuPont as one example.[2] Several years ago, DuPont undertook a major benchmarking study to examine best practices in the area of preventive maintenance in its plants. The focus was clear: The company believed it could significantly improve its costs by performing more preventive and scheduled maintenance and less unplanned maintenance requiring repairs of failed and injured systems. At the outset, DuPont performed about 30-percent planned maintenance and 70-percent unplanned maintenance. It set what it thought was a stretch target, suggesting its plants' aim to reverse that ratio, performing 70-percent planned and 30-percent unplanned maintenance.

In the course of a major benchmarking study, DuPont studied plant maintenance at 17 DuPont plants, eight similar plants run by other companies, and seven nonsimilar plants outside its industry. "Maintenance is maintenance," recalled Samuel Bookhart, formerly DuPont's benchmarking manager. "You can learn from anybody who does maintenance well." To ensure the success of the study, DuPont project managers enlisted the involvement of four maintenance managers from the highest performing DuPont plants. To DuPont team members' surprise, the best plants in the study resided outside DuPont and were performing 96-percent planned maintenance. This level of planned maintenance made DuPont's "stretch target" of 70-percent planned maintenance look puny.

The benchmarking study led to important findings and recommendations concerning maintenance planning, maintenance scheduling, and preventive maintenance. The recommendations held out the promise of impressive cost savings for DuPont's family of 17 plants. Unfortunately, those improvements proved elusive. Two years after the study was performed, the only plants achieving significant improvements were those that were home to the four managers who spearheaded the original study. The other DuPont plants made little or no improvement, continuing to perform only 30- to 35-percent planned maintenance.

A post-mortem analysis of the disappointing results revealed the following:

- Thirteen of the 17 plant managers were not meaningfully involved in the original studies.

- The 14 plants that did not participate directly in the project might be regarded as beneficiaries and they did not feel any ownership for the benchmarking project or its findings. On-going communications did not occur during the original study. Consequently, beneficiaries were never actively engaged in the project.

- The best practice recommendations were not embraced or accepted by beneficiaries who felt excluded from the original project.

Faced with first disappointing implementation results, DuPont showed exemplary discipline. Other companies might have dismissed the project as a failure, but DuPont was determined to get right what it had mishandled on first pass. The project champions set out to seek improvements in the remaining 14 plants. They started out by engaging the plants in the original project and its findings. To accomplish this goal, they interviewed managers at all the previously uninvolved plants, they sent out new questionnaires and undertook a more carefully planned recommunication process to demonstrate the benefits of implementing the project findings. With that extra implementation effort, all 17 DuPont plants improved significantly and the leading DuPont plant reached 95-percent planned maintenance levels.

DuPont's experience demonstrates several important principles: Project teams must carefully manage the dynamic relationship that exists among benchmarking constituents. Every major project seeking to achieve process change must have a sponsor supporting the benchmarking team. The team typically consists of two to 10 people and is frequently supported by a benchmarking advisor, who is an internal or external consultant experienced in benchmarking. The team then seeks to actively engage beneficiaries in the benchmarking process. *(Figure 5.1 illustrates the structure of a benchmark team and its beneficiaries.)*

Project Sponsorship

Involve all process owners in the project. If possible and feasible, place them on the benchmarking team. In many benchmarking projects, people from different functions and departments will participate. Some critical questions to ask at this juncture are:

- What process or function is the benchmarking focus area?

- What subprocesses within the function will be the specific targets for investigation?

- Who are the owner(s) of all processes and subprocesses?

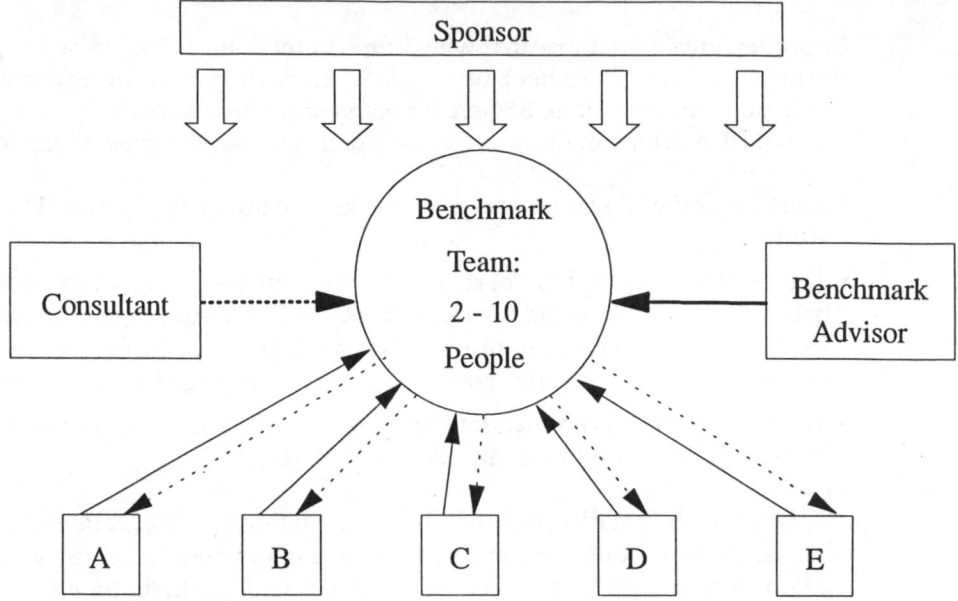

Figure 5.1. Benchmark team structure.

- Are process owners actively sponsoring and/or supporting the benchmark project?

Selecting Your Team

The selection of team members can at times prove difficult because more people may be involved in a function than can effectively serve on a benchmarking team. When selecting benchmarking team members, a few good rules of thumb apply. Team members should:

- Have line experience in the benchmark area.
- Be credible and competent in the process.
- Represent all primary functions or areas that own or influence the benchmark subject.
- Feel ownership and importance of the process.
- Have a designated leader.
- Be team-oriented and team players.

Roles and Responsibilities of Team Members

The roles and responsibilities of benchmarking constituents shift during the benchmarking project. Planning becomes especially challenging because of this dynamic. Like starters on a basketball team, players move in and out of lead shooting position.

The cumulative group effort—not the efforts of any one player—produces victory. The benchmarking sponsor, for instance, serves an active and primary role in the early project stages but remains on the sidelines during much of the information gathering and outreach. Other constituents, such as team members, advisors, consultants, or beneficiaries, play active and passive roles at different project stages. (*Figure 5.2 depicts the changing roles and responsibilities of benchmarking team members.*)

Gaining and Maintaining Beneficiaries' Support

Engaging beneficiaries in the benchmarking project can prove especially challenging. At times, the beneficiaries or their operating units may not be directly involved in the project. Consequently, benchmarking teams must take special care and be creative in building bridges. Some approaches that have proven effective in bringing beneficiaries into the benchmarking process include:

- Involving the beneficiaries and their employees on benchmarking teams, whenever possible.

- Communicating to all beneficiaries regular progress reports and important findings.

- Seeking feedback and counsel from beneficiaries to understand your organization's process capabilities in the benchmark target area.

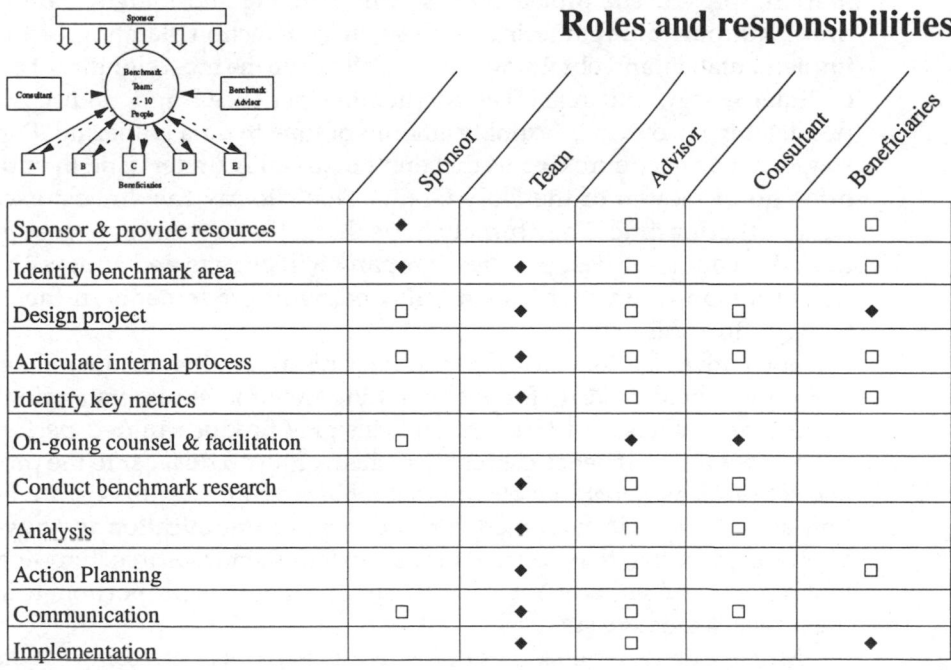

Roles and responsibilities

	Sponsor	Team	Advisor	Consultant	Beneficiaries
Sponsor & provide resources	◆		□		□
Identify benchmark area	◆	◆	□		□
Design project	□	◆	□	□	◆
Articulate internal process	□	◆	□	□	□
Identify key metrics		◆	□		□
On-going counsel & facilitation	□		◆	◆	
Conduct benchmark research		◆	□	□	
Analysis		◆	□	□	
Action Planning		◆	□		□
Communication	□	◆	□	□	
Implementation		◆	□		◆

Figure 5.2. Roles and responsibilities.

- Seeking feedback and counsel from beneficiaries who are developing question sets to evaluate other organizations' performance and capabilities in the benchmark target area.

- Piloting the question sets among employees in the beneficiary units to test their ease-of-use and to involve beneficiaries in the project.

- Seeking feedback from beneficiaries to develop and validate the list of benchmark partners.

- Sharing findings from site visits with the beneficiaries.

- Seeking early feedback from beneficiaries, concerning study findings before formal presentations.

- Making formal presentations to each beneficiary, tailoring the presentation to relate the findings and recommendations to the individual operating unit.

- Whenever possible, visiting beneficiaries at their facilities, branches, or offices. This is much more effective than inviting them to your office or communicating through memo.

The Benchmarking Learning Curve

Benchmarking, like other management initiatives, is subject to learning curve effects. Not surprisingly, benchmarking tends to grow easier, faster, and more effective with practice and experience. Consider the general sequence of a benchmarking project. The project moves from planning and understanding your own process through data gathering, analysis, insights, action planning, communication, implementation, and obtaining buy-ins. Where do the most significant expenditures of time, energy, and resources accrue? Inexperienced and undisciplined teams predictably devote an inordinate amount of time to data gathering. For many, this stage can be a quagmire. Teams that don't thoroughly understand their own process often go in "search of the Holy Grail." That's to say they invest excessively in data-gathering time. They futilely hope that if they accumulate enough information, then one clear best practice company will be revealed and, with it, insights will shine down on them, soon followed by improvements in their processes. Seldom does this occur.

Such investigations usually produce disappointing results. Thorough data gathering cannot make up for poor planning and process understanding. Moreover, important findings and breakthrough insights produce muted performance improvement results if beneficiaries reject them or turn a deaf ear to the project reports and action plans. Benchmark teams that achieve optimal results also invest adequate time and resources in early planning and in the communication and buy-in process. They recognize that these phases may seem less glamorous than hopping on a plane to visit another company. Yet these phases can have a disproportionate influence on the project's ultimate success.

Much of the experience curve effect is attained by compressing the time required to perform data gathering. As benchmarking teams grow more proficient, they perform data gathering more quickly. They also recognize that the actual data gathering is often less important than the way findings are implemented in the organization. Agonizing over which company is "the best" is futile. For most organizations, any company that is demonstrably better than your own can be a worthy role model. Since benchmarking is meant to be an "evergreen" process—it should be regularly updated—other companies may appear as benchmark partners in the future. Comparing your own operations over time to companies and industries with striking different approaches can be beneficial. As paradoxical as it may sound, comparing apples to oranges can at times produce the greatest insights.

(Figures 5.3–5.7 depict the time devoted to various benchmarking phases during failure-prone, typical, and optimal benchmarking projects.)

Designing your benchmarking projects for successful implementation is not especially difficult or complex. Careful planning, consideration of all constituencies, and attention to detail are all critical success factors. As the old saying long ago observed: "God lives in the details." Design for implementation recognizes this imperative to engage all constituencies and link their interests and efforts. Benchmark teams that embrace the DFI imperative can look forward to significant performance improvements. Those who forget or neglect this imperative will discover the dark side of the old management maxim: The devil also lives in the details.

References

1. "Surveying Industry's Benchmarking Practices," American Productivity & Quality Center's International Benchmarking Clearinghouse, Houston, Texas, 1992, p. 20.

2. Bookhart, Samuel, former benchmarking manager speaking at the Benchmarking Against The Best Conference, sponsored by the Institute for International Research on June 13, 1991.

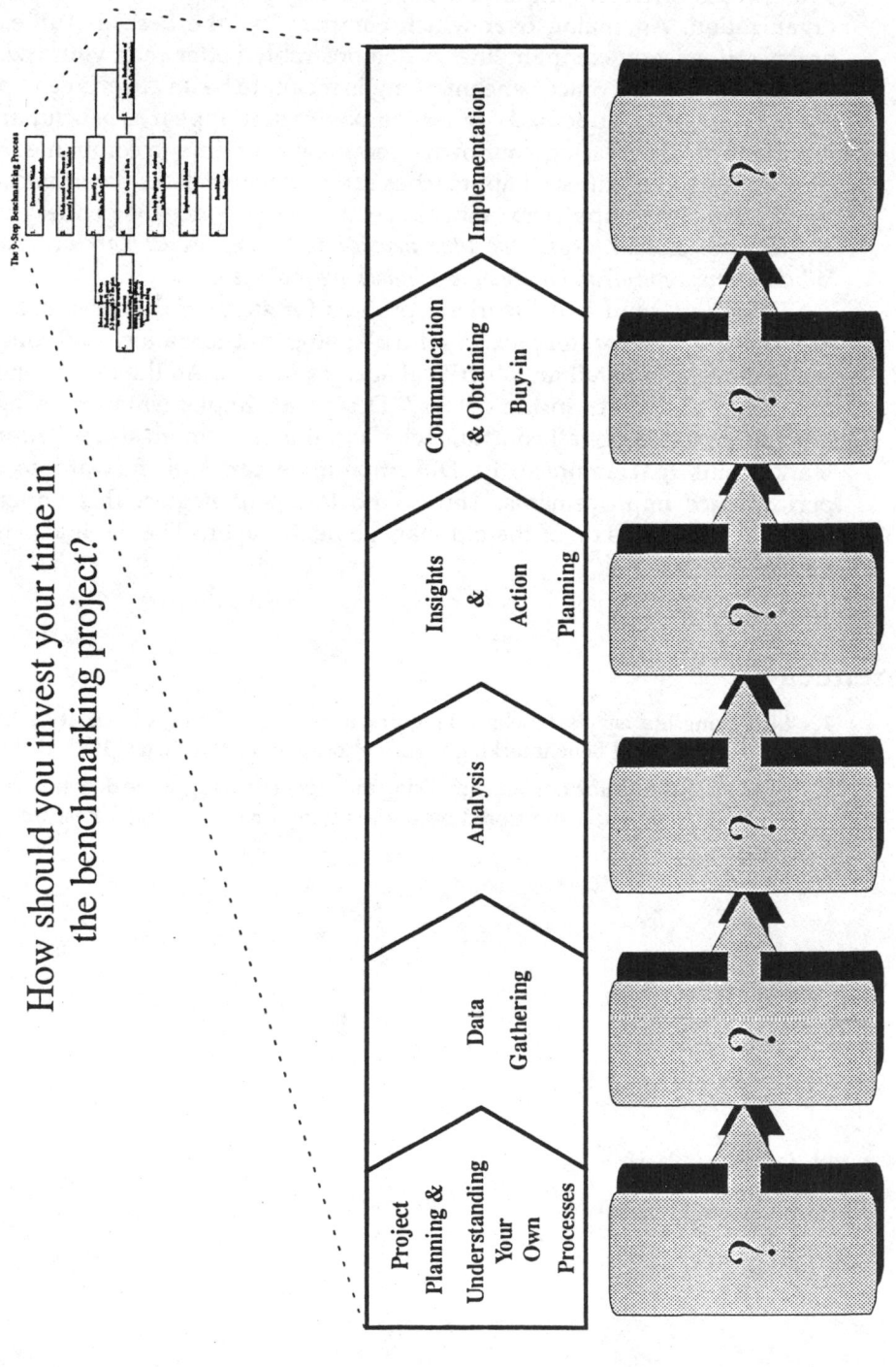

How should you invest your time in the benchmarking project?

Figure 5.3.

116

Failed benchmarking initiatives often make the same mistakes!

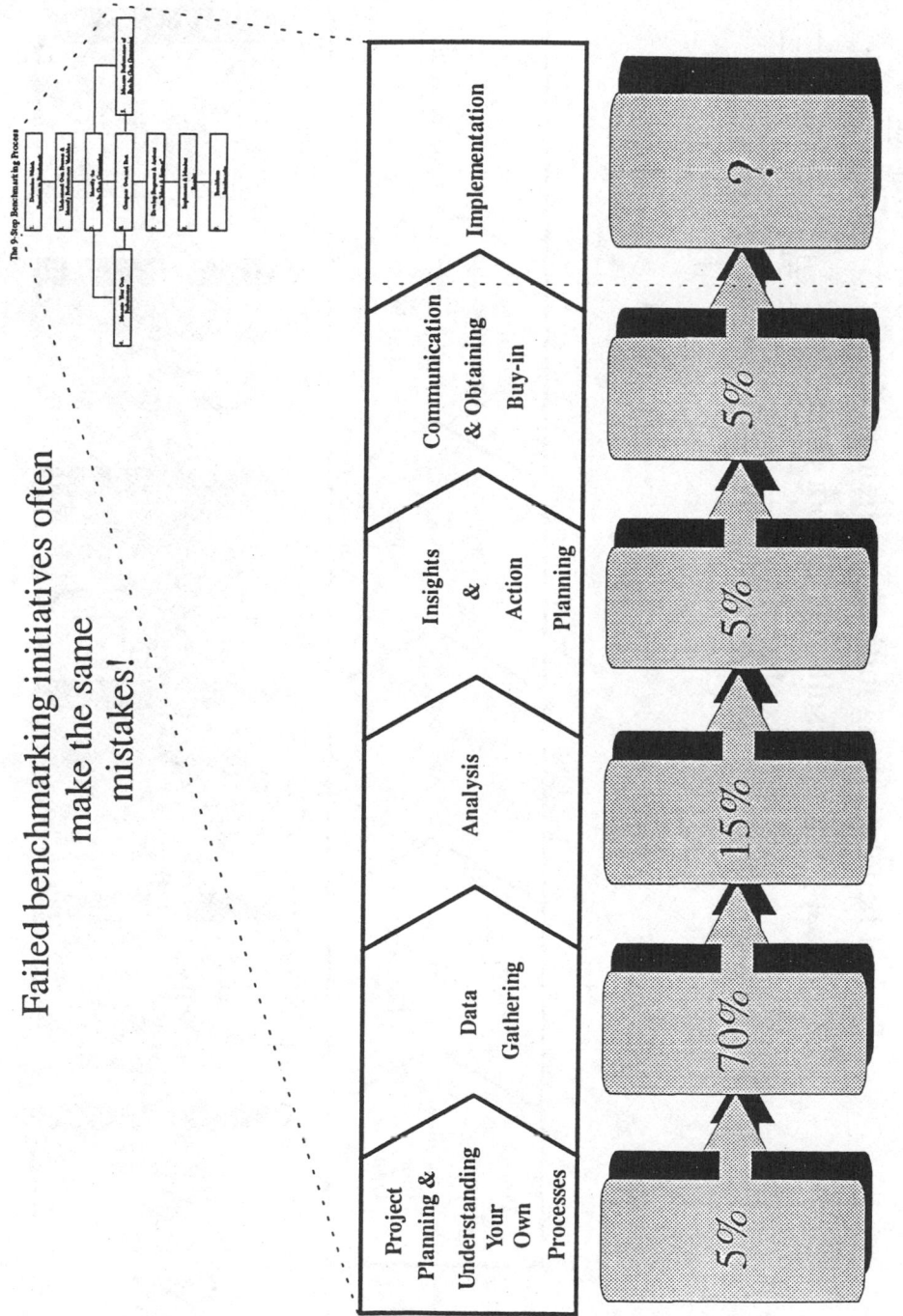

Figure 5.4.

Typical benchmarking initiatives often produce disappointing results due to poor time allocations!

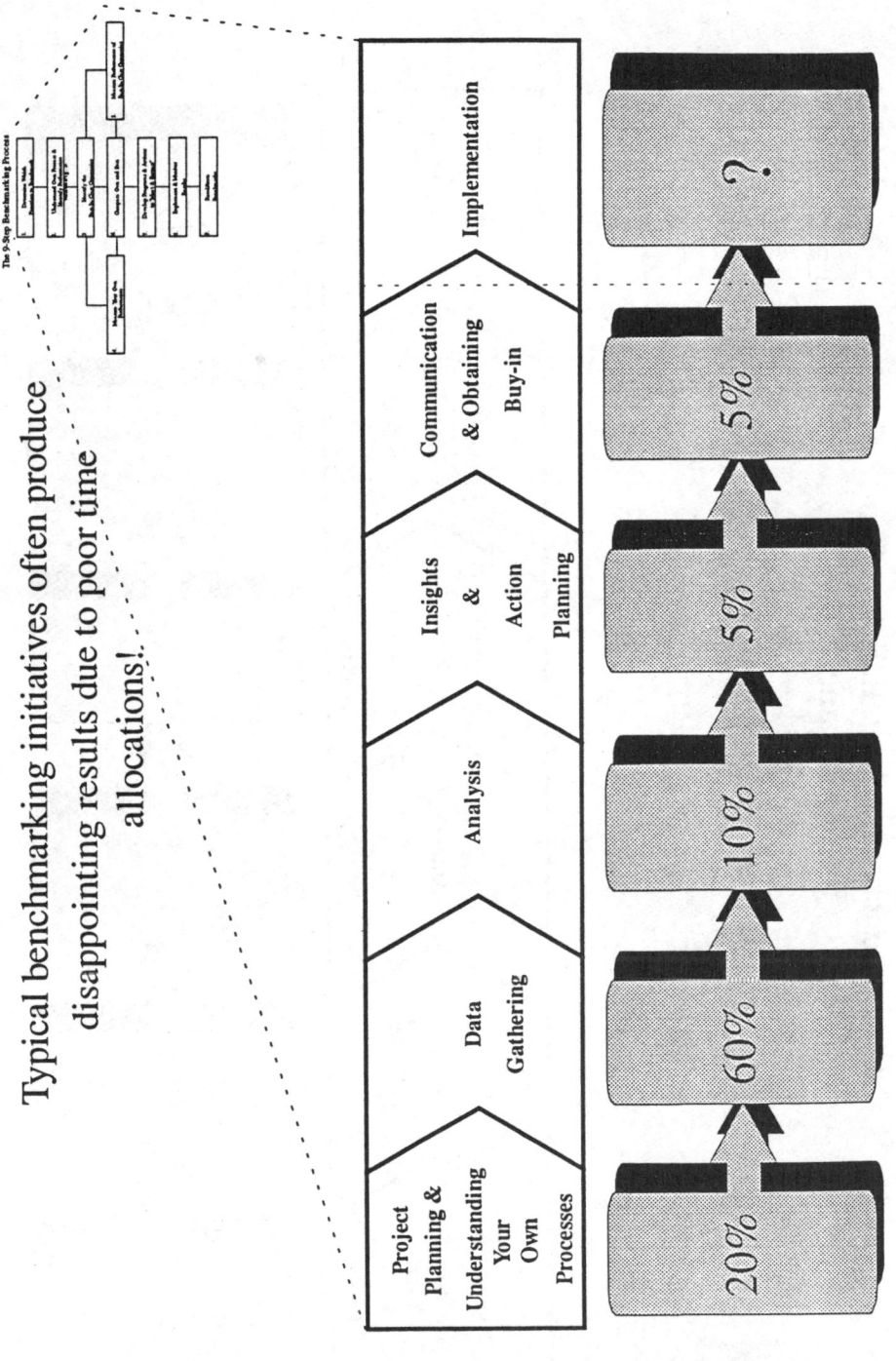

Figure 5.5.

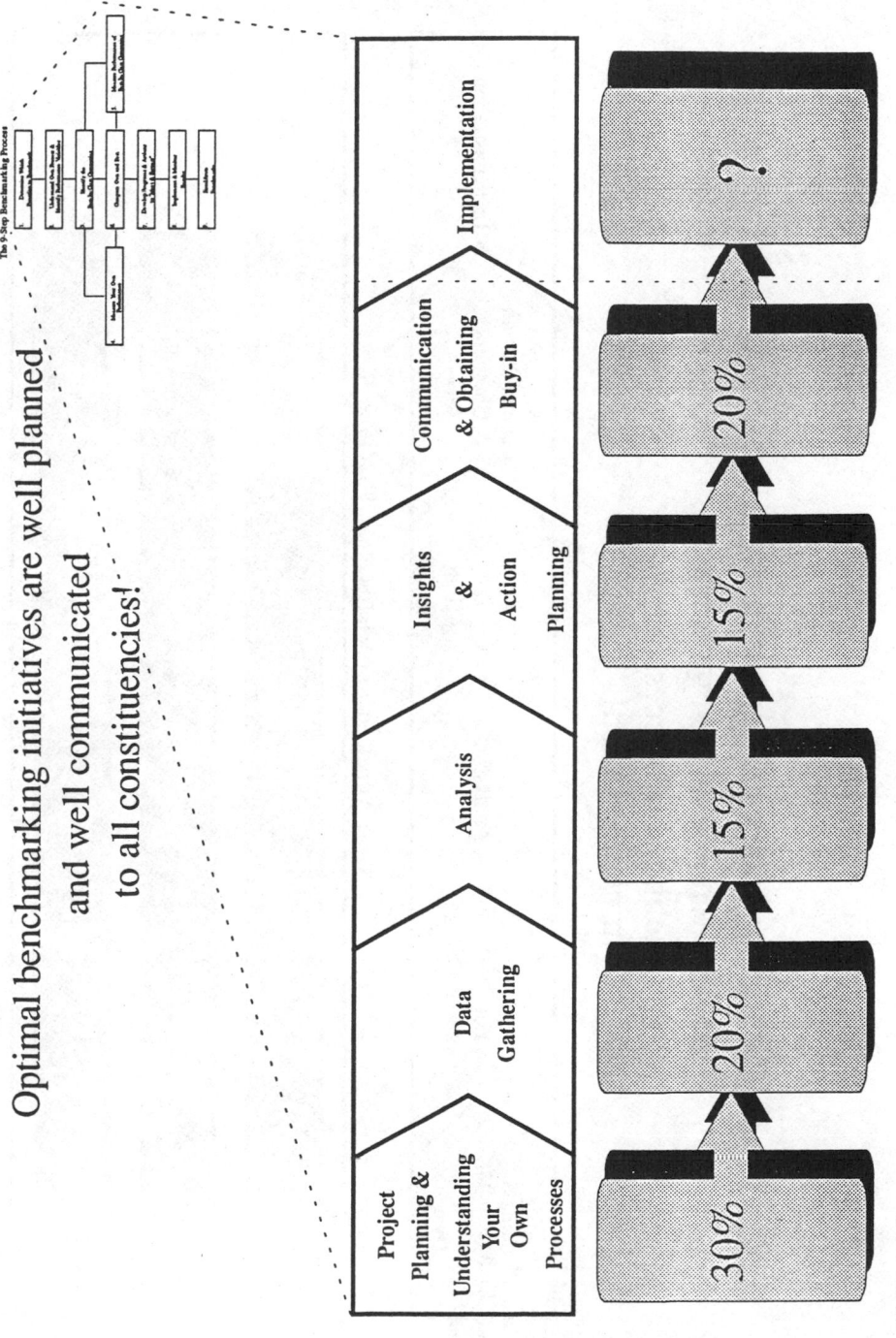

Optimal benchmarking initiatives are well planned and well communicated to all constituencies!

The 9-Step Benchmarking Process

Project Planning & Understanding Your Own Processes	Data Gathering	Analysis	Insights & Action Planning	Communication & Obtaining Buy-in	Implementation
30%	20%	15%	15%	20%	?

Figure 5.6.

119

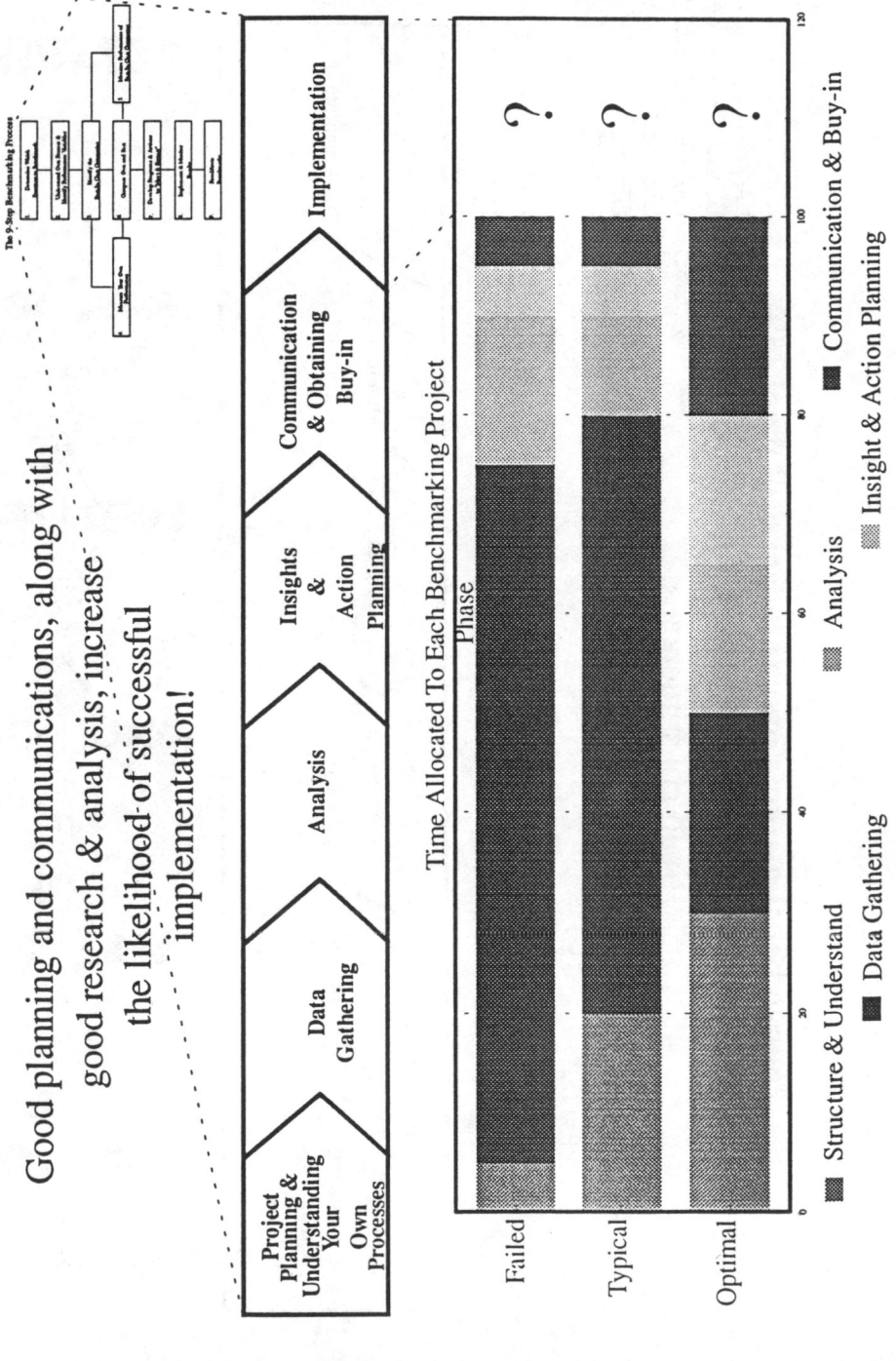

Good planning and communications, along with good research & analysis, increase the likelihood of successful implementation!

Figure 5.7.

6

Integrating Benchmarking into Your Organization

The president of a large chemical producer paused and then picked his words carefully: "I don't think we're ready to undertake benchmarking." The comment seemed oddly placed from the chief executive of a company with advanced quality systems and quality processes. Then he cut to the heart of his concern. "I don't want our employees flying off to Germany, Japan, or who knows where trying to find best practice companies. We have too much to do right here at home."

He made his point well. It can be counterproductive for teams to search for world best practices and global benchmarking partners when improvement opportunities abound at home. Nevertheless, benchmarking's basic skills are not only applied while in search of world best practices. When asked if it would be beneficial for this chemical company's employees to understand what constitutes best practice techniques within the U.S. chemical industry, he answered without hesitation: "Absolutely! We must keep our competitors under the microscope if we are to stay competitive. Margins are thin in our industry. We can't afford to fall behind."

When asked if it would be beneficial to understand the best operating practices within his company's many plants, he replied even more passionately: "Absolutely! I'm always urging our managers to share experience and ideas more actively. It drives me crazy when our people are always reinventing the wheel. We have an incredible untapped resource in our best people and our best ideas."

The search for best practices does not always require an odyssey to distant places. The identification of internal best practices and of industry best practices also hold great value. However, the awe-inspiring stories that circulate about best-in-world benchmarking have created more than a few misconceptions about what benchmarking is and should be. In the minds of many managers, benchmarking has *mistakenly* come to mean only the search for best-in-world practices. That's like defining "travel" as driving a Rolls Royce. Essential benchmarking skills can be applied in many situations and at all organizational levels. In this respect, benchmarking is like any word or concept of great currency in the English language. Webster's and other dictionaries offer multiple uses delineating all its different applications and meanings. Similarly, benchmarking has many different uses or contexts in which the basic skills have application. Benchmarking is not just one process with one context—that's to say it's not just the search for best-in-world practices. The search for effective operating practices takes many different forms.

"We definitely need to do *all* those things," observed the chemical company president when he understood the multiple contexts in which to study best practices. "I thought benchmarking only referred to best-in-world." This senior executive's misconception is not unusual. Benchmarking has become so widely invoked as a critical strategy for performance improvement that a Babel of meanings now proliferate. Many managers express confusion concerning precisely what benchmarking is, how to accomplish it, and when to apply it. This confusion was clearly evidenced when the American Productivity and Quality Center's International Benchmarking Clearinghouse surveyed 87 leading benchmarking companies concerning their benchmarking practices. Ninety-five percent of respondents agreed that "most companies don't know how to benchmark."[1]

Not surprisingly, many managers labor under the mistaken notion that benchmarking refers *only* to best-in-class—sometimes called best-in-world—benchmarking. Such investigations seek the best performers in a given function, without regard to the size, industry, or nationality of benchmark partners. Such investigations may yield significant insights—perhaps even major performance breakthroughs. Still, they often require many resources and consume much time. Best-in-class investigations are certainly not appropriate in all situations, functions, and organizations. If best-in-class benchmarking is your only definition for benchmarking, then every project produces a massive global or national search.

Not all companies need to undertake best-in-world benchmarking comparisons. Important factors such as size, quality, maturity, industry, and competitive issues may argue against such broad-reaching and far-flung investigations. The return on investment for such major projects may be low for some organizations. Derek Ransley, a senior quality consultant and benchmarking facilitator at the Chevron Research and Technology Company, likens best-in-class benchmarking to dining out at a five-star restaurant. The restaurant serves great food, presents a delightful ambiance, and charges a correspondingly high price. Consequently, most people will not dine out seven nights a week at the five-star restaurant. At times, a meal at McDonald's or some other local eatery makes better sense. On many evenings the increase in enjoyment will be marginal from a $10-meal eaten at McDonald's to a $200-meal enjoyed at the most expensive restaurant in town. The customer-perceived value may be much higher at the less expensive dining spot. Correspondingly, for most companies every benchmarking project does not need to be a best-in-world project. Learning to effectively apply benchmarking means learning to apply it appropriately in different situations.

Benchmarking "Under the Influence"

Even the most powerful business tools are counterproductive when misapplied. Benchmarking is no exception to this rule. "You can hurt yourself and other people with benchmarking," warns Fred Bowers, manager of corporate quality and benchmarking at Digital Equipment Corporation. "I'll go farther; you should have your license taken away for benchmarking under the influence. Under the influence of what? Of the euphoria of benchmarking. You want to go out and make visits."[2]

Research conducted by the American Quality Foundation and Ernst and Young provides further evidence of benchmarking's double-edge. For three years AQF and E&Y researchers surveyed 580 companies in four industries—autos, computers, banking, and health care. The purpose of this ambitious benchmark project was to identify the business practices that underlie the most successful organizations. Using frequency, regression, and other statistical analyses, the E&Y and AQF researchers tried to isolate business practices that correlate closely with high performance. They judged high performance by evaluating key financial and productivity measures such as return on assets and value-added per employee. In these studies, benchmarking rises as a business practice actively deployed by the most profitable and productive companies. For these world leading corporations, benchmarking core functions, such as product development and distribution and customer service, against the world's best results in a high payback.[3]

Interestingly, this research also suggested a dark side to benchmarking. If companies just beginning their quest for quality attempt to match the techniques used by world-class performers, they may actually set back their improvement efforts by trying to implement too much, too soon. "We believe there are at least two reasons why lower-performing organizations do not benefit greatly from benchmarking practices," observed the International Quality Study final report. "First, they are likely to be looking at inappropriate role models. The common practice in benchmarking is to examine the 'best of the best' or world-class organizations. Yet the IQS data have shown the practices that distinguish higher-performing organizations are often ineffective when adopted by lower-performing organizations." The IQS report continues: "Lower performers probably would find organizations that are on the threshold of medium performance, rather than world-class organizations, to be more helpful models. Second, the lower-performing organizations need to focus their resources on their core infrastructure. They should not diffuse their focus with the sophisticated practices they would see in the best of the best."[4]

The E&Y and AQF researchers suggested a better strategy for organizations just getting started or for those actively honing their skills in quality. They advise these less mature companies to first emulate worthy competitors and market leaders before trying to match the pace of the best-in-world functional leaders.

The wisdom of this observation is rooted in common sense. We must learn to walk before we run. Who, for instance, would attempt to drive a race car in the Indianapolis 500 before obtaining a driver's license? Correspondingly, a company should carefully select a type of benchmarking based on its organizational maturity and circumstances. Business units that are in the early stages of their quality-improvement process—or that lack the systems to perform their basic business functions—will find it counterproductive to go globetrotting in search of the world's best companies. Fledgling quality organizations that try to immediately mimic best-in-world performance systems—without first putting in place the requisite management foundations to support such systems—are a bit like yapping dogs that chase excitedly after passing cars: Should they be so lucky to catch them at an intersection, they are unprepared to do anything more than bark. Even with another company's best practices information in hand, some companies may not be able to implement these winning strategies—or the cost of implementation may prove prohibitive. Novice and intermediate quality-level companies often reap much greater rewards by studying internal best practices and best practice companies from within their regional markets and industries. The low-hanging fruit is often just as sweet as the fruit on the highest branches. Unfortunately, the essence of these broad-based statistical research findings has been lost or misunderstood by some managers, companies, and members of the media. For them, benchmarking has become a kind of Pavlovian bell that sends them searching far and wide for best-in-world comparisons without regard for whether the costs will be justified by the benefits.

The Seven Levels of Benchmarking

Effective benchmarking begins by understanding your position and your need. Not all benchmarking projects demand the same approach or resource investment. Experienced benchmarkers don't regard every benchmarking opportunity as necessarily a resource-intensive, best-in-world investigation. Instead, they evaluate which tools and skills to apply in the given situation and in what scope and measure to apply them.

As the principal tool to facilitate vicarious learning, benchmarking—or its fundamental investigative and comparative skills—can and should be applied in many ways. For instance, a manager reads about another company's winning strategies and this exposure may lead the company to adapt the other organization's proven ideas. Such informal benchmarking may seem undisciplined but it can be highly effective. Consider some proven practices of the Walt Disney Company, which has mastered the art of customer queuing in its theme parks. At Disney, large crowds frequently require park guests to wait in long lines—yet the guests seldom feel poorly entertained or frustrated while waiting. Disney entertains and communicates with its guests while they wait. They use basic design and communication techniques to effectively manage waiting. Disney provides its guests with constant sensory entertainment while they wait so that the waiting never feels like waiting. In lines, they play music; they set up video screens showing old movies; they pack every inch of every line with something interesting to look at. Disney designs exhibits (i.e., their work areas) so guests seldom see the length of the line. Disney cast members greet guests at each exhibit and advise the guests how long the wait will be. Reading about Disney's best practices in communicating with its customers to manage expectations and to ensure their satisfaction produces many insights.

These insights can be translated to apply to almost any organization that requires its customers to wait for delivery of its products or services.

Many effective managers also borrow ideas that they glean at conferences or through informal discussion with their peers at other companies. Not all forms of outreach require physical departures from the work place. Xerox Corporation's corporate law department, for instance, used benchmarking to help it improve its attorney recruitment and selection process. Like many corporations, Xerox suffered from suboptimal recruiting and retention of in-house lawyers. A quality-improvement team identified various causes and then applied benchmarking to help the law department seek solutions. Xerox's search led it to benchmark IBM, Pfizer, Aetna, PepsiCola, General Foods, and no less than seven other blue-chip companies from the Fortune 500.[5] Interestingly, Xerox accomplished its benchmarking project through telephone and mail surveys. Since Xerox conducts attorney selection criteria and law school recruitment off the premises of the corporation, the benefits of on-site benchmarking in this situation prove marginal.

In appropriate circumstances, visits to excellent companies can also provide excellent opportunities for vicarious learning. The Japanese excel at strategic study visits—searches for business insights rather than process-specific best practices. These tours may lack the structure of formal American-style benchmarking site visits. Nevertheless, the Japanese business visitors frequently return home with new ideas and insights that drive change and improvement in their organizations. *(Figure 6.1 illustrates the seven-level hierarchy for viewing benchmarking applications.)*

Benchmarking and Vicarious Learning

Some management experts define structured benchmarking as a single highly-evolved form of vicarious learning. Our view, however, inverts that model. We believe that benchmarking is the cornerstone concept of vicarious learning. In other words, a broad spectrum of benchmarking approaches exists, all of which result in beneficial vicarious learning. Each benchmarking approach brings a different level of rigor to the learning investigation. It is not constructive to classify only the most rigorous, resource-intensive examinations as "benchmarking." Less rigorous investigations involve the same essential skills of learning through external comparisons and outreach. They too often result in important insights that lead to significant organizational improvements. To ignore or disparage these less rigorous learning safaris is to ignore an important class of improvement opportunities.

Consider the example of the technology company that was ready to undertake a high-level benchmarking project investigating best-in-class approaches to the patent process. This major corporation recognized that its patent process performed less effectively than other global competitors. This weakness placed the company at a disadvantage in an industry where patent rights of its rich technology should have been a competitive advantage. A high-level task force prepared to undertake an in-depth six- to nine-month benchmarking process. Such a project could be expected to consume considerable resources, since the task force was staffed with several vice presidents.

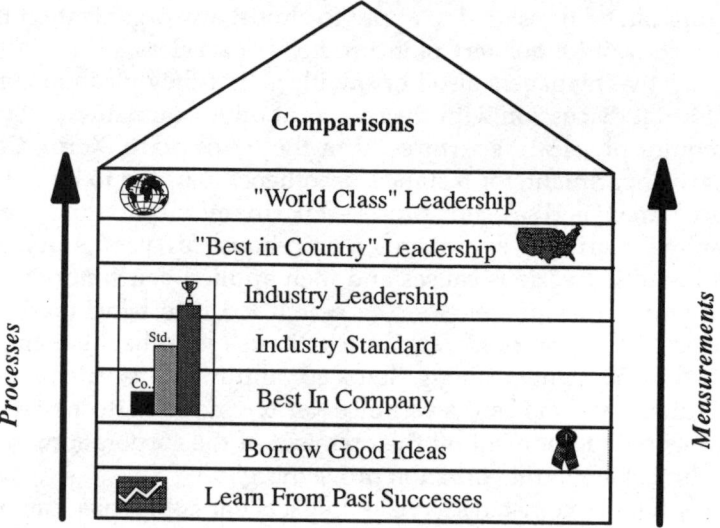

Figure 6.1a. Levels of performance benchmarking.

Before the benchmarking study kicked into full speed, a task force member asked who the corporation admired for its patent process. After some discussion, team members agreed IBM had an exemplary patent process. The question then arose: "As an interim strategy, can we study IBM's patent process and implement it as an interim improvement strategy?" The team hoped to improve the company's

Classified by Company or Business Unit Maturity		Novice	Intermediate	Master
"World Class" Leadership				✓
"Best in Country" Leadership				✓
Industry Leadership			✓	
Industry Standard		✓	✓	✓
Best in Company				
Borrow Good Ideas		✓		
Learn from Past Successes				

Figure 6.1b. Current practices.

Classified by Company or Business Unit Size		
	Small	Medium	Large
"World Class" Leadership 🌐			✓
"Best in Country" Leadership 🗺️		✓	✓
Industry Leadership	✓	✓	✓
Industry Standard	✓	✓	✓
Best in Company	✓	✓	✓
Borrow Good Ideas	✓	✓	✓
Learn from Past Successes	✓	✓	✓

Figure 6.1c. The ideal.

patent process without having to wait a year—the anticipated waiting time for the scheduled project—to begin improvement implementation. At first, people ridiculed the idea of borrowing the best features of IBM's patent process as impossible because the corporation at times competed with IBM. They concluded: "It will be a cold day in Hades before IBM shares its patent process with us." Taking another approach, a team member asked if the corporation employed anyone who had formerly worked at IBM and knew its patent process. A short investigation revealed that a current employee had worked at IBM and was intimately familiar with its patent process. This employee's knowledge enabled the company to quickly initiate interim improvements. The company later performed a more structured benchmarking project on the patent process. The early informal and unstructured process drove many improvements without spanning many months and several continents. Can one really term this pragmatic approach as *benchmarking*? Benchmarking purists will argue against calling it benchmarking. Yet, the organization performed effective outreach, learned vicariously, and imported external ideas that helped jump-start its process improvements. This *informal* benchmarking strategy proved fast and effective. It matters little that in this instance the outreach process diverged from a highly-structured, multistep, benchmarking process. In today's operating environment, time and resources are usually scarce. For the technology company's given situation and its circumstances, this *was* effective benchmarking.

Consider this issue in different terms. You are an investor choosing among two investment opportunities. One opportunity allows you to invest a maximum of $10,000 and doubles your investment in six months, giving you a quick 200-percent annualized return on your money. Contrast this opportunity with a $1 million investment that produces a $500,000 profit in 12 months, providing you with a 50-percent return. Which investment do you want to make? Both!

If the endpoint is performance improvement, then the dogma of any particular benchmarking approach seems unimportant. Benchmarking is a business practice or skill with countless forms and applications. Determining which form to apply in different situations is an important managerial decision. Every employee can become a continuous improvement manager. As such, all employees oversee a portfolio of potential incremental and major-magnitude improvements. In its many applications, benchmarking supports both types of improvement. Taken by itself, incremental change may not be sufficient to maintain market leadership. In turn, no one can live on a daily diet of breakthrough projects designed to produce continuous large-scale change; that results in chaos. Just like an investor, the benchmarker aims to achieve the greatest return with the lowest investment of resources. The benchmarker should use the approach that supports the highest return on investment.

Unfortunately, much confusion and uncertainty exist in the work place concerning the different benchmarking applications. We have developed a managerial model that describes a seven-level hierarchy of benchmarking opportunities and helps clarify these applications. This model organizes benchmarking according to approach and resource requirements. This integrated framework enables managers to understand that benchmarking can be formal or informal and highly structured or relatively unstructured. The framework recognizes that different benchmarking projects require different levels of rigor. It provides a tool by which to evaluate and select the appropriate approach and resource investment. The seven-level hierarchy embraces both informal and formal benchmarking. Briefly stated, the seven benchmarking levels include:

- Learning from past successes.
- Borrowing good ideas without regard to their corporate origin.
- Developing internal best practices.
- Matching industry average or standard practices.
- Establishing leadership through industry best practices.
- Targeting national best practices without regard for industry.
- Matching or exceeding world best practices.

Learning from Past Successes

The philosopher George Santayana captured centuries of wisdom when he observed that "those who cannot remember the past are condemned to repeat it." If managers and groups do not learn from past errors, mishaps, and failures, they will surely repeat them. This truth underlies a core concept of quality improvement. "Lessons learned," as Joseph Juran has called them, are critical to the improvement process. Effective managers always learn from experience. They evaluate historical and current performance and learn from both. Lessons learned are reflected in the improvements implemented throughout the organization. The Baldrige criteria, for instance, are girded with no less than 15 references to "evaluation and improvement" cycles. These cycles suggest how and where an organization has institutionalized the learning process in its operating systems. Since learning is the essence of continuous improvement, the Baldrige criteria probe to understand how deeply rooted are an organization's learning and improvement instincts.

One important exception exists to Santayana's axiom: Past successes—forgotten or remembered—typically will *not* be repeated unless they are understood. Yet most managers do not analyze past successes with the goal of distilling the critical factors that enabled the success. Schools do not frequently teach "success analysis" as a problem-solving approach. Most managers view themselves as "troubleshooters," "crisis managers," "problem solvers," or "fire fighters." They seek to identify problems, errors, mistakes—or other "out of control" events—and then focus resources and attention on reforming or correcting those problems. Few managers style themselves as "success shooters," "best practice managers," or "best practice hunters."

Identifying successful practices so that they can be understood and commonly repeated is a philosophically distinct management approach. When coupled with effective problem-solving techniques, success analysis offers rich rewards and returns. Moreover, it is easy, effective, and invigorating for employee morale. First generation benchmarking consists of sharing and studying success stories. In every success story lies the seed of a potential best practice. Pulling that precious seed from the husk of mundane experience is the challenge of managers focused on best practices.

The power of success analysis is evidenced in the following managerial situation: A large corporation in the hospitality industry had spent several hundred thousand dollars to investigate, study, and develop a roll-out strategy for a major company-wide service quality initiative. With an implementation plan drafted, the sponsoring team needed to win the support of an influential group of operating managers who would make the roll-out succeed or fail. Unfortunately, the presentation went poorly and stalled the implementation. In short, it became gridlocked in politics and problems among the leadership team. What promised to be a triumph for the sponsoring project team suddenly turned into a disaster. The senior and mid-level managers who had championed the initiative worried about their job security. The most senior project manager decided that the team should now abandon the initiative that they previously viewed as much needed.

Recovery meetings were informational but unproductive. Managers performed a post-game analysis of what went wrong. They generated a long list of implementation errors. Unfortunately, corporate politics caused most of these fatal flaws. They recognized repeated failures to influence key managers, repeated failures to win support of important departments, and repeated failures of individual executives to stay in touch with their colleagues. These post-mortem lessons seemed of little value, however, because they were difficult to translate into actions that would positively influence the situation—past, present, or future. Simply stated, the post-mortem determined that communication, trust, and cooperation had broken down among the leadership group.

A different tack subsequently proved far more helpful. Rather than analyze what went wrong, a small group went off-line of leadership politics and explored another subject. They contacted 15 influential managers from throughout the company and asked them to reflect on the following questions: "In the past five years, what company-wide initiatives have been uniquely successful?" "Could you please reflect on the critical success factors that enabled these initiatives to achieve such excellent results?" Much to the surprise of the group conducting this inquiry, the senior managers who were interviewed all identified the same two or three successful initiatives. Moreover, they quickly offered their thoughts and observations concerning what practices accounted for these critical few initiatives' unique success. When the 15 interviews were consolidated, they constituted a "white paper" on best practices to ensure successful implementation within the organization. These best practices were especially powerful because they were highly actionable and could be put into play to remedy the current situation and to better manage future implementation efforts. None of the practices was revolutionary. They simply represented a clear and repeatable set of steps and actions to ensure successful implementation. These best practices touched on issues such as setting clear goals, articulating the unit's strategic direction, sharing information in a clear, consistent, and timely fashion, involving operating officers in the decision process, and developing a systematic approach to communicate strategy and action initiatives to the field. These practices created the foundation of a recovery plan designed to put the service-quality project and its implementation plan back on track. By studying its past successes, this organization distilled various best practices that could help ensure future operating success. This learning was not easily accessible through traditional analysis of what went wrong in the failed effort. Systematical learning from past successes, therefore, begins when employees take time to share success stories. The goal is not to boast or trumpet one's own competency; the goal is to share successful strategies and practices that produce high performance. Successful quality initiatives build on the existing foundation of strengths in an organization. Total quality management involves more extensive actions than just finding and correcting weaknesses; it requires propagating and expanding existing strengths. Sharing success stories provides a look at organizational strengths that also reveal opportunities for performance improvements. This can be accomplished informally through work group discussion, or it can be approached with structured tools and analysis.

To this end, consultants at Best Practices Benchmarking and Consulting, Inc., have developed an approach applying a three-dimensional quality-assessment methodology to distill best practices and improvement opportunities from successful initiatives. They employ an analytic tool first developed by the Malcolm Baldrige National Quality Award program managers; this analytic tool can be used to examine the approach, the deployment, and the results of any successful initiative. Through the analysis, managers begin to isolate the key strengths and success factors of their successful initiatives. The approach-deployment-results framework, is a proven tool for evaluating the thoroughness, consistency, and sustainability of any performance initiative, not just a quality initiative.[6] Designed to help assess any performance initiative, this framework is especially useful in distilling best practices from organizational success stories. Best Practices Benchmarking and Consulting, Inc., has applied this tool with clients in many industries. In every instance, the management teams agreed that opportunities existed to further leverage their successes—or embryonic best practices—by more extensively deploying the successful approach or initiative in the organization. By their very nature, best practices are seldom fully deployed. Managers frequently find gold in the ground here.

Borrowing Good Ideas

The journey of a thousand miles begins with the first step, and the road to continuous improvement is paved one good idea at a time. Consequently, basic benchmarking skills embrace creative adaptation of good ideas, no matter how small, and they also support larger scale system and process improvements. Clearly, no individual or company can corner the market on good ideas. Innovative adaptation is therefore a high-yield strategy to supercharge continuous improvement. Milliken and Company, the Baldrige-award winning textile manufacturer, excels at involving employees in company improvements. Milliken celebrates the "steal shamelessly" ethic and adopts many good public-domain ideas—without regard to size or origin. Accordingly, improvements spring up from internal efforts and from external borrowing.

The genealogy of excellent borrowed ideas is long and distinguished. A case in point: The sticky hooked spines on the common burr inspired the man who invented Velcro. Nature's hairy seed pods create the perfect fastener, allowing them to stick effectively on animal fur, socks, shoe laces, and pants. By creatively adapting the burr's hooked spine structure, the inventor fashioned an all-purpose fabric fastener that has been applied in a thousand ways. What good ideas can you borrow—from nature or from your neighbor? The search for best practices begins with the search for good ideas that can move your organization, yard by yard, down the playing field of continuous improvement.

Small companies tend to be especially skillful at idea importation. Why? Starved of resources, small companies naturally develop a beg, borrow, and creatively imitate mentality that enables them to leverage others' experience and learning. Don Romine, president of Web Industries, a $20-million contract manufacturer based in Westborough, Massachusetts, puts it this way: "When we do plant-to-plant visits, it's in the spirit of friendly piracy." Romine and Web employees routinely visit other manufacturers and host other companies at their own facility. At one company the foreman described how the firm had collected 127 ideas from the factory floor the year before and had implemented more than 100 of them. "I thought, 'What a great idea,'" recalls Romine, "to measure not just product rate but idea generation." Though it takes time to set up and make the visits, Romine highly praises the benefits of the trips. The ideas he and others come back with, he says, are "little light bulbs going off: none in particular is necessarily revolutionary, but together they've made a very significant change in the way we do business."[7]

Despite their advanced technology and structured management systems, many large corporations can learn from smaller organizations. Exemplary operating practices can be mined within the fields of many entrepreneurial organizations that operate successfully in market niches. American Airlines, for instance, imports good ideas from small but progressive restauranteurs. To improve food service on its flights, American consults regularly with its "Chef's Conclave"—an elite advisory group made up of 14 of today's top U.S. restaurant chefs. Their ranks include some top names in U.S. cuisine—Wolfgang Puck of Spago in Los Angeles, Alice Waters from Chez Panisse in Berkeley, California, Stephan Pyles of Baby Routh in Dallas, and Larry Forgione of New York's An American Place.

"The chefs have given us sound advice," observes American Airlines CEO Robert L. Crandall. "For starters, noting that customers in their own restaurants tend to order their favorite meals again and again, they suggested that we offer fewer selections. They recommended we pick a few outstanding dishes and strive to prepare them perfectly every time," says Crandall. "So we have cut way back on the nearly 500 different menus we used to offer. Having fewer menus also makes it easier to maintain the quality of our offerings, which are prepared in 175 flight kitchens around the world. It also reduces our costs."

Crandall continues: "The chefs also came up with lots of good ideas to improve value. Where we once offered freshly cut fruit salads, for example, our fruit appetizers now consist of melon slices, which give our customers more fruit—and save lots of cutting and peeling in the kitchen. And on some flights," says Crandall, "we now offer such imaginative entrees as roast chicken with chili pesto sauce and salmon with tomato basil fettucini."[8]

Steal This Idea®

The reservoir of good ideas runs deep in virtually every organization. Nevertheless, teaching employees and managers to plumb that idea well can be difficult. In many organizations, there are years of built up resistance to borrowing. To demonstrate how easy it can be to tap into these untapped resources, Best Practices Benchmarking and Consulting, Inc., has developed a set of exercises designed to encourage employees to creatively borrow good ideas. Called *Steal This Idea®*, these exercises underscore how much more fertile are the grounds of adaptive learning and innovative imitation than the rocky grounds of pure invention. Steal This Idea®, exercises encourage employees to brainstorm good ideas they have observed without regard to the idea's origin. Best Practices' consultants have conducted this exercise with hundreds of managers and executives. The identification rate of good ideas is always impressive. In a typical session, individuals will identify between three and eight ideas in five minutes of solitary brainstorming. In other words, a group of 20 people will identify 60 to 80 ideas in five minutes. Admittedly, some ideas may have small merit, and some may be redundant. But even if half the ideas are unactionable, that still leaves 40 ideas worthy of implementation consideration. For most organizations, that's a great return for a five-minute investment.

After managers have identified ideas they admire, Steal This Ideas® asks them to evaluate the origins of these exemplary ideas. Usually they borrowed 90 to 95 percent of these ideas and tailored them to fit their situations. Only 10 or 15 percent of the time do managers propose ideas invented by individuals on their teams. Steal This Idea® dramatizes this borrowing by using a scoring system that mimics the reward and recognition systems of most organizations.

Team members first total their ideas and then score them, weighting invented ideas more favorably than borrowed ideas. During this scenario, pure inventions receive a score five times greater than innovative adaptations. This traditional score reflects the reaction of many American organizations—they reward invention more than innovative adaptation.

Performance improvement, however, is blind to the lineage of good ideas. Creative adaptation can drive improvement just as forcefully—perhaps more forcefully—than invention. A second idea grading system underscores this deep-rooted prejudice for invention. When innovative adaptations receive equal weight as inventions, team scores soar. Employees become intoxicated with the fact that they can champion creativity. Of course, the Steal This Idea exercise leads employees to puzzle why their organizations consciously or unconsciously favor pure inventions so much over applied creativity and innovation. The exercise also produces a tangible result: It generates a bumper crop of improvement ideas. They have worked elsewhere and are ready for import into the participants' own organizations.

The habits that Steal This Idea encourage are the habits that drive many excellent organizations. They are inspired with a "we-can-learn-from-everyone" attitude. These organizations accelerate learning and change by developing good ideas from both internal and external sources. Then they tailor the ideas to suit different situations and needs. These high-performing organizations have escaped the "not-invented-here" syndrome. Their culture and reward and recognition systems support borrowing ideas as much as creating ideas.[9] Consider a sampling of innovative ideas and adaptations that have percolated up in Steal This Idea sessions.

Customer Service Telephone Waiting Queues.   WordPerfect Corporation and Lotus Development Corporation both have borrowed features from drive-time radio disc jockeys to make waiting on their customer service 800 telephone lines less annoying and more interesting. WordPerfect and Lotus both employ live "hold jockeys" who announce the number of callers in your queue and the average waiting time. At Lotus, hold jockeys announce various Lotus products and play music. (Both software makers seem to favor jazz and soft rock.) At Lotus, there is also a feature to determine your number in line and the estimated wait. *What can you do to make waiting for your customers less nettlesome and even more enjoyable?*

Flowers in the Rest Rooms.   The president of a Midwest-based independent supermarket chain observed that many of his daytime customers were women with young children. During shopping trips, these moms frequently needed to park their shopping carts and change diapers in the store's rest rooms. When this executive's anchor store remodeled, management expanded the size of the rest rooms and—borrowing an idea from Stew Leonard's Dairy in Connecticut—placed flowers in the rest rooms to make the environment more inviting. It was a small touch that won frequent praise from patrons. *What areas in your building have similar importance to your customers and could be enhanced by flowers?*

Make a New Product Wish.   Microsoft Corporation understands the importance of a steady flow of new products. The computer software colossus dedicates a new-product-ideas telephone number for customers who have ideas concerning new features and applications. "Microsoft provides several avenues to give product feature requests," notes Microsoft's directory of products and services. "The following is a voice number for Microsoft Excel and Word. You can offer suggestions on other products by calling the dedicated support number at Microsoft Product Support Services. We welcome your ideas for future versions. Microsoft Wish - 206-936-WISH." This idea has been borrowed by other companies that want to hear from their customers concerning what new products, services, and features they should be providing. *How can your company apply this idea to encourage its customers to share their new product wishes with you?*

Demonstrating Value to Your Customers.   Staples doesn't just advertise low prices; the Massachusetts-based office supply chain demonstrates value at point-of-sale. To reinforce the value it provides its customers, Staples prints out the actual discounts on customer cash register receipts; this feature highlights customer savings and reinforces the company's low-price and high-value image in the minds of customers. Customer receipts itemize all purchases and then automatically calculate your savings off retail prices. "At List Prices, Your Purchase Would Cost $216.67," declares a typical Staples receipt. "At Staples' Low Prices, You Paid $82.46 And Saved $134.21." *Where can you provide discount information to your customers, helping them appreciate the value and savings they receive by purchasing products and services from your company?*

Mobile Headsets for Customer Service Representatives.   Weaver's Business Interiors, a Wisconsin-based furniture design company, observed McDonald's drive-through window employees using cordless microphones and headsets. These mobile communication units allow the restaurant chain's employees to remain in constant contact with customers, even when they step away from the drive-through window to fill an order. It also solved a persistent problem of Weaver's and other customer-focused companies. Sales people and customer service representatives no longer miss calls when they step away from their desks. Consequently, Weaver's imported the idea from the fast food industry into its customer service department. *How might mobile communication technology keep your company in better touch with your customers?*

Adopt a Common Space.   The CEO of a Midwest hospital borrowed an idea she heard on Minnesota Public Radio and creatively adapted it to improve operations in her hospital. The public radio story reported about an innovative approach to public green space beautification. To overcome the persistent problem of litter-strewn islands on public streets, Minnesota encouraged citizens to assume ownership for keeping clean these common green areas in their neighborhoods. The program had a felicitous effect. In this green-space program, the hospital CEO found an elegant and simple solution for a nagging problem in the hospital. Common areas, such as hallways and patient waiting areas, lay in no particular department's jurisdiction. Consequently, they became littered, wall-marked, and floor-scuffed without any hospital employee initiating immediate cleanup. These areas were also often the ones most visible to patients. By adapting the Minnesota green-space program to the hospital, the CEO solved a persistent problem for her facility and improved conditions for both employees and patients. *What common area could be better maintained if your organization were to launch an "adopt a green space" program?*[10]

By themselves, none of these borrowed ideas is momentous. But each solved long-standing problems for the organizations importing them. Moreover, they solved these problems quickly and inexpensively and left plenty of opportunity for other innovative adaptations. These improvements benefited the organizations twice over: They improved performance and regenerated employee enthusiasm.

Developing Internal Best Practices

Gold lies in the ground of your organization. It rests in the internal best practices that reside within your best performing individuals, teams, departments, and operating units. However, in many corporations this natural wealth remains unmined. Studying one's brothers and sisters to identify internal best practices is decidedly unexciting, especially when compared to a road trip to the foreign soil of another corporation. Though it may not seem as exciting or beneficial as visiting another corporation, developing and deploying internal best practices provides a simple and powerful approach to drive performance improvement. Milwaukee-based Johnson Controls Battery Group is a case in point. During the 12-month period following initial implementation of the Battery Group's *Best Business Practices Program* in 1992, dramatic improvements were achieved in productivity, quality, safety, transportation, inventory management, and profits. This performance improvement was especially impressive because it was achieved during a weak economy that produced declining sales for the Battery Group.

The first seeds for Johnson Controls' best practices effort were planted in 1991 when a group of managers set out to develop a set of performance measures that would enable the company to compare its 12 plants. The team began by equalizing differences in product mix and customer segments among plants. Even after factoring out these variations outside plant management control, significant performance differences remained. For years, managers had presumed that performance differences could be explained by these factors. In fact, they did not.

"When we forced ourselves to look in the mirror, we had to ask: 'Why do certain plants consistently outperform the others?'" explains Keith Wandell, Vice President of Operations for Johnson Controls Battery Group. "Product and customer mix alone could not explain the differences. The differences simply came down to plant performance. We had to find out what made the best plants perform and translate those practices to all plants."

To identify its best business practices, Johnson Controls Battery Group brought together 42 top managers and supervisors from all 12 plants and all functions and assigned them to five teams. Together they identified and consolidated the division's best practices. The goal was simple, explains Wandell: "We wanted to succinctly state in a few pages the things that made the best plants successful."

The team found that fundamental management practices, not specific activities or programs, separated the best from other plants. The best plants focused on quality, safety, effective delegation, teamwork, employee involvement, and structured problem solving. "The management process was the foundation," observed Stan Schulz, who coordinated the division's efforts. "If you don't build the right management process, you will forever search for ways to tweak the system and only achieve modest improvement. If you build the management team and the basic management process, and involve everyone in it, you get tremendous leverage."

During its best practices identification project, the division developed a set of 88 monthly performance measures falling into five critical management areas: financial control, production management, quality, transportation, and health and safety. These 88 measures allow employees at any level, including plant floor employees, to compare their performance against any other plant and against benchmarks achieved by the best plants. If someone lags behind operating targets, it is that individual's responsibility to seek help from peers at the best plants. "The natural reaction when faced with a difficult problem is to try to solve the problem yourself," notes Schulz. "What we're doing is encouraging our people not to solve every problem by themselves; we're teaching them to pick up the phone and call someone else who has already solved the problem."

"The mind set for several generations of managers had been to have the plants compete against each other," adds Schulz. "Plant managers were very protective of any small advantage they developed. The system promoted behaviors that didn't facilitate improvement at the pace we needed and now require."

The process has unleashed tremendous innovation and creativity among employees. "We're getting ideas from employees on the floor of our plants that engineers and staff people in Milwaukee never even dreamed of," marvels Schulz. "They are simple, pragmatic solutions to important problems. New problem-solving techniques and new language are being applied to test these solutions. Employees everywhere are talking pareto analysis, statistical process control, and design of experiments. There is tremendous innovation going on at the line level."

Operations Vice President Wandell requires all 12 plants to achieve aggressive goals developed as part of Best Business Practices Measurements, but he declined to declare the best business practices themselves mandatory. "We know the processes defined by our Best Practices Manual work, but we need to encourage innovation. If results can be achieved without using the processes defined in our Best Business Practices, all we ask is that the new practices are shared and integrated into the manual," which is updated about twice a year and now runs about 70 pages.

Coupled with strong leadership, clear goals, and a culture dedicated to continuous quality improvement, Johnson Controls Battery Group's best practices strategy has helped supercharge the unit's rapid performance improvements. The program's success has even spurred visits from enthusiastic customers—such as Honda, which wants to replicate the program at its other suppliers.

Despite successes such as this one, many corporations—arguably most corporations—do not systematically study and learn from their best performers. In a recent survey of the top 80 U.S.-based executives in a multinational corporation, most answered that they "sometimes" or "infrequently" studied what other company units were doing. Without detailed knowledge of sister units' areas of excellence, it's difficult to import or leverage the most effective operating practices and strategies of each unit. This is ironic because a company can more easily research and leverage its internal best practices than external best practices. This multinational corporation study found that the companies examined their internal best practices less than they performed external benchmarking studies. As one senior executive in charge of development in this highly-decentralized corporation observed: "It's axiomatic that we don't pay any attention to what each other is doing."

The "if-it's-not-invented-here, it-can't-be-any-good" syndrome has an equally limiting corollary: "If it *is* invented here, it can't be any good." Rooted deeply in human nature, this principle represents the flip side of managerial parochialism, which presumes the world is no larger than our particular view of it. Parochial organizations distrust, disregard, or denigrate outside views. The opposite instinct resides in many nonparochial organizations; they presume instead that good ideas never blossom in their own gardens. For these organizations, the grass always appears greener outside. People presume that colleagues down the hall or in other plants and branches must conduct business similarly to them. Consequently, they underestimate the benefits of studying other colleagues' approaches. The actuality could not be farther afield. Great variation frequently springs up among operating groups as their work practices and innovation rates evolve at different speeds and in different directions.

Internal best practices prove fruitful for many reasons:

- Participants share the same culture and speak the same operations language, making comparisons fast, easy, and meaningful.

- In most cases, participants share the same measurement system, again making comparisons more meaningful.

- Participants frequently use similar technology, processes, and systems. Consequently, small practice differences can have a significant impact on performance.

- Antitrust and competitor barriers do not exist when sharing best practices within an organization.

- It is easier to identify benchmark partners and to establish working relationships with professional peers within one's own organization than across companies.

Establishing an internal best practices strategy is akin to conducting external benchmarking. It proceeds in a structured fashion, but usually at a faster pace. Companies can more readily access information and best practitioners can more easily be engaged in these internal projects. Consider Table 6.1, which is an example of Mutual Life Insurance Company of New York's (MONY) establishment of a best practices strategy for its sales process—a core activity in the life insurance industry. The MONY strategy proceeded through a multi-step process that mirrors most external benchmarking projects.

MONY is one of a rapidly growing number of organizations implementing internal best practices strategies. Other organizations that have begun to leverage

Table 6.1. Implementing Best Practices in Sales at MONY

Implementation phase	Objective	Approach
1. Determine process steps to be examined (e.g., recruiting, interviewing, selection, fact finding, etc.)	To focus on areas that have greatest impact on the sales success.	Use quantitative and qualitative data to determine highest impact areas.
2. Establish criteria to be used to determine who are the best.	To ensure credibility of selection and best practice findings.	Use quantitative data (sales volume, sales numbers, persistency, etc.) for past three years along with management observations.
3. Select best managers and sales people.	To get the "best of the best."	Senior management selection based on above nominating data.
4. Conduct best practice sharing rallies.	To identify and understand critical success factors and individual best practices at each step of the sales process.	Steal This Idea®, Best Practices exercises, brainstorming, process mapping, fishbone diagramming, pareto analysis, general discussion.
5. Develop best practices information.	Develop finer detail to best practices information.	Surveys, small group discussions, one-on-one interviews, and field observation.
6. Consolidate best practices from the best of the best.	Consolidate the "class of best practices," which is more comprehensive than any one practitioner's best approaches.	Synthesize information, use written guidelines, process maps, comparison matrices, various charts, and other presentation tools.
7. Pilot the best practices standards and training with a championship unit.	Staged implementation designed to achieve immediate success and support from key influential players.	Multi-site service organization implementation strategies.
8. Refine based on pilot experience.	Continuous improvement of practices and implementation.	Continued internal and external benchmarking.

the best of their best include: Marriott, GE, Johnson Controls, Chevron, and Avon Products. The list can be expected to grow.

Developing External Best Practices

"The universe," as Thoreau long ago noted, "is larger than our views of it." Consequently, many organizations discover that their quest for performance excellence through best practices leads them outside their own organizations. Brave new worlds exist outside our own work places. Innovation may occur in the most unlikely places. It is only when you venture outside your own practices that your perspective shifts and your context expands. This journey outside often leads organizations to small incremental improvements, leapfrog jumps, and—at times— to radical redesign and reengineering of their entire systems.

Competitive benchmarking is the first level above internal best practices. It focuses on the practices of direct and indirect competitors within one's industry. There are many stories and examples testifying to the benefits of competitive benchmarking. One compelling example of competitive benchmarking occurred when a senior vice president requested a capital expenditure that would help his manufacturing company improve quality and performance by 20 percent in a critical area of its business. The improvement looked impressive. The president innocently inquired where the chief competitor stood on this operating issue. The vice president who requested the expenditure did not know the answer, but pledged he would find out. The president postponed the decision until the next meeting when the vice president could report back with the competitor information.

Members of the organization began asking customers and suppliers how they compared to the competition. Inquiries quickly revealed that the primary competitor had also targeted this function as an improvement area and had set major stretch goals to focus improvement efforts. The competitor was aiming at a 200-percent improvement. Back at headquarters, the organization received this news from the field with skepticism. The response back to the field was curt: "Those numbers can't be right. Keep digging." When a sales representative drove by the competitor's headquarters, he was surprised to see a corporate banner, emblazoned with "200%," hanging from the flagpole. This stretch goal had become a rallying cry for improvement and the focus of organizational efforts. Additional inquiries among suppliers to both companies brought more alarming news. Apparently, tension was high at the competitor because operating teams were currently performing at 150-percent improvement rates, 50 percent behind their target! This story ends on a humorous note. When the senior vice president reported back to the executive committee, he soberly informed his colleagues that there had been a typographical error in the previous meeting's handouts. A zero had been inadvertently dropped from a key operating figure. The stated improvement target he was projecting was 200 percent, not 20 percent.

The point in this story is dramatic. How long can any company—industry leader or laggard—operate at 20-percent improvement rates while its primary competitor aims at 200-percent improvement gains and quickly accelerates to 150 percent of target? The answer is obvious: "Not very long!"

Competitive benchmarking has been refined into an art by companies like Xerox and Motorola. For them, it is a comprehensive skill that begins with competitor intelligence monitoring. It grows to include techniques such as reverse engineering, and then evolves to embrace full-scale benchmarking. Advanced benchmarking activities may focus on strategy, product and service features, performance levels, or core process and technology capabilities.

Reverse Engineering

Traditionally regarded as a manufacturing approach, reverse engineering can be a useful competitive benchmarking tool for manufacturers and service providers alike. Auto makers routinely buy each other's cars and take them apart to see how they are made. Xerox does the same with copiers. Even a company like L.L. Bean applies this tool. L.L. Bean routinely orders packages from itself and from its competitors to determine how L.L. Bean, when viewed from the customer's perspective, stacks up against the competition.

When applied broadly, reverse engineering can embrace the entire sale, order, delivery, billing, use, and after sale process. Reverse engineering does not merely include the technical process of disassembling another product; reverse engineering may scrutinize the entire customer path of a competitor. Imagine you are studying the service capabilities of a service provider, such as a mail order company. The reverse engineering process can begin with a review of the competitor's catalog, its prices, product line, layouts, and niche marketing approaches. Next you place your order, observing the speed of response, customer contact training, access to information, flexibility in indexing customer location, credit card verification, and ability in providing product information. All these activities speak volumes about the competitors' information, human resource, market research, and product capabilities. How do they compare with your own?

The next stage in comprehensive reverse engineering scrutinizes delivery, packaging, and billing capabilities. Delivery offers information concerning dispatch speed, distribution sites, transport means, and vendor relations. Does the competitor use Airborne, Federal Express, UPS, or the Postal Service? What are the cost and capability implications for each? The packaging can reflect many operational and cost issues as well. Billing capabilities also reveal operational details concerning terms, speed, ease of use, and value-added information provided through the bills that might create some type of competitive advantage.

Classic reverse engineering focused primarily on the examination of the competitor's product. Disassembly allowed an engineering team to examine issues of design, number of parts, ease of assembly and repair, and other engineering-related issues. Viewed more broadly, reverse engineering also yields important competitor information by scrutinizing documentation, warranty and guarantee terms, and even after-sale service. When a company reverse engineers its competitor's entire product or service process, it develops an important knowledge about the competitor's general strategy, cost position, and its process, product, and service capabilities. All prove critical in assessing one's own strengths and improvement opportunities. This information drives resource investments and strategic planning. (*Figure 6.2 illustrates how reverse engineering techniques can be employed by benchmarking teams.*)

Hitch Your Wagon to the Stars

Competitive comparisons within the industry can occur at many levels. Frequently, though, organizations make the mistake of setting their sights on easily accessible

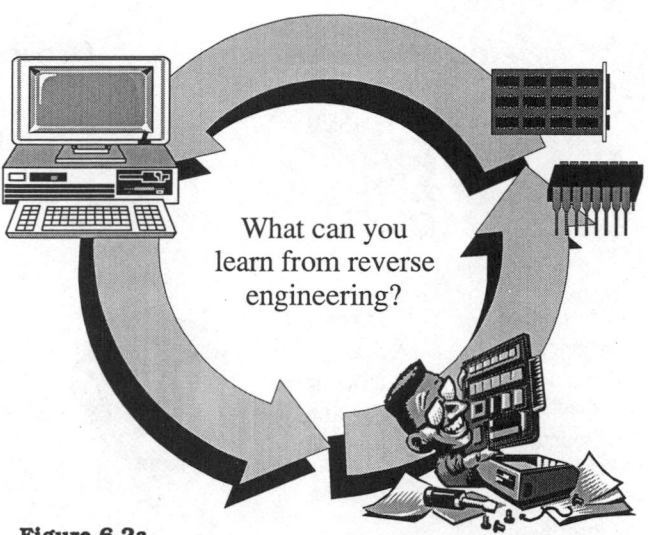

What can you learn from reverse engineering?

Figure 6.2a.

benchmarks such as industry norm, industry average, or industry standards. Trade associations conveniently track these valuable reference points. Yet setting targets on the average, norm, or standard, condemns the organization to mediocrity. These measures represent the melding of all performers into one middle-of-the-road or acceptable point. For benchmarking to have value, you must—in general—set your sights on the leadership position. That is where the real gold and real gains lie. Leadership can and should be interpreted according to your situation. For early stage companies, their competitive benchmarks might consist of local market competitors that have excelled. At later stages, leadership may call for targeting the

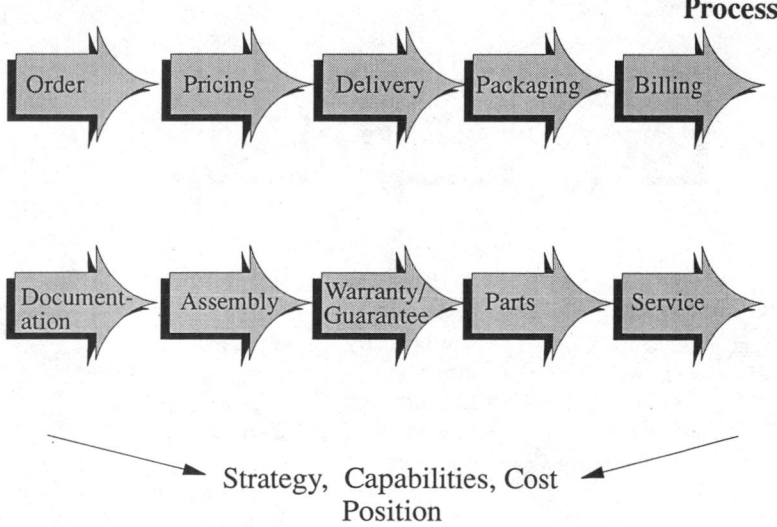

Figure 6.2b. Process.

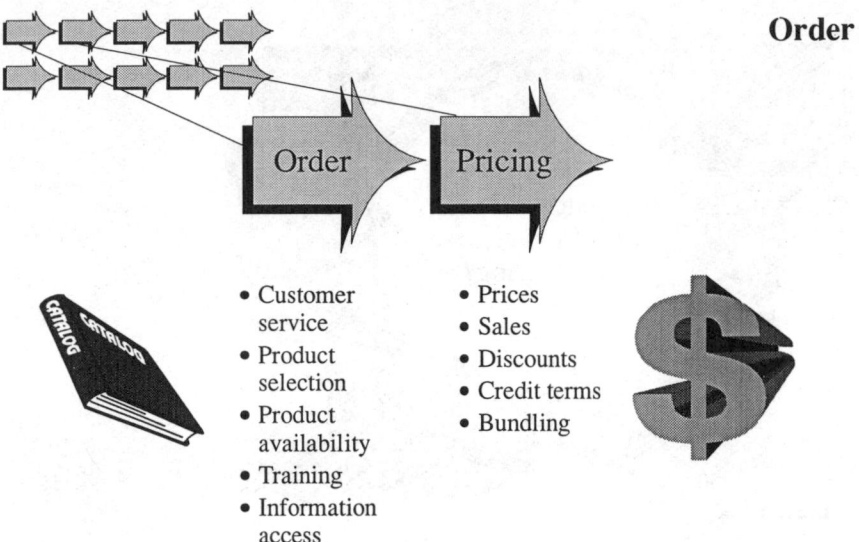

Order

- Customer service
- Product selection
- Product availability
- Training
- Information access

- Prices
- Sales
- Discounts
- Credit terms
- Bundling

best in your industry, without regard for their location. A common benchmarking pitfall is the desire of many benchmarking teams to go in search of the holy grail. Distance, of course, has no correlation with excellence. Those who chase after the single company that is the absolute best will search forever. "Don't let your search for the great idea blind you to the merely good idea," advises inventor Bob Metcalfe.

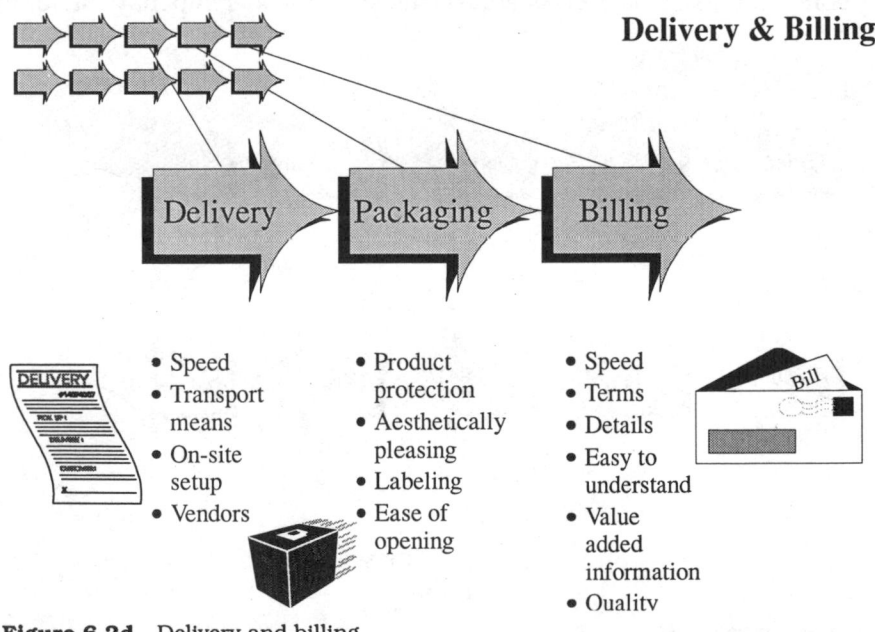

Delivery & Billing

- Speed
- Transport means
- On-site setup
- Vendors

- Product protection
- Aesthetically pleasing
- Labeling
- Ease of opening

- Speed
- Terms
- Details
- Easy to understand
- Value added information
- Quality

Figure 6.2d. Delivery and billing.

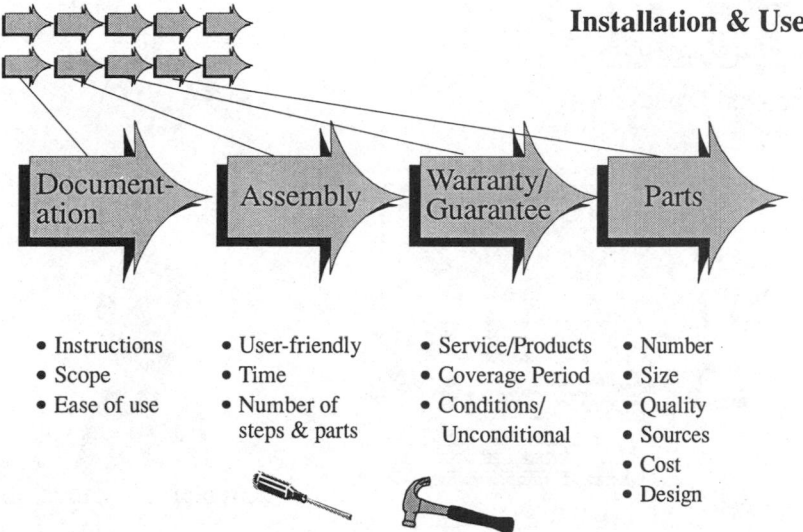

Installation & Use

- Instructions
- Scope
- Ease of use

- User-friendly
- Time
- Number of steps & parts

- Service/Products
- Coverage Period
- Conditions/ Unconditional

- Number
- Size
- Quality
- Sources
- Cost
- Design

Figure 6.2e. Installation and use.

"Reject everything except for the very best and you'll end up with nothing." Educator Donald Kennedy has similar feelings: "A lot of disappointed people have been left standing on the street corner waiting for the bus marked *perfection*."[11] Hardly ever does a single company embody all best practices in a given function. The holy grail-like search for one best practitioner will make you an "industrial tourist" but not an effective benchmarker.

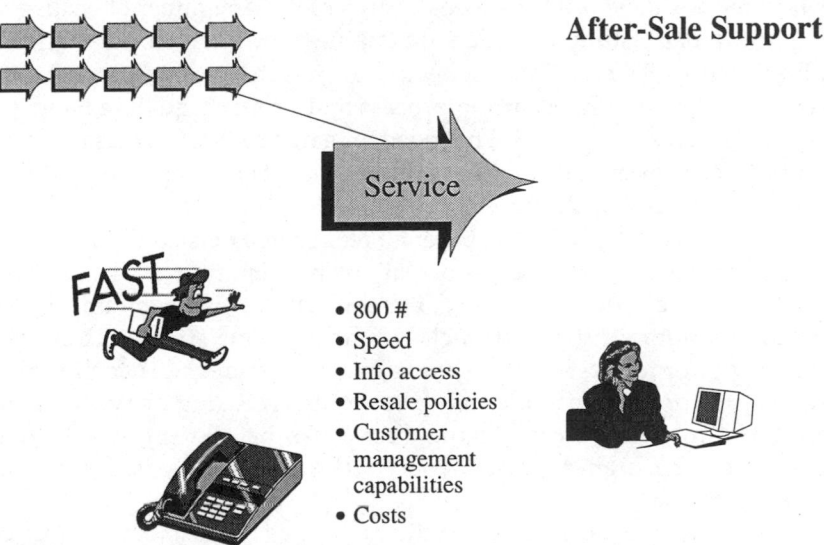

After-Sale Support

- 800 #
- Speed
- Info access
- Resale policies
- Customer management capabilities
- Costs

Figure 6.2f. After-sale support.

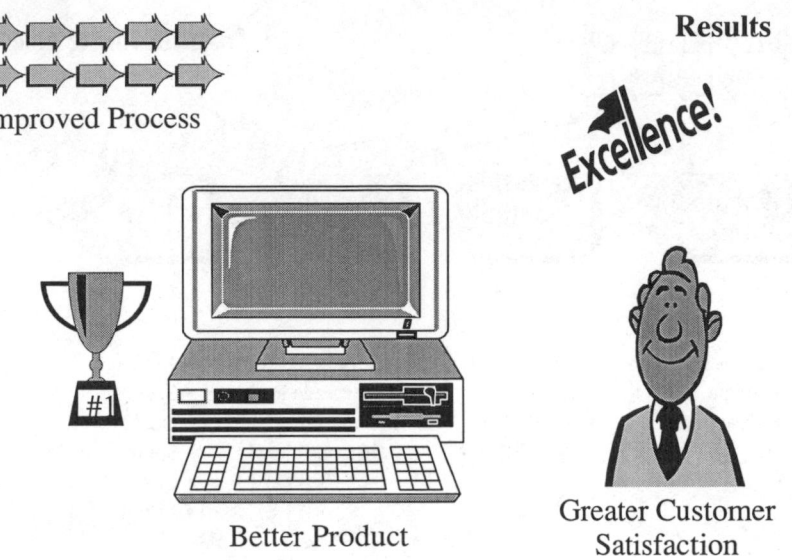

Improved Process

Results

Excellence!

Better Product

Greater Customer Satisfaction

Figure 6.2g. Results.

Best-in-Class Benchmarking

In baseball, the American penchant is to focus on the long-ball hitter. American sports fans revel in home runs. This predisposition to home runs—opposed to RBI (runs batted in)—probably explains why so many managers are ordering up best-in-class benchmarking studies without regard to cost, need, or return. At a regional financial services company, a senior vice president ordered a best-in-class benchmarking study on loan foreclosures. Dutifully, a team of assistant vice presidents and staff people deployed to conduct such a study. The project quickly—and fortunately—stalled within a week or two of the assignment because the team did not know what to study. It needed to examine more than an operating metric relating to foreclosures. The real interest lay in how to achieve low foreclosures through an excellent sales and underwriting process that properly qualified and screened loan candidates. This company did not need to search for best-in-class because it still had plenty to learn from other regional competitors. It also lacked any budget to support a national or worldwide search.

At day's end, the winning baseball team cares mostly about how many runs it has placed on the scoreboard—not about how many runs were earned through home runs. The same holds true of benchmarking. Not every improvement project requires a multiindustry best-in-class benchmarking study. Not every company is ready to perform a best-in-class investigation. Best-in-class benchmarking seeks the highest performers without regard to industry. Rising above industry comparisons, best-in-class benchmarking first seeks best-in-country benchmark partners. One rung higher on the benchmarking hierarchy is best-in-world benchmark partners.

The need must be great to benchmark at this highest level. Before commissioning a best-in-class study, consider whether you truly want and need to search for best-in-class companies. They are quite distinct and fewer than merely "really good" companies from which you may also learn much. Answer the following questions before sending your troops out into the field:

- *Are your operations mature enough to emulate a world-class performer?* Most companies need to learn to walk before they run. If your company is small or mid-sized, it may not be realistic to spend millions of dollars on information technology that will enable you to match the systems of a best-in-class multinational corporation. Learn first from industry leaders or other excellent companies.

- *Do you compete internationally or globally?* Organizations operating profitably in regional niches don't necessarily need to look outside the country for effective practices that will make them more competitive. Companies that compete globally need to look farther afield. If you are competing in the global marketplace, you must know what constitutes the best among your foreign competitors.

- *Are you willing to devote adequate time, resources, and capital to undertake a best-in-class search?* True best-in-class searches are by their nature very time-consuming, expensive, and logistically complex. Do not undertake such a study without adequate resources. You will condemn the team to frustration and likely failure.

- *Is your project area strategically important enough to warrant a best-in-class search?* Best-in-class studies should have a payoff that is commensurate with the effort. Target them at operating areas that have significant impact on the business' critical success factors. Such studies should have an adequate strategic, cost, revenue, or customer satisfaction impact to promise a satisfactory return on investment.

If you answer "no" to any of these questions, carefully consider if best-in-class benchmarking may exceed your current need. Studying excellent companies from other industries, without agonizing over whether they are truly the best-in-class, can deliver many results. A growing number of organizations report significant payoffs from simply studying other good companies outside their indigenous industries. Stepping outside one's own industry is a useful first step in breaking the cultural mold that confines one's thinking. "If you want to create a breakthrough," counsels Digital Equipment Corporation's Benchmarking Manager, Fred D. Bowers, "measure apples to oranges." At Digital and other corporations, such apples-to-oranges measures usually require the benchmark team to jump the fence and explore other industry practices.

AT&T Benchmarks Customer Complaint Handling[12]

What does AT&T have in common with a mail carrier, a hotel chain, and an airline? All have customers that—at times—complain. Dedicated to improving its complaint handling process, AT&T's operator services benchmarked three companies from other industries. All are highly effective at complaint handling. With 15,000 AT&T operators handling six to seven million calls per day, AT&T quickly recognized that complaint handling excellence was important to developing and maintaining customer loyalty. Moreover, it understood that the experience of leading companies in other industries might hold valuable lessons for AT&T's operator services.

When operator services' customers want to voice a complaint, they have three options: They can write a letter to AT&T CEO Robert Allen, call "00" to speak to an operator or the service center supervisor, or complain to a manager. Complaints addressed to senior management or not satisfied at the lowest level go to an executive complaint handling group that decided how to respond. Until 1984, AT&T had a monopoly on telephone services and did not worry much about losing customers. However, with the rise of competitors such as Sprint and MCI, AT&T elevated customer satisfaction and retention as a major corporate theme. Operator services possessed enough self-awareness to recognize it handled complaints suboptimally.

Consequently, a team of four complaint-handling experts and one benchmarking manager set out to examine AT&T's complaint handling process and to improve it by identifying best practices among leading companies from several industries. The team began by brainstorming problem areas that might help them focus their benchmarking and improvement efforts. For example, AT&T did not teach operator services' employees how to project a "friendly, helpful image." Discovering techniques that would enable employees to learn how to project this image would be of great benefit. Establishing high-priority problem areas also helped the team determine the scope and goals of the study. These included setting reasonable time frames for handling complaints, developing flexible methods and procedures for complaint handling, providing statistical information for reports to management, tracking complaint status, and following-up after complaint handling to ensure customer satisfaction.

Ultimately, the team identified 11 areas that they used as evaluation criteria for assessing other companies' complaint handling capabilities. These evaluation criteria included elements such as CEO participation in the complaint process, use of statistical reports, use of computers, use of award systems to encourage customer service, a focus on employee empowerment, and the way in which customer calls are received (through an 800 number or through collect calls). The AT&T library staff assisted in the effort by identifying background articles on the companies. Profiles of each prospective benchmarking partner and its standing on the evaluation criteria were constructed from this research and summarized in a selection matrix, which enabled the team to make a data-driven decision on benchmarking partners to visit.

During its research, the benchmark team determined that customers complained about different problems in different industries, but team members observed that the complaint handling process was similar across most industries. Consequently, they decided to seek best complaint handling practices without regard to industry. Ultimately, the team agreed to visit three companies—a mail carrier, a hotel chain, and an airline. A detailed question set was developed and mailed to the benchmark partner in advance so that they understood the purpose of the visit and could have the right people in the room.

During site visits, the team worked with a carefully constructed agenda that placed the most important topics first on the schedule. One person was selected to open the meeting with a short presentation overview about AT&T, Operator Services, and the Complaint Handling Function. The partners reciprocated with similar presentations.

Each team member was assigned a section of the question set that they were responsible for asking as well as answering if the partner wanted to learn about AT&T procedures.

Although AT&T was willing to share information during its site visits, the partners did most of the talking.

Judy Horsfield, the AT&T benchmarking manager, took careful notes during the benchmarking visits. At each site visit's conclusion, the AT&T team immediately debriefed and consolidated its findings in a detailed trip report. The trip report was later shared with the rest of AT&T through the corporate benchmarking database. Horsfield says the AT&T team assessed its findings from the 3 visits and identified the following critical success factors to improve customer complaint handling in operator services:

1. The manager who is champion of the complaint handling process must participate in and provide financial support to the complaint handling group.

2. The company vision for the future, company culture, company attitude, and policy toward its customers must be understood by all complaint representatives.

3. The complaint handling job requires full-time dedicated employees whose sole responsibility is to process customer complaints, manage the day-to-day operation, and report customer information to the rest of the company.

4. Each complaint representative must be connected to the computerized complaint tracking system for easy access to past and present customer data.

5. The complaint handling group's organization reporting level must be closely tied to the process champion, (i.e., there should be limited management levels between the two).

6. The complaint handling group must be centralized to:
   - Prevent breaks in the process flow.
   - Allow easy access to complaint information.
   - Create a family atmosphere and prevent burnout.
   - Eliminate duplication.
   - Centralize data collection and reporting.
   - Keep communication flowing.

7. Each member of the complaint handling group must be empowered to satisfy a customer within reason and according to AT&T policy.

8. The three main focuses of the complaint handling group should be:
   - People—its employees.

- Service—to its customers.
- Profit—for the company.

9. The mission of the complaint handling group is "to satisfy the customer."

10. A formal training program must be in place. A buddy system should be created to connect new employees with more experienced representatives until the new employee has reached a designated confidence level.

11. Customers must have easy access to the complaint handling group, by phone, mail, or computer.

12. Strong employee commitment is necessary to:
    - Create team spirit and a family atmosphere
    - Encourage peer support

13. Well-established methods and procedures must be in place describing an organized flow of the complaint process (from complaint inception through customer satisfaction).

14. Each complaint representative must have a high level of skills acquired through many years' experience in the field and college training.

15. Measurements and metrics are compiled by the complaint handling group to ensure quality.
    - Daily statistics:
      - ✓ The number of complaints received.
      - ✓ The number of complaints resolved.
      - ✓ Turn-around time.
    - Trend analysis to determine customer complaint trends over time.
    - Baseline statistics to compare AT&T's complaint data to those of best-in-class, industry averages, and the FCC.

16. Internal competition is eliminated among complaint representatives. Each employee is judged as an individual.

17. A reporting mechanism is in place to report complaint information to upper level management. This will allow management to keep on top of customer issues, desires, and satisfaction levels.

18. Work loads are equally distributed across complaint representatives. Each representative is accountable for their customer base.

19. Complaints are coded, categorized, and entered in the computer tracking system to be used for statistical purposes.

20. Contact information is maintained with the names and phone numbers of process owners across the company who can assist in resolving customer complaints.

21. The complaint handling process should push resolution to the lowest level.

22. Only experienced complaint representatives have the authority to sign for the champion and executive management on letters to customers.

23. Standards are set for offering symbolic atonement to customers (e.g., "We are sorry for the inconvenience" gifts).

24. The sharing of information is encouraged among complaint representatives (e.g., a pool of effective letters to customers, a bulletin board posting ideas that work well, a suggestion program to change the process).

25. Continuous improvement.
    - Consultants are used to improve the process.
    - Industry benchmarking is on-going.

With the initial benchmarking study completed, AT&T then launched the difficult process of implementation, which Horsfield predicts will be a multiyear process in view of the scope of the benchmarking findings. Already the company has implemented two major improvements: It installed a database to track complaints and it centralized the complaint handling group. Moving forward, AT&T plans to adapt the other critical success factors it observed among its benchmark partners. The result is a dramatically improved and far more effective complaint handling system than could ever have been internally engineered.

Select the Appropriate Benchmarking Approach

Best-in-class benchmarking can produce dramatic breakthroughs. The insights that AT&T gained when studying three national leaders from the express mail industry, the airline industry, and the hotel industry produced insights that AT&T never would have developed by studying other telephone companies. Not every company could or should undertake such an extensive national investigation. Best-in-class benchmarking should be pursued with careful planning and discretion. Hitting the long ball is not the only way to score runs. Sometimes getting a base hit will win the ball game. Managers can learn to integrate benchmarking as a flexible skill into various improvement and outreach approaches. They must evaluate each opportunity to determine the appropriate project scope and investigative rigor required by the benchmarking study. Different projects and situations require different approaches. Each company must work from its own experience and maturity base. Benchmarking, like any important management task, requires judgment, skill, and practice. *(Figure 6.3 depicts different approaches to benchmarking and reflects the varying resource investments required of each approach.)*

References

1. "Surveying Industry's Benchmarking Practices," APQC International Benchmarking Clearing House, Houston, Texas, 1992, p. 9.

2. Bowers, Fred, "Benchmarking Against The Best," proceedings from the International Research Institute conference on benchmarking, May 1991.

3. "Quality: Small and Midsize Companies Seize the Challenge—Not A Moment Too Soon," A Special Report by the Staff of Business Week, November 30, 1992, pp. 66-67.

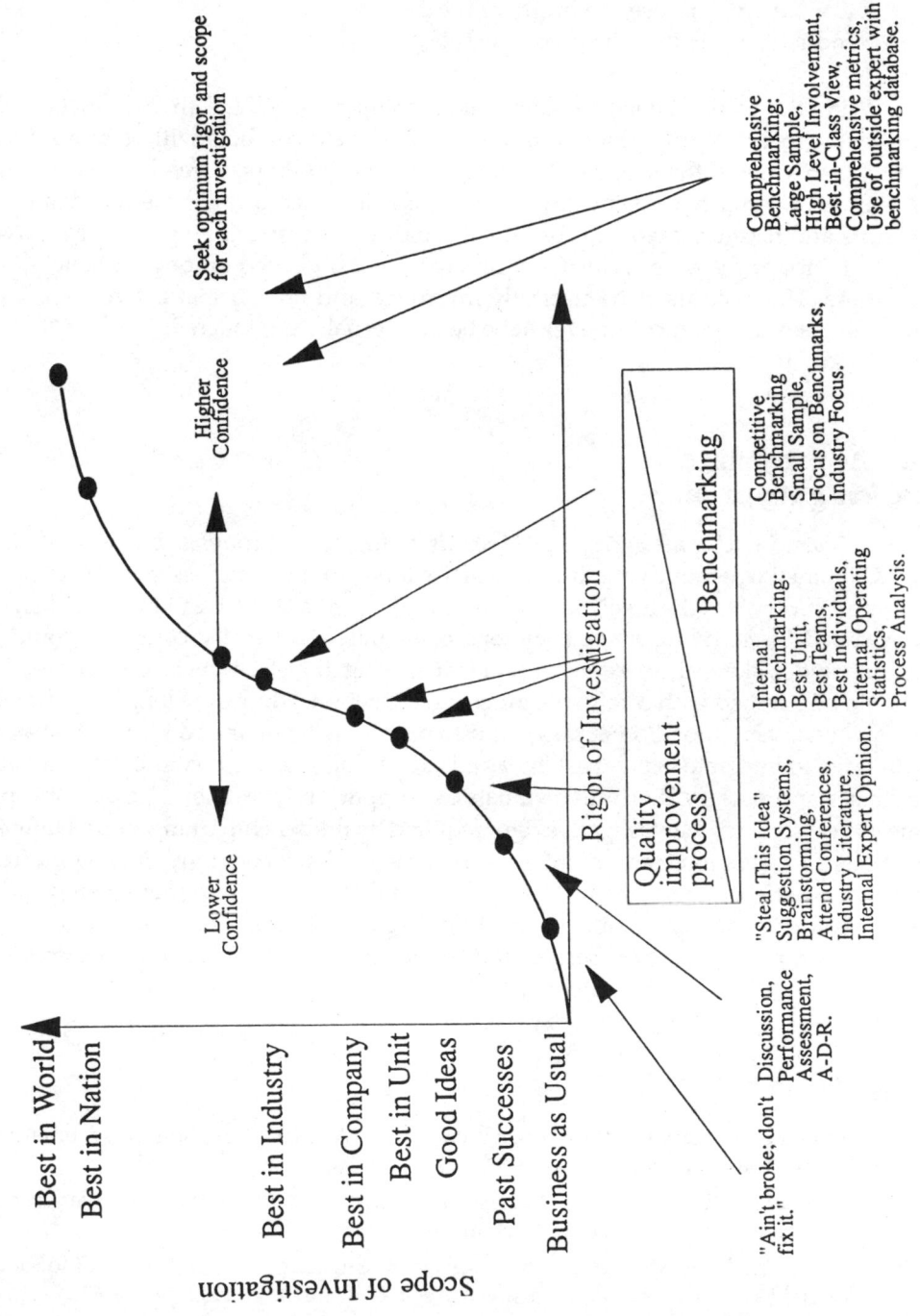

Figure 6.3. Concept of scope and rigor.

4. "The International Quality Study[SM], Best Practices Report: An Analysis of Management Practices that Impact Performance," a joint project of Ernst & Young and the American Quality Foundation, Cleveland, Ohio, 1992, p. 38.

5. Weise, Richard H., "Representing the Corporate Client: Designs for Quality," Prentice Hall Law & Business, Englewood Cliffs, N.J., pp. 10-57, 10-74.

6. For a full discussion of this approach, see "The Baldrige: What It Is, How It's Won, How To Use It To Improve Quality In Your Company," by Christopher Bogan and Christopher Hart, McGraw-Hill, 1992, pp. 208-214.

7. Brokaw, Leslie, "Where Great Ideas Come From," *INC.*, January 1993, p. 74.

8. Crandall, Robert L., Chairman's Letter to readers of American Airlines in-flight magazine, called *American Way*, March 1993, p. 10.

9. For a full description of Steal This Idea™ and a library of some of the more ingenious ideas that have been benchmarked through it, consult the executive resource library.

10. See additional ideas in the Steal This Idea™ appendix.

11. von Oech, Roger, "*Creative Wack Pack*," 1989, Box 7354 Menlo Park, California 94026, Card #48 (green).

12. Judy Horsfield, AT&T Benchmarking manager, served on the AT&T complaint handling benchmark team and graciously shared her experiences for this case study.

Putting Benchmarking to Work in the Executive Office and Boardroom

Benchmarking and Leadership

In performance improvement no communications tool is more powerful than the actions of leadership. Accordingly, many corporations are challenging their senior executives to model the behaviors, attitudes, values, skills, and actions that they advocate for the organization. In other words, American leadership is being asked to *walk the talk.* Consequently, the time has come for benchmarking to be put to work in the executive office and boardroom. Already a powerful tool of front-line operating managers, professional staff, and front-line teams, benchmarking has far-reaching potential for senior executives, CEOs, and corporate directors.

Borrowing best practices from market leaders and world-class performers is an effective way to manage change and to supercharge continuous improvement. A clear and steady focus on best practices can provide the impetus for change, can help identify what must change, and can provide a vivid picture of the end point after change. Benchmarking should, therefore, be common practice among every organization's leadership team. Best practices benchmarking has *at least* five critical, high-level applications from which senior executives and board members can benefit:

- Benchmarking enables executive officers and directors to know where their corporations stand versus the best.

- Benchmarking is a practical tool for creating a fast-learning organizational culture that's dedicated to continuous improvement. (See Chap. 2.)

- Benchmarking is a *sine qua non* for *strategic* planning to attain market leadership. (See Chap. 9.)

- Benchmarking is a breakthrough technique to accomplish organizational restructuring, reengineering, or redesign. (See Chap. 10.)

- Benchmarking is a valuable tool to assist board members in performing their duties of corporate governance and oversight.

Understanding Where Your Corporation Stands Versus the Best

Benchmarking emerges as a leadership responsibility of first-order importance. To be effective leaders, managers must be aware of other corporations' strategies, tactics, and operating successes. Senior executives, directors, and their corporations need to "know where they stand" relative to competitors and best-practice performing organizations. Enlightened leaders recognize that they can learn valuable lessons from outside their corporations and outside their industries. It often helps to look to others for different perspectives. That's why the search for best practices—inside and outside one's own organization and industry—is a fundamental part of total quality management. After internal improvement efforts get underway, fast-learning organizations inevitably search outside themselves to other units, companies, and industries; they seek to accelerate their own performance improvements by studying others' best practices. The goal of all quality and benchmarking initiatives is the same: to target and achieve total performance excellence. Translated, that means benchmarking serves to continuously improve products, services, sales, profits, cost management, customer satisfaction, employee satisfaction, and all other measures of financial and nonfinancial success. As a performance improvement tool, therefore, benchmarking is a vital instrument of leadership.

The evolution of the Malcolm Baldrige National Quality Award criteria highlights this fact. In 1988, when the Baldrige performance-assessment criteria were first published, no direct references to benchmarking appeared in the leadership category. In 1989, however, the Baldrige administrators updated the criteria and then first probed senior management's role in "learning about quality of domestic and international competitors." In 1992, the Baldrige criteria specifically inquired about senior executive leadership activities that "might also include ... benchmarking." That same year the National Quality Award guidelines suggested that senior executives "describe indicators, benchmarks, or other bases for evaluating and improving organizational structure." By 1993, the Baldrige leadership guidelines stated that "activities of senior executives might also include leading and/or receiving ... benchmarking." (For a complete discussion of benchmarking and the Baldrige criteria, see Chap. 17—Best Practices Toolchest for Executives.)

Some companies have carried the benchmarking banner even further into their managerial ranks. At one regional trucking company, every member of the executive committee—right up to the president and CEO—have agreed personally to perform benchmarking visits to other leading companies in their core responsibility areas. Such actions ensure that senior management is well informed of its competitive position, enabling the leadership team to make better decisions and to more effectively chart the company's course through the fast waters of rapidly changing markets.

An Essential Leadership Skill

Bellwether reports from the corporate workplace demonstrate the growing importance of benchmarking within the executive office. DuPont, for instance, has written benchmarking into the job descriptions of all vice president-level officers. DuPont and other organizations regard benchmarking as an essential skill for all competent managers. Other corporations, such as Ameritech, AT&T, Hewlett-Packard, Pacific Bell, Digital Equipment Corporation, Eastman Kodak, Xerox, and IBM, have created executive-level benchmarking jobs to oversee and spearhead corporate benchmarking efforts. Texas Instruments Defense Systems and Electronics Group, a 1992 Baldrige Award winner, understands the broad-reaching relevance of benchmarking as a managerial tool. At TI, benchmarking is one of six "leadership thrusts" employed to help the organization achieve its fundamental objective of "customer satisfaction through total quality."[1] The manager in charge of broadly spearheading benchmarking at TI has the sterling title of "Benchmarking Champion for World Operations."

Universal Card Services, AT&T's credit card company, regards benchmarking as a critical success factor for senior managers. Another 1992 Baldrige Award winner, UCS classifies benchmarking as one of 12 leadership activities. Senior executives—including the CEO—are expected to participate in benchmarking visits to other companies. In one such study, the UCS senior executive team visited Walt Disney. Their goal was to understand how Disney effectively deployed strong customer-focused operating values throughout a large and diverse work force. UCS executives were impressed by the way Disney began talking about the company's values in the preemployment interview process. This practice and others were imported into AT&T Universal Card Service and they have helped UCS establish a strong corporate culture dedicated to "customer delight."

In addition to conducting benchmarking visits to other world-class companies, AT&T UCS *requires* its senior executives to stay tuned to its market by meeting with customers, listening to customers' calls, reviewing daily process measures, meeting quarterly with suppliers, reviewing and using customer feedback, reviewing the program management process monthly, co-chairing monthly customer listening post meetings, hosting team sharing rallies, conducting "meeting of the minds" sessions, holding quarterly employee meetings, and "owning" Baldrige self-assessment categories. Many of these activities give rise to informal or unstructured benchmarking opportunities.[2]

Leaders Set the Culture

"Good is the enemy of best and best is the enemy of better," says Roger Milliken, CEO of Milliken & Company. With this attitude, Milliken has created a "we-can-learn-from-everyone" corporate culture at the South Carolina textile company. Innovative adaptation or creative imitation is the operative concept encouraging Milliken & Company's employees to look outside their work groups, business units, and corporation for ideas that can help them continuously improve performance. Roger Milliken supports this active learning culture with personal actions that reinforce the stated values of a fast-learning organization. Milliken's CEO is known to stay late at the office and on occasion he works late with associates on detailed projects. He has no doors on his office, eschews corporate parking spaces, executive dining rooms, and other perks that might separate him from the team that he leads. These largely symbolic gestures encourage idea exchange, information sharing, and accessibility among employees—including the CEO.

Every senior manager would do well to consider what effect his or her own actions have in developing an operating environment conducive to benchmarking, innovative adaptation, and fast learning. In its broad-reaching inquiries into all aspects of an organization's performance improvement systems, the Baldrige criteria systematically probe leadership's role in supporting benchmarking as an enabler of continuous improvement. Try answering the following questions; all probe issues that Baldrige examiners must assess when evaluating a company's total quality systems:

- Do you visibly support benchmarking as an activity that fosters quality excellence?

- Do you make an effort to learn about the quality of domestic and international competitors?

- Do you study what other companies are doing so that your unit can import their most effective operating practices and strategies?

- Does your leadership employ benchmarking information to enable it to evaluate your organization's strategy?

- Does your senior executive team carefully study what your competitors are doing before making major business decisions?

The first reaction among many executives is to say, "Yes, we support benchmarking." However, a gap often exists between managerial intentions and management's supporting actions. In a Best Practices Benchmarking and Consulting, Inc. survey of the 80 most senior executives of a multinational corporation, all operating executives indicated that they *frequently* attempt to learn about competitors, and most executives said that they "visibly support benchmarking as an activity that fosters quality excellence." Despite executives' personal commitment to benchmarking, other survey questions revealed that "the walk does not always match the talk." In many instances, managerial systems and actions did not support or encourage benchmarking and active learning through comparisons and outreach. For instance, benchmarking primarily concerned price and product feature comparisons. It did not examine a wider range of operating issues, such as process capability, quality, new product development success, or speed to market. Moreover, managerial systems, such as reward and recognition, performance appraisals, information sharing networks, and the strategic planning process, did not support benchmarking.

In this corporation and others like it, senior executives are struggling to understand what their roles should be in championing benchmarking efforts within their organization. They are challenged to develop managerial systems that support an organizational culture dedicated to innovative adaptation. Responding to this new role, senior executives have significantly increased their numbers at benchmarking events such as the Malcolm Baldrige Quest For Excellence conferences, where newly announced Baldrige winners share their successful strategies. The Quest For Excellence conferences are the epitome of conference-style informal benchmarking; those attending can compare their companies' performance management approaches with those of a select few companies chosen to receive America's most prestigious award for quality excellence. In 1993, approximately 20 percent of attendees at Quest For Excellence V were CEOs, presidents, vice presidents, or other senior executive officers in their organizations.

Benchmarking Project Champions

Benchmarking presents many organizations with a management paradox: Benchmarking studies are typically most effective when conducted by front-line managers and employees who actually operate the systems and processes under examination; yet, benchmarking projects produce suboptimal results if the senior leadership doesn't champion these projects. Establishing the right mix of front-line involvement and senior executive involvement can be tricky. Clearly, senior executives play a leading role in determining the success of benchmarking projects. Senior executive influence is often disproportionately larger than the actual time and direct involvement they may dedicate to any given project. This is especially true of complex process benchmarking projects.

In the American Productivity and Quality Center's 1992 survey of 87 companies active in benchmarking, 61 percent of respondents stated that "no top-management support" was a "great or very great" factor influencing unsuccessful benchmarking efforts. Senior management support ranked second, behind poor planning, as the most frequent reason a benchmarking study fails.

In the same survey, 98 percent of benchmark respondents also observed that process owner involvement was of "great or very great importance" in a successful benchmarking study. This involvement may extend to senior executive team members who are responsible for championing individual processes that are critical to the organization's success. Moreover, 83 percent of benchmark survey respondents indicated that top-management commitment was of "great or very great importance" in encouraging benchmarking within the respondents' organizations. In fact, top-management commitment was the most frequently cited factor encouraging benchmarking in the surveyed organizations.[3]

Boil down all these survey findings to their essence, and it's self-evident that senior executives play a critical role—both direct and indirect—in enabling benchmarking investigations and best practice improvement strategies to succeed.

Monitoring the Competition

Benchmarking prompts senior managers to focus systematically on the competition. Moreover, it expands the typical competitive scrutiny beyond price and product feature comparisons. "Power benchmarking," which embraces process comparisons, performance comparisons, and strategic comparisons, embraces the widest range of operating areas. It includes—but is not limited to—examinations of process capability, organizational structure, quality, reliability, warranty costs, cycle times, human resource management, technology management, and many other important competitive factors.

Some industries and organizations have long understood the importance of effective competitive intelligence. Authors Noel M. Tichy and Stratford Sherman describe the intense scrutiny on competitors at General Electric:

To promote clear thinking and fast decision-making, the CEO often makes operating executives prepare a few simple slides that describe the essence of their business situations. The slides show the answers to basic questions such as these:

- What does your global competitive environment look like?
- In the last three years, what have your competitors done?
- In the same period, what have you done to them?
- How might they attack you in the future?
- What are your plans to leapfrog them?

Benchmarking provides senior executives a structured approach by which to develop competitor information. The information may be process-oriented, performance-oriented, or strategic in nature. More importantly, best practices benchmarking provides a systematic process for evaluating the adequacy of one's own position and course relative to competitors. Whether you manage a multinational corporation or a professional sports team, awareness of one's competitors is critical for successful leadership. Pat Riley, coach of the New York Knicks, carried Sun Tzu's *The Art of War* with him throughout the 1993 basketball season. When asked why he kept this military strategy book, written nearly 2500 years ago, at his bedside during the playoffs, Riley simply quoted Sun Tzu on the importance of competitive intelligence: "If you know your enemy and know yourself, you need not fear the result of a hundred battles."[5] Senior executives that engage in benchmarking root their leadership in fact-based performance information. No wonder their success rates surpass those of other managers who operate without so systematically considering competitor capabilities and marketplace realities.

Corporate Governance and Oversight

Now that benchmarking is being profitably applied to organizational structure, strategy, performance, processes, products, and services, it's time to put it consistently to work in the boardroom. The potential benefits are many for corporate directors. In performing their duties of corporate oversight and governance, board members must routinely evaluate the adequacy of senior management's goals, plans, and overall strategy; benchmarking information provides essential reference points that make these evaluations much more objective and reflective of the current marketplace.

"There is no other factor in a company that is as important as knowing where you stand against your competitors. I'd lay that out against any company in any industry," observes James L. "Rocky" Johnson, the Chairman Emeritus of GTE Corporation. Johnson, who sits on many boards, says he always requests competitive benchmarking information from management, for it helps him assess the organization's current performance and future prospects. "If I had to pick one thing that I thought was most responsible for GTE's success it would be benchmarking ourselves against our competitors and then going about trying to be the best to compete in all those markets," adds Johnson.

The cautionary tales emerging from boardrooms across the nation, highlighted by forced leadership changes at corporations such as General Motors, American Express, Compaq Computer and Digital Equipment, speak of a new activism among boards. Increasingly, board members are no longer content to serve as passive corporate advisors; many are actively reviewing performance and operations, unwilling to accept the prognostications of management without independently testing them against competitive realities. In this regard, benchmarking information represents a valuable source of objective information by which corporate directors can better oversee the affairs of the companies which they serve. Committee work involving executive compensation, organizational structure, succession planning, pension trust, public policy, and strategic planning is made excellent by the use of benchmarking.

Imagine how much more quickly a long list of directors might have acted if they had personally perused competitive comparisons and benchmark studies that provided a market-based perspective to the organization's goals, strategies, and performance. How long would a corporate board sit quietly if it observed management setting 7- to 8-percent productivity improvement targets when competitors were setting and achieving 50- to 60-percent improvement goals? Not long! Benchmarking information provides an important diagnostic test of the organization's current and future health.

"Whatever can help board members assess where the company is and what's the likelihood of its success, that would be invaluable. That is the value of benchmarking," observes Vincent Cannella, senior partner in charge of Quality at KPMG Peat Marwick. "...Especially if I'm a part-time board member, it would be nice to have benchmark information to help evaluate how this management is doing ... how does it match up with its peers? What's the likelihood of this idea or enterprise working? I need to be equipped as a board member to make sure my name isn't soiled; I don't want to be part of a company that ran amok."

Ironically, the deep-rooted cultural barriers that often discourage front-line employees from borrowing from the best are equally obstructive among directors. When preparing for her first assignment on one of several boards, a young director said she began by benchmarking the role of effective board members so that she could undertake and meet her responsibilities. An older director and CEO, who also sits on several boards, confided how envious he was of her having the opportunity to review and learn from the experience of what other highly-effective boards were doing. He noted that people simply expected him to know how to conduct affairs as a board member and therefore he never dared openly inquire what his role should be. What a shame it will be if directors become encumbered by the same cultural obstacles that can handicap unempowered front-line employees.

A powerful and flexible new business tool, benchmarking is not conceptually difficult to understand or apply. Operating managers are increasingly putting it to good use—and so are growing numbers of senior managers. There is no reason why boards should not benefit as well. In both the conduct of general operations and of corporate governance and oversight, benchmarking for best practices is a sound business strategy.

Leadership Actions

When General Electric CEO Jack Welch returns from a visit to Wal-Mart raving about the practices of the world's largest retailer, his actions strongly communicate the acceptability of adopting and adapting others' excellent practices to advance the organization. Then, when Welch writes about his experiences in his annual letter to shareholders in GE's annual report, employees begin to believe he means what he's saying:

> In 1991, we shared best practices with a number of great companies. We learned something everywhere, but nowhere did we learn as much as at Wal-Mart. Sam Walton and his strong team are something very special. Many of our management teams spent time there observing the speed, the bias for action, the utter customer fixation that drives Wal-Mart; and despite our progress, we came back feeling a bit plodding and ponderous, a little envious, but, ultimately, fiercely determined that we're going to do whatever it takes to get that fast.[6]

What can senior executives and board members do to put benchmarking to work in the executive office and boardroom? How can an organization's leadership foster creative adaptation throughout the corporation? Here are some action strategies that we have observed in different companies stretching across diverse industries. Consider how you might deploy these action strategies in your organization:

- Employ competitive benchmarks to provide an external context for operating reviews.
- Require benchmarks as part of the strategic planning process.

- Use benchmarking information to establish aggressive goals and standards.

- Study the strategies and core processes of other excellent companies.

- Use benchmarking information to evaluate and improve organizational structure.

- Review the products, services, and practices of other companies and visit their facilities whenever possible.

- Invest in employee education to bring outside ideas into your organization.

- Create lending libraries that focus on competitors' and other high performers' winning strategies and systems.

- Reward and recognize creative adaptation or the borrowing of good ideas and systems.

- Invite customers, suppliers, and other outside speakers to share their ideas and strategies with your employees.

- Make best practice information sharing and innovative adaptation a skill evaluated in the performance review and promotion process.

- Actively promote and publicize the benefits of borrowing from the best.

- Sponsor regular meetings and discussions where employees and managers exchange ideas and explore best practices. Place on the agenda of executive meetings a regular period of time to explore external ideas, systems, and managerial approaches that might be profitably imported into the company.

- Make benchmarking a responsibility delineated in senior-level job descriptions.

- Employ best practice information in the problem-solving and continuous-improvement process.

- Link compensation to improvement against benchmarks in core operating functions.

- Focus organizational resources on identifying and deploying best practices, as well as on other quality-improvement strategies, such as defect reduction and error removal.

- Study what competitors and other leading companies are doing before making major business decisions and capital investments.

- Lay out work areas to encourage idea sharing and information exchange.

None of these action strategies alone is momentous. However, when taken together, they present a strong and consistent message that is articulated through the leadership's words and actions. No single action or pronouncement by an organization's leadership will ensure benchmarking's success throughout the company. Yet consistent words and actions lay a strong foundation for success. Some well-worn management wisdom observes that "the speed of the leader determines the rate of the pack." Certainly the speed with which senior executives and directors embrace benchmarking in their words, actions, and management systems will determine the rate at which the line organization employs best practice strategies to drive innovation and improvement.

References

1. Hayes, Hank, President of Texas Instruments Defense Systems & Electronics Group, speaking at the Quest For Excellence V Conference in Washington, D.C., February 15-17, 1993.

2. Kahn, Paul, former CEO of AT&T Universal Card Services, speaking at Quest For Excellence V Conference in Washington, D.C. February 15-17, 1993.

3. Surveying Industry's Benchmarking Practices, APQC International Benchmarking Clearing House, Houston, Texas, 1992, pp. 15-20.

4. Tichy, Noel M. and Sherman, Stratford, *Control Your Destiny or Someone Else Will*, Doubleday, New York, N.Y., 1993, p. 26.

5. Riley, Pat, interviewed by NBC Sports during the 1993 National Basketball Association semifinals with the Chicago Bulls.

6. Welch, John F., Jr. and Hood, Edward E., Jr., "To Our Share Owners" letter in GE's 1991 Annual Report, February 14, 1992, pp. 1-5.

8

Benchmarking and Strategic Planning

Benchmarking has not traditionally been an integral part of strategy setting and strategic planning. It should be, though. Without some external reference points by which to evaluate and validate an organization's strategies, plans, and goals, management is flying in the dark. Corporate histories have recorded some notable crashes by fine companies that tried to fly without looking out for what others were doing and accomplishing. If they had gathered benchmarking information in advance, many of these failed efforts might have been avoided.

The Phoenix-like resurrection of Compaq Computer Corp. is one of the most dramatic examples of how easily accessible benchmarking information helped a Fortune 500 company redefine its strategy and regain market share that was lost when the personal computer market suddenly turned treacherously price sensitive in the early 1990s.

The rapid market shift transformed the personal computer industry from a battleground where leading-edge technology and quality won significant price premiums and market share, to a near commodities marketplace where advanced technology and quality are requirements of doing business and where low prices win sales. This 180-degree shift turned Compaq's originally successful strategy—provide superior technology and engineering and command premium prices for it—into a recipe for slow growth and declining market share.

What makes the Compaq strategic benchmarking story so compelling is that it was initially commissioned by Compaq Chairman Benjamin M. Rosen behind the back of former Compaq chief executive and co-founder Joseph R. Canion. Mr. Rosen, a venture capitalist who had experienced several high-tech booms and busts, was increasingly convinced that the PC business was fundamentally changing and that Compaq had to begin competing directly with the low-cost PC clone makers. But Mr. Canion's senior-management team was slow to shift strategy and meet this new competition head on.

Frustrated with the senior management team's resistance, Mr. Rosen quietly recruited two mid-level Compaq managers as "secret operatives" to seek benchmark information pertaining to the cost and availability of computer components. The secret operatives went to work without the knowledge of the CEO and his lieutenants when they attended the 1991 Comdex computer trade show in Las Vegas with the chairman. Strolling around as "nobodies," wandering from booth to booth, they were amazed to discover that some suppliers offered them component prices that were lower than what $3.3 billion Compaq was then obtaining. The benchmark team quickly assembled two demos that convinced Mr. Rosen that Compaq could launch new entry-level computers much faster and more cheaply than management had been telling him.[1] Later, when Compaq engineers began studying some of the low-priced PC clones, they discovered they were made much better than Compaq had ever suspected. All this information led to the unceremonious ouster of Mr. Canion and resulted in the rapid development of a new strategy for Compaq.

This seminal benchmarking study led to one of the most dramatic and dazzling turn-arounds in recent history. Still, one wonders how much sooner the company might have shifted course and avoided the battering of its stock and co-founder if strategic benchmarking had been a regular part of Compaq's planning process.

With just this understanding, a growing number of companies have begun to use benchmarking as a critical step in the strategic planning process. By reviewing the products, prices, practices, strategies, structures, and services of competitors and other companies, managers can evaluate the adequacy of their own goals, plans, and strategies. At Mutual Life Insurance Company of New York, for instance, all executives are required to develop benchmarking information on their primary and secondary competitors as part of the company's newly revamped planning process. Says MONY Quality Officer Jan Howard: "Planning without awareness of what your competitors are doing is like flying a plane over the Alps in heavy fog without any instrument controls." (*Figure 8.1 illustrates benchmarking's role in strategic planning.*)

Benchmarking adds a very useful navigational compass to the strategic planning process. By identifying and studying others, strategic benchmarking is especially useful in the following areas:

- Determining where your organization stands versus the competition and best performers outside your industry.
- Validating the adequacy of short-term and long-term goals.
- Setting and refining corporate strategy that has the highest likelihood of succeeding.
- Ensuring core processes that are critical to the organization's success are competitive with the rest of the market.
- Ensuring the company's use of technology is adequate to help it maintain its position within its chosen markets.
- Ensuring that critical issues such as structure, price, performance, products, and services are adequate to succeed against competition in your chosen markets.

Figure 8.1. Strategic plan.

BEST PRACTICES

- Ensuring your supplier capabilities are adequate to enable your company to succeed in its chosen markets.

- Identifying key factors to attain market leadership.

For a growing number of companies, including the likes of Johnson & Johnson, AT&T, and Digital Equipment, benchmarking has become a critical part of the strategic planning process. "We see benchmarking as an enterprise, not an operational function," says Robert H. Ogle, director of worldwide quality improvement at J&J. "I would never run a company without using this again."[2]

The findings of strategic benchmarking may be painful to the egos of managers who pride themselves on running world-class operations. Discovering that another company has developed a better and more effective approach can be humbling. But the sting of such discoveries when there is still time to take advantage of the learning is much less painful than prolonged market share shrinkage, profit erosion, plant closings, and repeated work force reductions.

The Strategic Thinking Process

In his book *Strategy Pure and Simple: How CEOs Outthink Their Competition*,[3] Michel Robert argues persuasively that traditional formulas for competitive analysis and strategy setting are flawed. Based on 15 years of experience with CEOs from 300 corporations, Robert contends that setting a business strategy is not merely performing competitive analysis; an organization must strategize and plan before developing competitive tactics. Robert identifies eight steps that an organization passes through during the strategic thinking process, and benchmarking can serve as an enabling tool to ensure the success of each step in the strategic planning process. *(Figure 8.2 depicts the strategic planning process as conceptualized by Robert.)*

Step 1: Analyze the organization's current profile. What is the organization's current state? The answer is an organizational profile that includes current performance trends for all products and services, growth in all geographic areas, market share standing, the organization's core business concept or driving force, and areas of operating excellence.

Step 2: Analyze internal and external performance variables. Examine the internal and external operating environments that positively or negatively influence future performance. Internal performance variables include key features of products, markets, customer segments, operating strengths, weaknesses, and growth opportunities. External performance variables include various competitive considerations, strategic opportunities and strategic threats, or operating vulnerabilities.

Step 3: Explore driving forces and strategic options. What skills, resources, or advantages can the organization leverage in order to be successful in the marketplace? The leadership team might answer this question by formulating two or three high-leverage areas on which to build its strategy. Ultimately, management must choose the

Strategic Thinking Process

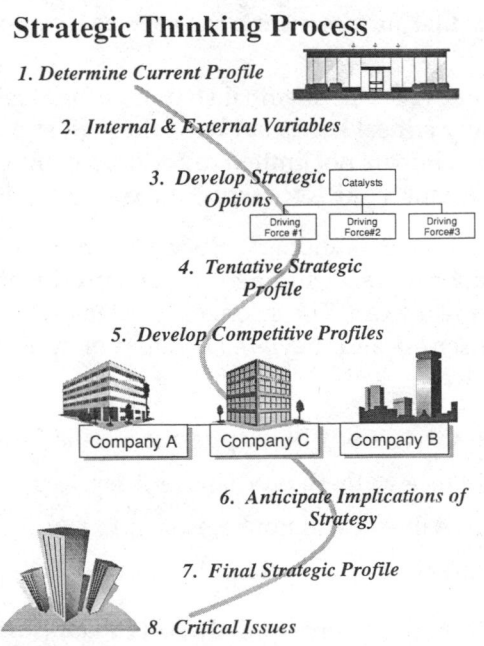

1. *Determine Current Profile*

2. *Internal & External Variables*

3. *Develop Strategic Options*

4. *Tentative Strategic Profile*

5. *Develop Competitive Profiles*

6. *Anticipate Implications of Strategy*

7. *Final Strategic Profile*

8. *Critical Issues*

Figure 8.2. Strategic thinking process.

best one or two strategic leverage points or driving forces and translate them into an integrated strategic mission, vision, or operating plan.

Step 4: Develop a tentative strategic profile for the organization. The strategic profile articulates the fundamental business concept, including areas of excellence that will exceed competitors' performance. It also delineates key milestones or implementation time frames and it may pinpoint future products to be emphasized and deemphasized. The profile identifies target markets and customer segments, defines corporate beliefs and values, delineates market size, growth rates, and expected investment returns and profit margins.

Step 5: Develop competitive profiles. The organization's management assembles similar profiles for three or four major competitors, scrutinizing the competition's products, services, pricing, and market selections. By examining competitor's profiles over a period of years, management can better anticipate the course each competitor is likely to pursue in the short-term future.

Step 6: Test and evaluate the implications of the organization's proposed strategic course. By testing the organization's current profile, management can better assess the extent of change required to fulfill its strategic vision. The leadership wants to be certain that important operating strengths are fully exploited, weaknesses are minimized, key beliefs are not violated, major opportunities are exploited, and major threats are avoided. Management also wants to match its strategic profile against competitor profiles to ensure that the organization's adopted strategy accurately reflects the competitors' strengths, weaknesses, and current and future actions. These strategic profile tests suggest how the organization's strategy can be modified and strengthened.

Step 7: Finalize the strategic profile. At this step, management reshapes and fine-tunes the strategy by addressing issues that surface during the various tests conducted in Step six.

Step 8: Resolve any remaining critical issues. In this final step of the strategic planning process, management resolves any critical issues surfaced through strategic profile analysis. Such *critical issues* include but are not limited to considerations of organizational structure, core operating systems, work force skills, and compensation systems.

Strategic thinking is iterative or evolutionary. Robert notes that it typically reflects the contributions of many people, including the CEO, president, and other members of the senior-management team. The strongest strategic profiles are frequently developed through consensus and they are anchored on four cornerstone decision areas:

1. Pinpoint which products and services will be emphasized in the future
2. Identify which customers will receive these products and services
3. Define which market segment will and will not be pursued
4. Outline which geographic markets will and will not be developed

As a business process, benchmarking consequently adds an all-purpose tool to the strategic planning process.

Strategic Benchmarks and Competitive Projections

Before setting strategy, the organization must establish its current position. Managers can do this most effectively by developing baseline measures or benchmarks that help define where their organizations stand compared to the competition. Strategic variables, such as market share, customer satisfaction standing, product and service performance (such as defect rates, cycle time, and warranty data), productivity, and cost, all help delineate an organization's competitive position. Other benchmark measures, such as growth rates, profitability, return on assets, or return on equity, may also be useful in evaluating your organization's relative performance standing.

Analyzing strategic performance variables is straightforward. Assemble benchmark profiles for three or four primary competitors. Then compare their benchmark measures with your organization's to identify your relative strengths and weaknesses and to spot competitive opportunities or threats. Do your organization's perceived strengths seem consistent and intact after careful comparison with competitors? Under scrutiny, do your strengths remain the basis for a sound, implementable strategy? If not, then revise the strategic foundation on which the organization is building its performance plans and goals.

Consider the example of GTE Telephone Operations. Competitive benchmarks and best practices played a leading role in GTE's development of its five-year strategic plan in 1993. Benchmark analyses helped GTE first identify competitive cost gaps that were traced to broken, outdated, or inefficient processes. Best practice benchmarking then helped the company diagnose weaknesses in its core processes. In most cases, those process or operating system weaknesses were the root causes of uncompetitive costs. GTE senior management chose to reengineer all major processes that were deemed uncompetitive and which hindered the organization from achieving its strategic plans and goals. During this process, benchmarking was also the tool that helped reengineering teams identify the best practices that would drive successful redesign efforts in its processes and systems.

Performance benchmarks are also useful for creating competitive forecasts. The Z-chart is one of the benchmarker's favorite tools for evaluating operating performance. The Z-chart maps or projects the performance of two companies—the host organization and a benchmark or competitor organization—over a multiyear time horizon. The Z-chart's power lies in its graphic depiction of what rates of improvement will be required to match or surpass another benchmark organization. The Z-chart dramatizes the fact that organizations compete in full motion—not standing still. That's to say current competitor performance positions are sure to change because all worthy competitors are also engaged in constant improvement efforts. If your organization takes a year to match a competitor's current position, expect that the competitor will have advanced by the time your organization reaches the competitor's original position. Many companies have seen their strategic plans fracture because they judged a competitor's capabilities based only on current position, rather than by determining where the competition would move based on its current position and the velocity of historical improvement rate. In this respect, successful planning can be likened to shooting skeet: Managers must aim ahead of their targets, plotting a performance improvement trajectory that accelerates faster than their opponents' and aims into the future, if they are to best their competitors in the marketplace. *(Figure 8.3 demonstrates the creation of a Z-chart.)* The company that fails to embed this "moving target" logic into its planning will forever remain a laggard. Consider the case of Company A. In 1990, Company A conducted benchmark analysis and found its unit cost ($150) was 16.7 percent higher than Competitor B, whose unit cost was $125; this put Company A at a major competitive disadvantage. To eliminate this gap, Company A undertook a cost-reduction regimen that targeted roughly 6-percent annual cost improvements in order to equal Competitor B by 1993. However, Competitor B continued targeting 15-percent annual cost improvements and ended up achieving 11-percent improvements during the next three years. By 1993, Competitor B's unit costs were down to $83.75, while Company A attained its original goal of besting $125 per unit. However, the competitive gap had actually widened from 16.7 to 32.9 percent. Despite Company A's average improvements of 5.6 percent for each of the three years, Company A actually fell further behind. It failed to view its competitor as a moving target!

For this very reason, a growing number of companies, such as Johnson & Johnson, AT&T, GTE, Xerox, MONY, and Digital Equipment Corporation, have established benchmarking as a critical part of the strategic planning process. Indeed, skillful managers and strategic planners have long understood this "moving target" dynamic of successful competitive strategy, regardless of whether they employed formal planning tools such as a Z-chart. For instance, to promote strategic thinking and planning at General Electric, CEO Jack Welch requires operations executives to prepare a few simple slides describing the essence of their business situations. Benchmarking helps them address these questions about competitive dynamics:

The Z-Chart

Creating a Z-Chart to perform gap analysis

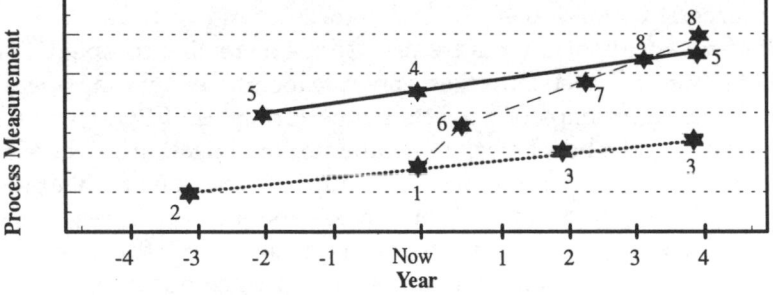

1. Our current "baseline" measure is here.
2. Review past performance to assess the trend line.
3. "Business as usual" improvement continues the same trend line.
4. Identify the current level of your benchmark or competitor.
5. Project competitor's improvement rate from past & present performance.
6. One time incremental change is tactical & may only partially close the gap.
7. Projected new "process" improvements represent strategic change.
8. Strategic changes enhance your improvement rate, enabling you to
    meet or surpass the benchmark's performance.

Figure 8.3. The Z-chart.

- What does your global competitive environment look like?
- In the last three years, what have your competitors done?
- In the same period, what have you done to them?
- How might they attack you in the future?
- What are your plans to leapfrog them?[4]

Benchmarking and Improvement Planning

Benchmarking can inspire incremental improvements and major magnitude leapfrog advances. Both applications empower ongoing improvement planning that girds every company's strategic planning process. The current Baldrige Criteria speak directly to this linkage among benchmarks, benchmarking, and continuous improvement plans and objectives. The current Baldrige Criteria observe the following about the dynamic role performance indicators and benchmarks play in driving both incremental and breakthrough improvements:

> Nonprescriptive, results-oriented Criteria and key measures and indicators focus on *what* needs to be improved. This approach helps to ensure that improvements throughout the organization contribute to the organization's overall objectives. In addition to fostering creativity in approach and organization, results-oriented Criteria and key measures and indicators encourage "breakthrough thinking"— openness to the possibility for major improvements as well as to incremental ones. However, if key measures and indicators are tied too directly to existing work methods, processes, and organizations, breakthrough changes may be discouraged. For this reason, analysis of operations, processes, and progress should focus on the selection of and the value of measures and indicators themselves. This will help to ensure that measure and indicator selection does not stifle creativity and prevent beneficial changes ("reengineering") in organization or work processes.
>
> Benchmarks may also serve a useful purpose in stimulating breakthrough thinking. Benchmarks offer the opportunity to achieve significant improvements based on adoption or adaptation of current best practice. In addition, benchmarks help encourage creativity through exposure to alternative approaches. Also, benchmarks represent a clear challenge to "beat the best," thus stimulating the search for major improvements rather than only incremental refinements of existing approaches. As with key measures and indicators, benchmark selection is critical, and benchmarks should be reviewed periodically for appropriateness.

Two examples illustrate the power of benchmarks and benchmarking to stimulate both incremental and breakthrough improvements. Consider first how WordPerfect, a leading developer of word processing software, has applied benchmarking to drive innovation and continuous incremental improvements in its customer-service support system, a core process area for WordPerfect. WordPerfect competes head-to-head with Microsoft, a much larger software company that also commands significant market share in the word-processing market. To compete against Microsoft, the colossus of the software industry, privately-owned WordPerfect has targeted customer-service support as a core competency area. Each year

WordPerfect spends upwards of $24 million on long-distance phone bills generated by customers calling into WordPerfect's help desk, which extends numerous toll-free 800 phone numbers to its customers. WordPerfect is so successful at managing its customer support functions that the Utah-based concern has emerged as a leading benchmark in call-center management and customer-help lines. Indeed, WordPerfect routinely handles 17,000 to 18,000 calls a day and the company's special phone switching systems allows it to handle up to 1300 calls at any moment. To effectively manage such large call volumes, WordPerfect has benchmarked the practice of radio disc jockeys communicating with car-bound listeners during high-volume drive-time commuter periods; WordPerfect creatively adapted the concept and employs "call jockeys" to play music, give WordPerfect user tips, and discuss forthcoming releases with customers who are waiting for a Help Desk associate to answer their call. The call jockey concept allows WordPerfect customers to avoid listening to recorded messages, which customer feedback revealed was especially irritating to callers. Such continuous improvements and innovative adaptations have firmly established WordPerfect at the head of its industry in customer-service support. In a *PC Magazine* 1992 survey, for instance, WordPerfect's word processing customer support was rated the industry's best, with an 8.5 customer score on a scale of 10. None of its rivals—not even giant Microsoft—scored higher than a 7.5.[5]

At Motorola, benchmarking has led to breakthrough improvements, as well as to incremental improvements. One such breakthrough was achieved by a Motorola team working to dramatically reduce the time it took for the operating units to close their books each month and prepare management reports. A corporate-wide stretch goal to reduce key cycle times by 50 percent spurred the team's improvement efforts. However, one operating obstacle that was rooted in the traditional system threatened to thwart the team's efforts: Many Motorola front-line employees recorded their work hours on traditional time-stamp clocks. The information on the time cards was then keyed into a computer—often with many small typographical errors—that later caused the entire process to slow down. The result: Sometimes Motorola took nearly three weeks to close their books and the improvement team was targeting less than one week. Team members concluded that the time-clock system would never be adequate for them to meet its ambitious cycle-time goals. Consequently, team members looked outside the company's traditional time-keeping systems to achieve a breakthrough. In supermarkets and other industries they observed the widespread and effective use of bar coding technology to quickly process similar types of information. By creatively adapting this existing technology to the time-keeping and payroll process, Motorola improved its overall financial reporting process, reduced system-clogging errors and met its ambitious cycle-time-reduction goals. Without the innovative adaptation of techniques applied in other industries, though, the Motorola team may never have achieved such dramatic performance improvements.

Benchmarking and the Strategic Use of Technology

Technology is the source of great excitement and frustration for many organizations. Blue-chip companies such as American Airlines, Federal Express, PepsiCo's Frito-

Lay, Wal-Mart, and USAA have used technology to help them achieve competitive advantages against competitors. Many other companies, though, have been discouraged by the disappointing investment and performance returns that major technological investments have produced. When well managed, technology can enhance value to customers, reduce cycle times, simplify complex processes, and improve costs and productivity—all to the strategic advantage of the corporation. When poorly managed, technology can annoy customers, hobble cycle times, lard the corporation with egregious costs, and damage productivity—all to the disadvantage of the organization. Consequently, benchmarking has emerged as a fundamental tool helping managers better evaluate the features, functionality, benefits, roles, and costs of technology. Various benchmarks are a familiar part of the information technology landscape. Moreover, increasing numbers of "I.T." professionals are evaluating the best practices of different systems and procedures in the world of computing, telecommunications, and other broad-based technologies. Telecommunications companies, for instance, have used benchmarks and benchmarking to help them better evaluate digital technology versus analog technology. The regional telephone companies use sophisticated switching systems to route calls to and from interexchange carriers, such as AT&T, MCI, and Sprint. In fact, switching equipment constitutes one of a phone companies' single largest capital investments. In the early 1980s, therefore, telephone companies gathered comprehensive benchmarks comparing both analog and digital technologies to help them evaluate under what circumstances each technology was most advantageous and which vendors were best at supplying each technology. Benchmarks included one-time costs, maintenance costs per line, minutes of downtime per line per month, and various performance measures for processing time and failures. Initially, the benchmark data substantiated the purchase of digital equipment only in central offices of 100,000 or more lines. Subsequently, though, software upgrades and new equipment models drove down digital technology costs and drove up digital performance; then, with few exceptions, digital equipment became the best technology in nearly all instances. The ongoing benchmark studies helped telecommunication planners know when they could optimally shift their strategic plans from relying on the one technology to the other. Again, the 1990s, benchmark studies are leading telephone companies to similar conclusions about the cost effectiveness of fiber optics over coaxial or copper cable.

Evaluating the best technologies is only one of benchmarking's many uses in the I.T. area. As a management tool, benchmarking provides valuable information concerning the most effective practices and procedures to utilize existing technologies in support of an organization's strategic plan. Consequently, best practice studies have mushroomed in technology-intensive areas that are strategic cornerstones for many organizations. Computer help desk management, customer call center management, computer capacity planning, computer preventive maintenance, computer training and education, technology-intensive logistics operations, technology research and development, network management, and the technology patent process are all sources of recent best practice studies among a broad cross-section of major American companies. Many organizational processes that support the effective use of technology are linchpins in effective strategy implementation.

Take, for example, a company like Toshiba, which competes broadly in many high-technology markets throughout the world. Toshiba is a best practice company for patent and intellectual property management. Toshiba's skills in this support process are especially important because the company excels at applied research and process improvements, rather than in the area of industry-shattering breakthroughs, new technology development, and pure research. Like other Japanese companies, Toshiba overcomes any potential weakness in developing brand new patentable technologies by securing myriads of patents that cover the very processes required to manufacture existing and future products. "Even if there is a pioneering patent, we can improve on the process so even the basic patent holder must use our process patent," observes Kensuke Norichika, general manager of Toshiba's intellectual property division.[6] So successful has Toshiba been in this strategy that Tokyo-based Toshiba obtained the highest number of U.S. patents in 1991. Toshiba directly supports its strategy by carefully monitoring and broadly benchmarking the patents and technologies of other companies. "The basic function of the intellectual property staff is to assist in the invention activities of the researchers," adds Norichika. In this way, Toshiba's intellectual property department and its highly-effective patent process support the company's overall business strategy. Benchmarking helps the company validate its core competencies and stay abreast of—or ahead of—the competition.

Benchmarking and Human Resource Management

Managing people is the "soft side" of most business operations. But what organization can achieve its long-term goals without integrating the people-side of operations into its strategic plans? Human resource development and management are so essential to successful strategy implementation that the Malcolm Baldrige National Quality Award assessment criteria dedicate an entire category to the people side of quality. Moreover, the Baldrige guidelines begin their exploration of a company's human resource system by examining "Human Resource Planning and Management." Under this item, Baldrige applicants describe how they integrate their human resource plans and practices with their overall strategic and operational performance goals and plans. To this end, applicants are challenged to describe how their organizations address the needs and development of the entire work force.

Consider the experience and observations of Winston Chen, chairman of 1991 Baldrige-winner Solectron, a manufacturer of computer circuit boards and subsystems. He describes how continuous renewal of Solectron's human assets is a cornerstone of this world-class manufacturer's business strategy:

The only way to keep up—and prepare for the future—is to invest in internal development. We must encourage employees to take ownership and share responsibility for the company's financial results. If you want to compete just on minimum costs, you can do this in a lot of places in the world. But in the United States, employees' intellectual capacity must play a bigger part in business.

It's important to tell employees what the future will look like. Training is the crucial cornerstone of any business philosophy. People are the biggest asset in a company. We have buildings and parts, but without employees, we have nothing.

Solectron employees spend 100 hours a year, or 5.5 percent of their working hours, training in new technology, leadership, motivation and other business topics. In high-tech industries, 20 percent of an employee's knowledge becomes obsolete each year. We have to continue training or our skills will be gone in five years.[7]

Benchmarking is a principal tool to assist managers in their efforts to integrate human resource plans and practices with the strategic business plan. For Solectron and other companies that view human resource management as a source of competitive advantage, placing the right people with the right skills in the right jobs and then continuously developing and retaining those people are strategic imperatives. Operating benchmarks help managers evaluate their unit's position and performance on critical work force factors, such as retention, training, learning, and productivity. Best practice benchmarking then illuminates the most effective practices and procedures for implementing successful human resource plans and actions.

Ritz-Carlton Hotels, a 1992 Baldrige Award winner, has integrated a unique system of benchmarks into the very fabric of its human resource recruitment and selection system. This luxury hotel chain invests significant time, energy, staff, and capital to nurture a corporate culture that is passionately devoted to customer service, and these human resource benchmarks enable managers to make better hiring decisions. Ritz-Carlton has performed comparative studies to identify and profile the personality traits of the best performing employees in every major job classification within its hotels. This world-class service company has then developed a 55-question preliminary interview process that is designed to profile job applicants along the same dimensions. A job candidate's responses to these benchmark questions provide Ritz-Carlton with a personality profile that can be compared with the profiles of Ritz-Carlton's highest performers in respective job functions. In this way, Ritz-Carlton can better evaluate if a job candidate's personality and skill profile are likely to make the candidate successful in the job for which he or she is applying. Used as a preliminary performance indicator, the Ritz-Carlton personality trait selection system has proven highly effective, enabling Ritz-Carlton hotels to hire better and retain employees longer than other competitors.

Human resource benchmarking is additionally compelling because many generic processes apply to human resource functions across many industries and organizations; generic processes, such as selection and recruiting, orientation, training, development, reward and recognition, and performance evaluations, apply to nearly all industries and organizations. This fact makes cross-industry comparisons and learning much easier and more powerful. Human resource benchmarking has been especially active in the following six operating areas:

1. *Recruitment and selection.* Benchmark data enables management to determine how well its selection process performs when compared with the selection systems of competitors and other companies operating in its recruitment markets. Key benchmarks such as candidate acceptance rates, employee turnover, and recruit-

ment cycle times quickly signal the effectiveness of an organization's selection process. In turn, best practice information can help management fine tune or restructure its selection process and strategy. Best practice benchmarking has led some companies to abandon traditional college recruitment campaigns in favor of lateral hiring from other organizations, and it has encouraged some organizations to pull back from help-wanted classified advertising in favor of seeking new hire nominations from current employees. Benchmarking can instruct management how to keep pace with industry trends and marketplace changes. Armed with this information, management can fine tune its short-term recruitment strategies and anticipate longer-term structural changes in work force supply, capabilities, and diversity.

2. *Employee education and development.* Solectron Chairman Winston Chen contends that in his industry 20 percent of an employee's knowledge becomes obsolete each year. Consequently, continuous training and employee development are the fuel to sustain ongoing competitiveness. Benchmarking is a primary instrument helping human resource managers to evaluate the adequacy, quality, and relative standing of employee development areas such as education, training, retraining, employee performance reviews, career path planning, and empowerment. Benchmark-generated baselines describe where the work force stands today and they serve as guideposts by which to navigate into the future.

3. *Employee relations.* A sage observer of organizations long ago observed that "the customer gets what management deserves." Accordingly, customer service is sometimes great and sometimes wretched. Either way, employee relations and morale are bellwethers of an organization's customer service quality. Companies that value their employees and effectively communicate the organizational values and vision tend to deliver better products and services to external customers. Benchmark analysis helps an organization evaluate critical internal performance indicators, such as employee satisfaction levels and trends, grievances, suggestion rates, employee involvement levels, turnover rates, and accident statistics. Best practice studies help the organization see how to fortify its areas of strength and vulnerability.

4. *Reward, recognition, benefits, and compensation systems.* The wave of restructurings, downsizing, and corporate makeovers in the 1990s seems to share at least one common denominator: Managers and employees are being asked to accomplish more with less. In today's "flatter," more customer-focused organizations, front-line employees take on greater authority, decision rights, job responsibility, and risks. Managers are expected to grow top-line sales performance with fewer resources in order to deliver better bottom-line profit performance. To accomplish such herculean feats, an organization's systems for motivating and maintaining employees must be suitable to the challenge. Consequently, organizational reward, recognition, benefits, and compensation systems have taken on greater importance than ever before. Benchmarking provides performance data and best practice information to help managers evaluate the adequacy of their organization's employee motivation systems and to help managers improve these motivation systems in order to support organizational strategy.

5. *High-performance work teams, self-directed work teams, employee involvement and empowerment.* The continuous quest for improving productivity has led many organizations to explore new structures, strategies, principles, and approaches to work-force management. The result is a revolution in the way people work together. Employees at all levels are taking on new roles and responsibilities, and they increasingly are working in teams that possess expanding management liberties, such as the right to hire, fire, reward, recognize, set budgets and operating targets, and share financially in the performance improvements they generate. Benchmarking is a principal tool helping organizations monitor and explore this bold new frontier of work structures, strategies, and collaborative formats.

6. *Work force composition, diversity, scheduling and well-being.* Structural changes in the workplace are mirrored by sweeping changes in work-force demographics and individual employee expectations. Organizations increasingly wrestle with ways to open themselves to greater employee diversity. They recognize that happy employees perform more effectively for the organization. Consequently, the successful organization of the 1990s is adapting to societal changes, such as growing numbers of women and minorities in the work force, and it is addressing employee issues, such as the need to secure day care, elder care, or flexible work hours. Benchmarking helps management stay abreast of the best work-force deployment strategies so the organization can operate most effectively with a diverse work force in diverse regional markets.

Benchmarking and Suppliers' Capabilities

"If pure water flows from the upper stream," observed Hiroto Kagami, senior manager of quality assurance for Canon, Inc., "there is no need to purify it farther downstream." This wisdom is the bedrock on which world-class organizations increasingly link their operating strategies to the direct capabilities of their suppliers. An organization cannot be world-class if the companies supplying its materials, parts, labor, or other components are third-rate. "If we are going to be a benchmark supplier of equipment, our suppliers need to be benchmark suppliers of parts," says Anthony Pollock, Xerox's manager of North American commodity operations. "It just does not work any other way."[8] Xerox and other leading companies have reconceived traditional customer-supplier relationships. Price negotiation is no longer the only or most important dimension of these relationships. Xerox, for example, regards suppliers as partners or joint venture associates in its business; consequently, Xerox has cut its supplier base from more than 5000 in the early 1980s to 420 in 1993. Xerox works closely with this smaller cadre of suppliers—awarding them multiyear contracts, sharing business plans and operating information, inviting them into the new product design and development process, providing them with clear quality specifications and with training so they can achieve these aggressive standards.

When viewed in such a light—as operating partners—suppliers assume much greater importance in the strategic planning process. Consequently, developing capable suppliers is of strategic importance for organizations such as Xerox and a growing number of other leading companies. With capable suppliers, an organization has a better chance of succeeding in its chosen markets. Benchmarking, in turn, helps the organization measure and assess its suppliers' relative capabilities in a fast-moving and competitive business environment. Such information is vital for working with suppliers to define, develop, and communicate quality standards and to set meaningful improvement plans.

Consider how the role of supplier management has evolved at Motorola. In 1985, Motorola benchmarked Xerox's supplier management practices. Motorola's former director of materials and purchasing described that meeting with Xerox as "A religious conversion."[9] Motorola learned quickly and increasingly placed great importance on its relations with its supplier. Motorola's goal was to produce near defect-free products; to achieve its goals, Motorola wanted its suppliers to commit equally to ambitious quality-improvement efforts. Consequently, in early 1989 Motorola notified all its suppliers by letter that it was requiring them to announce their intentions to apply for the Malcolm Baldrige National Quality Award within five years. This notification came within six months of Motorola's winning the Baldrige Award in the manufacturing category. Why did Motorola impose this requirement? Motorola recognized that its defect-free quality goals depended on its suppliers' ability to deliver defect-free parts, equipment, components, and raw materials. No matter how good Motorola's internal manufacturing processes, the company could not achieve its quality goals without supplies of comparable quality. Motorola executives also knew from experience just how much a company learned about the quality of its operations by performing a Baldrige total quality assessment. Motorola therefore used the Baldrige application process as a learning device for encouraging process improvement among all its suppliers. Some vendors complained that Motorola was wielding this learning instrument like a hickory stick. Motorola's management pointed out that suppliers did not *have* to go through the Baldrige process, but such vendors would likely lose Motorola's business.

As a result of Motorola's demand, its suppliers have shown steady improvement as measured by Motorola's vendor management system. Nearly 10 years after Motorola benchmarked Xerox's supplier management system, Motorola's supplier certification and management process is now regarded as a world-class system worthy of benchmarking. Both Xerox's and Motorola's systems ensure that suppliers are enablers, not disablers, of the companies' strategic plans.

Borrowing Strategies and Projecting Success

All successful companies are captives of their culture, prisoners of the traditional paradigms through which they have previously competed. A survey of winning organizations around the world quickly reveals there are many strategies on which they can successfully compete. Consequently, a growing number of organizations are studying the strategies of other successful enterprises and adapting the best aspects of battle-tested strategies. Consider the case of Bath Iron Works, the fourth-largest shipyard in the United States. In the 1980s, Bath Iron Works had allowed itself to become a one-customer company completely beholden to the U.S. Navy.

When the Cold War ended abruptly in 1991, the Navy no longer needed to build large combat ships. This about-face in customer needs threatened the very existence of Bath Iron Works. Bath CEO Duane D. "Buzz" Fitzgerald determined that his company had to devise a new strategic plan. Rather than start from scratch, though, Fitzgerald sent a four-person team of vice presidents to perform strategic benchmarking at ten shipyards in Holland, which had already undergone a defense conversion. Bath benchmarkers found Dutch shipbuilders applying their core competencies or capabilities to design and construct boilers, nuclear containment vessels, and bridges—in addition to shipbuilding. These strategic benchmarking explorations enabled Bath Iron Works to reconceptualize its essential strengths and its future strategic course. Bath CEO Fitzgerald eventually concluded: "Maybe our core competency is not shipbuilding but rather the design, manufacture, integration, and testing of complex structures. Certainly that would include ships, but it's much broader than just ships."[10]

Rather than develop a strategy from scratch, Bath Iron Works used benchmarking to examine the winning strategies and approaches of companies that were phasing themselves out of or diversifying within the combat shipbuilding business. Benchmarking encouraged Bath to abandon its narrowly conceived product orientation and become much more market oriented. Benchmarking helped Bath reconceive its traditional strengths and reinvent its future options. Other companies' successes, failures, approaches, and lessons learned—all gleaned through benchmarking—provide valuable evidence by which to formulate and predict the potential success of a strategic plan.

Summary

Benchmarking is a power tool of strategy. It fortifies the planning process and enables strategic thinking. Benchmarking applications are expanding rapidly in the area of strategy; an executive summary of the most important uses includes the following:

- The development of benchmarks or comparative performance indicators for important business factors such as market share, products, services, productivity, customer satisfaction, and costs are critical for planning purposes.
- Benchmarking helps an organization project its competitors' future performance; such competitive projections are instrumental when evaluating the adequacy of short-term and long-term goals or targets.
- Benchmarking helps an organization leverage internal and external resources for market leadership.
- Benchmarking stimulates long-term planning to ensure core business processes remain competitive.
- Benchmarking identifies best practices for employing technology to support long-term strategy.

- Benchmarking tests an organization's structure and operating systems to assess whether they are adequate to successfully execute strategy.

- Benchmarking establishes the best human resource practices to fully develop employees and realize work force potential in support of the organization's strategy.

- Benchmarking helps an organization evaluate its suppliers' capabilities and manage its supplier relations to support strategy.

- Benchmarking enables an organization to develop, refine, or improve its strategy by observing and adapting the competitive approaches of other successful enterprises.

Benchmarking is not an afterthought of organizations that are highly skilled at strategic planning. It is not a one-time event to fulfill a reporting requirement of the budget cycle. Quite the contrary, benchmarking is a hallmark of effective strategy development. It is an on-going enabler of strategic design, strategic planning, and strategic thinking.

References

1. Allen, Michael, "Developing New Line of Low-Priced PCs Shakes Up Compaq: The Chairman Went Behind Chief Executive's Back To Target Cheap Clones," *The Wall Street Journal*, Monday, June 15, 1992, pp. A1, A8.

2. Biesada, Alexandra, "Strategic Benchmarking," *Financial World*, September 29, 1992, pp. 30-36.

3. Robert, Michel, *Strategy Pure and Simple: How CEOs Outthink Their Competition*, McGraw-Hill, New York, N.Y., 1993.

4. Tichy, Noel M. and Sherman, Stratford, "Control Your Destiny or Someone Else Will," Doubleday, New York, N.Y., 1993, p. 26.

5. Dubashi, Jagannath. "Customer Service," *Financial World*, September 29, 1992, p. 58.

6. Meyer, Richard, "Patent Management," *Financial World*, September 29, 1992, p. 56.

7. Chen, Winston, "Lead By Example: Use High Personal Standards, Not Words, to Motivate Performance," from Tom Peters' *On Achieving Excellence* management newsletter, August 1993, Volume 8, Issue 8, pp. 5-6.

8. David, Gregory E., "Supplier Management—Xerox," *Financial World*, September 28, 1993, p. 2.

9. Ibid.

10. Biesada, Alexandra. "Strategic Benchmarking," *Financial World*, 29 September 1992, pp. 30-31.

9

Benchmarking and Business Process Reengineering

"We are going to break the cycle of mediocre performance. … How are we going to do it? We're going to do it by not accepting mediocre returns year after year. We're going to do it by reinventing the way we do business, by reengineering our processes and procedures for doing business, by revitalizing the people who conduct those businesses. And that's the first point I want to get across: these things are going to happen."

RON COMPTON, PRESIDENT
Aetna Life & Casualty

Four Performance-Improvement Approaches

Sort through the rhetoric of modern management practice, and four fundamental approaches exist for improving performance. These four cornerstones of improvement are continuous improvement, managed reform, organization restructuring, and process reengineering. Benchmarking is noteworthy in this context because it is the single component to all four approaches; it enables both incremental change and major magnitude breakthroughs.

Benchmarking turbocharges all forms of performance improvement by leveraging best practice experience inside and outside the organization; its power is greatest when integrated into an enterprise's core processes and infused into all the organization's performance-improvement efforts. By acting as a catalyst for learning, benchmarking accelerates the speed of improvement and frequently generates improvement approaches that might otherwise lie outside the thinking, experience, or grasp of internal improvement teams.

Continuous improvement can focus on any one or combination of the following performance dimensions:

1. Value—enhancing value to customers through the products and services received.

2. Defects—reducing errors, defects, and waste.

3. Fast response—quickening responsiveness and/or shortening cycle times.

4. Productivity—more productively and/or effectively utilizing resources.

Different companies tend to emphasize different dimensions in their improvement efforts. Nevertheless, the four fundamental approaches to continuous performance improvement can then be represented on a simple matrix or conceptual map. The axes of this map relate the *degree of change* (i.e., tactical or strategic) with the *pace and longevity of change* (i.e., immediate or extended). (Table 9.1 illustrates the four fundamental performance-improvement approaches.)

Tactical change usually represents small-scale improvements, such as a process enhancement that reduces defect rates or improves cycle time. Strategic change embraces enhancements, modifications, and improvements on a much larger scale, such as improving product quality by 10-fold or reducing new product development times by 50 percent. Change can occur at different rates or velocities. Immediate change occurs quickly, within a one- to three-month time frame. Extended

Table 9.1. Four Fundamental Performance-Improvement Approaches

Degree of change		Velocity of change	
		Immediate	Extended
	Tactical	Continuous improvement	Managed reform
	Strategic	Organization restructuring	Process reform

change unfolds during the course of many months and often plays out for several years. Reengineering projects, which are extended change events, may require two, three, or more years to completely deploy and embed within the fabric of a large organization. These two dimensions of change yield four possibilities for performance improvement:

- Immediate tactical change (i.e., continuous improvement)
- Immediate strategic change (i.e., organizational restructuring)
- Extended tactical change (i.e., managed reform)
- Extended strategic change (i.e., process reengineering)

Immediate Tactical Change (ITC)

ITC or *continuous improvement* involves daily, weekly, or monthly small-scale improvements. Managers use performance indicators and regular cycles of planning, implementation, and evaluation to drive change, but this type of change is evolutionary because it occurs in many small, incremental improvements, none of which may look or feel revolutionary. Over time these many small improvements powerfully advance the organization's performance systems. Benchmarks and best practice benchmarking are invaluable tools of continuous improvement. They point the way to improvement and frequently trigger operating insights that produce operating gains. A service company, for instance, was wrestling with how to improve its internal mail distribution system. Operating benchmarks revealed the system was slow and inefficient; large piles of unsorted mail signaled the first-in, first-out process was not highly efficient. Why? One reason was that small numbers of problem mail that were poorly labeled or addressed slowed down the entire system. An employee who drove to work on a toll-road proposed an elegant partial solution. The mailroom adapted the concept of an exact change lane, where cars move quickly through a processing system. Now, the mailroom separated nonproblem mail with clear, accurate address labels (akin to exact change vehicles) and problem mail that was unclearly labeled and would require research to deliver. This immediate tactical change, inspired by informal benchmarking, improved the speed and work flow of the entire system. Process-improvement teams still would have to wrestle with the problem of how to keep poorly addressed mail from entering the system, but they achieved important short-term improvements before solving the more difficult and complex problem.

Immediate Strategic Change (ISC)

ISC or *organizational restructuring* includes many different approaches to rearranging work within an organization: Functional transfers, structural reorganizations, downsizing, and realignments all represent common types of ISCs. These changes often target the creation of "synergies" where the new structure is intended to be more effective and less costly than the sum of its parts. For example, a Fortune 500 organization had a large headquarters and 10 operating companies, each of which provided various forms of management education. Obviously, the education and training in these 11 units were potentially redundant. The organization benchmarked the training practices of five other companies to learn how they managed executive education at half the relative costs. By shutting down six of the 11 training groups and transforming the other five (including headquarters) into specialized organization-wide education centers for financial, marketing, technology, engineering, and general management training, the organization eliminated the redundancy and maintained functional expertise. The organization accomplished the change within a few months.

Extended Tactical Change (ETC)

ETC or *managed reform* is a catch-all category embracing a wide array of both reactive and proactive management actions. Such actions are intended over time to remove errors, problems, faults or abuses, to reverse negative trends, or to improve capabilities that will ensure operations meet management objectives. These changes may require several years to roll out, refine, integrate, and complete. A cable television company, for example, found that its signal transmission quality was substandard in several of its service communities in Texas. Service performance indicators initially showed flat results; then subscriber complaints suddenly tripled and performance trends plunged. Management reacted by assembling an emergency team to evaluate the situation, perform root-cause analysis, and make immediate recommendations. The key action steps included: Stop using one supplier's coaxial cable which, during the last year, performed poorly and was frequently found to be defective; replace 15 sections of faulty cable going to 3000 customers and rebate those customers for their trouble; provide new test equipment to 30 technicians for troubleshooting problems at other locations. All told, the recommended steps took nine months (and $1 million) to complete—typical of the cycle time and cost of extended tactical change.

AT&T, Chevron, Caterpillar, Corning, Digital, DuPont, Florida Power & Light, GTE, Kodak, Phillips Petroleum, Texas Instruments, Xerox and many other organizations have successfully applied benchmarking to enable all four types of performance improvement. When it comes to process reengineering, though, benchmarking is an absolute necessity. Benchmarking is arguably the fastest, most resource effective, and most powerful approach to helping an organization reconceive and redesign the way its core processes operate. Benchmarking and process reengineering have therefore become the focus of tremendous interest and excitement throughout the private and public sector. They warrant the most detailed discussion of all.

Business Process Reengineering

Business process reengineering or *extended strategic change* (ESC) is a relatively recent and radical performance-improvement approach. Reengineering is *ex-*

tended because it can take years to fully implement, and it is *strategic* because it reinvents the way an organization conducts its business. Companies as varied as AT&T, Aetna, American Express, Ameritech, Chemical Bank, Deere & Company, Ford Motor Company, GTE, Texas Instruments, and Westinghouse are actively reengineering.

Reengineering starts with a clean slate and with the underlying assumption that there are *no sacred cows*. Reengineering strives to reinvent how work is done; consequently, it promises dramatic breakthroughs with spectacular performance improvements of 30 to 90 percent. Through process reengineering, for instance, IBM Credit reduced cycle time to a tenth of what it was for its loan processing while producing ten times the number of loan agreements. Thanks to reengineering, GTE Telephone Operations increased from 1 to 60 percent the number of telephone service problems resolved within 5 minutes when their customers call for repairs. Ford Motor Company adopted Mazda's invoice-less accounting system, and as a result reduced its head count for accounts payable by 75 percent from 500 to 125, while also reducing cycle time and the number of errors per transaction.

Best practice benchmarking, the "power tool" for leveraging other leading organizations' experience, learning, and innovation, is a principal instrument for achieving breakthroughs that produce such spectacular results. Benchmarking is like the fabled looking glass that took Alice into Wonderland; it provides the mechanism for improvement teams to access the larger world outside their organizations. Benchmarking springs open the door to the collective treasure of highly-effective practices, approaches, and systems perfected in other industries and operating cultures. These experiences and insights frequently provide the key or inspiration for reengineering teams to completely reinvent their organization's own core processes and systems. Process reengineering's goal is the development of a sustainable competitive advantage—achieved through substantially increased value to customers, improved economics through lower cost, and/or expanded output capabilities, and improved cycle times, quality, and productivity. The defining characteristics of process reengineering are summarized in Table 9.2 and the following discussion.

The following sections elaborate some of the more complex characteristics of process reengineering.

One-Time Change Frequency.   Process reengineering is undertaken as a one-time project that focuses on major magnitude performance improvement. Such large-scale change efforts are seldom accomplished quickly or easily. Moreover, they create initial discomfort for many people involved, who will be asked to learn to work in new ways. Consequently, constant reengineering is not an appropriate steady diet for most organizations. Yet, in a paradoxical sense, reengineering may be complete but it is never done. Reengineered processes must be fine tuned and steadily advanced; here reengineering eventually crosses the border back into the province of continuous improvement. Best practice benchmarking moves easily across these borders, serving equally as a tool of leapfrog improvements and incremental change.

Table 9.2. Characteristics of Business Process Reengineering

Time factors:	✓ One-time change frequency ✓ Long-term time requirements ✓ Change measures in year(s)
Risks:	✓ Moderate to high
People factors:	✓ Top-down participation ✓ Integration of high-level teams ✓ Top-down, bottom-up communication flow
Primary enablers:	✓ Senior management support ✓ Process owner concept ✓ Vision of future state ✓ "No sacred cows" ✓ Best practice identification
Range:	✓ Clean-slate starting point ✓ Broad, cross-functional scope ✓ Improvement change through radical breakthrough
Tools:	✓ Data collection techniques ✓ Process mapping techniques ✓ Teamwork ✓ Breakthrough thinking techniques ✓ Information technology ✓ Best practice benchmarking
Benchmarking:	✓ Assess reengineering need ✓ Stimulate breakthrough thinking ✓ Migrate best practices into organization ✓ Manage culture change ✓ Manage reengineered process refinement and continuous improvement

Moderate to High Risk. Reengineering *is* revolutionary and revolutions are risky by definition. They embrace change and change is uncomfortable for those who enjoy the status quo. Process reengineering is likely to embrace process redesign, job redefinition, new applications of information technology, employee skill enhancement, and revision of managerial systems such as selection, compensation, performance evaluation, and career development. Such radical changes often reshape an organization's very culture. When an enterprise radically reinvents the way it does business, the organization requires people to stretch as much as its systems. Fear, anxiety, and resistance are no uncommon early reactions. Kent Foster, President of GTE Telephone Operations, has likened process reengineering to remodeling your house over three years while you and your whole family continue to live in it. The discomfort is real, and at times inhabitants will question whether such massive renovations are attainable while everyone continues to dwell and work in the existing structure.

Indeed, not all who undertake reengineering efforts achieve their goals and visions. A 1991 CSC Index survey found that 25 percent of about 300 North American companies involved in reengineering reported that they were not meeting their goals. Other reengineering experts place the failure rate much higher, estimating 70 percent of organizations undertaking reengineering do not fully achieve their stated goals.[1] Nevertheless, the prospect of creating three-fold to ten-fold performance improvements in a core operating process remains so alluring to many organizations that they embrace reengineering despite its high risks. Benchmarking then proves an important risk management tool. By providing reengineering participants with a clear and vivid picture of the endpoint of their change efforts, the reengineering process becomes less fearful and more manageable.

A Clean-Slate Starting Point.   Unlike continuous improvement teams that build on a foundation already in place, reengineering teams deliberately start with a clean slate. They set out to reinvent the process; they do not intend to massage it toward incremental improvement. Bold breakthroughs often require the reengineering team to step outside customary work approaches and business models. Existing work systems are frequently organized to reflect specific functional strengths that reside in individual departments. Work, however, flows across departments; it is horizontally oriented. Yet strong functional organizations tend to be vertically oriented. They organize performance systems according to functional "silos" that force work into vertical structures. Most functional groups—Marketing, Sales, Engineering, Procurement, Manufacturing, Customer Service, Finance, and Human Resources—have created their own independent silo structures, as well as their own self-contained work processing paths. These functional silos act as corrals for sacred cows; that's to say they harbor suboptimal work systems that are self-perpetuating. The people who perform the work protect these systems from large-scale improvements because stretch improvements might diminish individual authority and importance. Ironically, then, the effective functional organization of a work system can cripple its optimization through cross-functional reorganization. To overcome the root cause of this endemic problem, reengineering teams begin work as if no system preexisted. They work from a clean slate.

Senior Management Support.   Process reengineering is a radical approach to performance improvement. Senior management's active support and involvement are therefore essential in overcoming the natural resistance to such broad-reaching change. Table 9.3 illustrates the role of leadership in lending authority to the four fundamental approaches to performance improvement. Senior leadership, usually the CEO or COO, play a leading role in process reengineering.

Broad, Cross-Functional Scope.   Work tends to flow horizontally across departments; consequently, functional boundaries blur when designing a system to perform work quickly, efficiently, and with customer requirements in mind. Organizations can be viewed as gigantic cross-functional processing systems: customer support producers and frontline processes transform materials, information, and

Table 9.3.

Performance improvement mode	Extent of resistance	Implementation authority	Authority to overcome resistance
Process reengineering	Extreme	COO or CEO	Absolute
Focused restructuring	Moderate	Vice president	Specific
Focused process improvement	Limited	Director	General
Continuous improvement	Somewhat	Process owner	Variable

other "inputs" into products or services that are delivered through frontline processes to customers in the market.

Figure 9.1 provides an example illustrating what a local telephone exchange carrier looks like when its operations are viewed as a gigantic processing system. The customer sits at the top of this model, elevated to the equivalent of the driver's seat. Frontline processes deliver products and services through distribution chan-

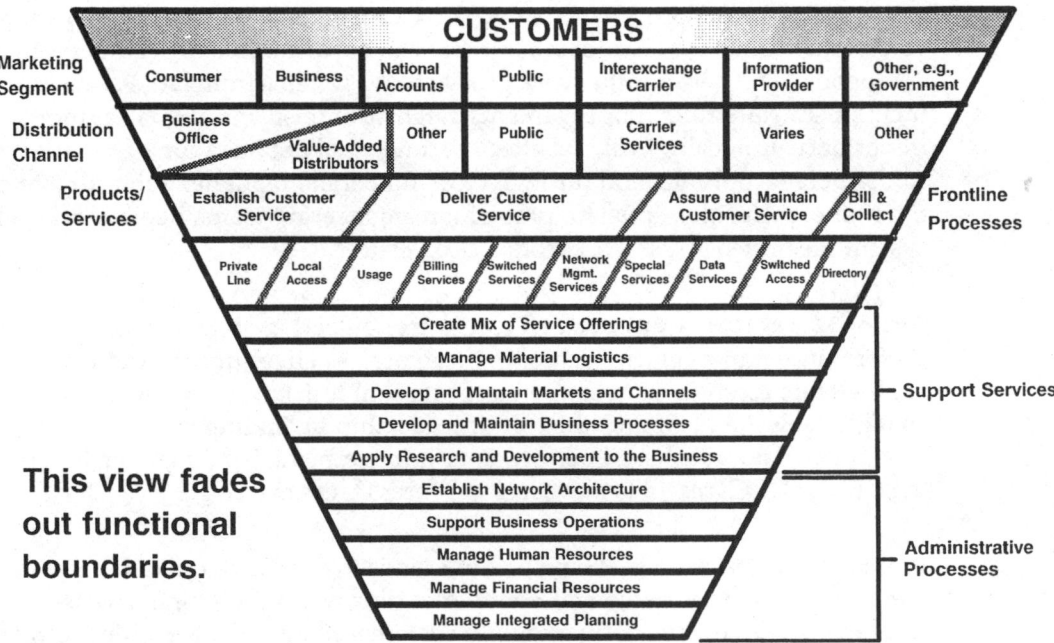

Figure 9.1. Viewing a company as a gigantic processing system.

nels into customer marketing segments. Frontline processes are sustained by support processes, which are, in turn, supported by administrative services. This model enables the reengineering team to focus on those primary processes which are most essential to the success of the organization. By applying process benchmarking, the reengineering team can objectively evaluate the adequacy of its current process system and then envision ways to reinvent the system.

Choose-Use-Pay

Within every organization, a few processes are primary. They are the ones that deliver the greatest value to the customer. During reengineering, organizations must therefore separate primary sequence processes from support service and administrative processes. Primary sequence processes are typically the time-based, order-to-delivery processes. Support services back up primary sequence processes and produce things needed by the primary sequence. In turn, administrative services plan, regulate, and integrate the primary sequence processes and the support service processes. To help reengineering team members identify high value-added, primary sequence processes, GTE reengineering teams developed the *choose-use-pay* process model. It is a simple but powerful concept that classifies all customer-driven, cross-functional services as being part of a three-phase customer sequence: The customer selects its service provider in phase 1; the customer uses the service in phase 2, and in phase 3 the customer pays.[2]

The litmus test regarding any subprocess' importance is to establish its role in this three-phase customer-focused sequence. (*Note:* The customer-oriented outside-in view for services is typically "choose-use-pay"; however, the customer view for products is often "choose-pay-use.") Reengineering strives to redesign primary sequence processes to create the greatest value for the customer and the organization. Support and administrative service processes may also be redesigned to provide greater value to the primary sequence. (*Figure 9.2* illustrates the Choose-

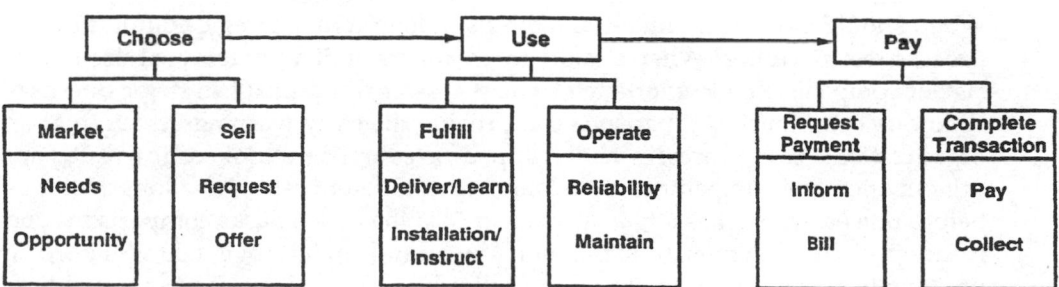

Figure 9.2. Outside-in and inside-out views to main sequence processes.

Use-Pay Model as applied by GTE. This information graphic also provides outside-in and inside-out views to the primary sequence processes. Virtually every process that appears on this process map is a candidate for best practice benchmarking during reengineering or afterwards in the continuous improvement stage.)

Though reengineering is still in its infancy as a structured performance improvement practice, core operating principles are rapidly evolving to describe some generic approaches to redesigning effective work systems. Table 9.4 reports seven "Principles of Business Process Reengineering" observed by Michael Hammer, one of reengineering's most elegant and passionate proponents.

The reengineering experiences of GTE and other organizations are quickly adding to this body of knowledge concerning what works and what doesn't. Table 9.5 presents two additional principles that have emerged from GTE's experiences with reengineering.

Information Technology

Information technology is often a master key allowing companies to open the door to radically new and improved work system designs. The advent of powerful computers and applications software enable companies to simplify once complex processes, to bring information previously available only to diverse specialists into the grasp of frontline employees with access to powerful databases, and to enable once linear processes to parallel process. The potential end results are improved economics for the business, improved cycle times for the process, and enhanced value for customers. Thomas H. Davenport, author of *Process Innovation: Reengineering Work through Information Technology* and a noted reengineering expert, identifies no less than nine different ways in which information technology enables innovation in work processes. All nine impact areas can be observed by benchmarking innovative applications of information technology to improve work processes in different industries such as insurance, telemarketing, paper manufacturing, express mail, automobile, snack foods, and computer manufacturing. (Table 9.6 describes I.T. impact on work systems.)

Deciding When to Reengineer

When should a company undertake the risky, long-term strategic commitment of process reengineering? When should an organization literally reinvent itself? Persistent competitive benchmarking can help answer this difficult strategic question. Rigorous benchmark comparisons can provide the early warning system that an organization's core processes have slipped or completely broken down. If your organization and competitors are advancing at different rates, how long will it be before one enterprise has a fatal advantage? Do the benchmarks comparisons and future projections demonstrate that only quantum-leap improvement will keep an organization from brutal quality or cost disadvantages? If the answer is yes, then process reengineering is probably the process to prescribe.

Table 9.4. Seven Principles of Business Process Reengineering.
*(Principles of Business Process Reengineering. Source: Michael Hammer,
"Reengineering Work: Don't Automate, Obliterate," Harvard Business
Review, July-August 1990: pp. 108-12.)*

Compress linear processes	1. Organize around people performing outcomes, not tasks. Have one person perform all the steps in a process and design the job around an objective or outcome instead of tasks. 2. Have those who use the output of the process perform the process. Use computer-based data and expertise to reengineer processes so that people who need the result of a process can do it themselves.
Move work	3. Add information-processing work into the real work that produces the information. Organizations that produce information now need to also process it.
Develop benefits of scale	4. Treat geographically dispersed resources as though they were centralized. Use databases, telecommunications networks, and standardized processing systems to get the benefits of scale and coordination while maintaining the benefits of flexibility.
Link parallel activities	5. Link parallel activities instead of integrating their results. Forge links between parallel functions and coordinate them while their activities are in progress rather than after they are completed. Information technology brings independent groups together going forward.
Empower and flatten the organization	6. Put the decision point where the work is performed, and build control into the process. The people who do the work should make the decisions, and the process itself can have built-in controls.
Eliminate redundancy	7. Capture information once at the source. Bar coding, relational databases, and electronic data interchange (EDI) make it easy to collect, store, and transmit information.

Reengineer core processes that, if competitively weak, are an Achilles heel to the organization's strategy, market position, and future success. Large differences in quality, productivity, and cost are telltale signs that the organization is vulnerable. After performing regular benchmark comparisons with the best world competitors

Table 9.5. Two Added Principles for Reengineering from GTE

Two-touch compression	8. As an adjunct to Principle #1, have no more than two people touch a main sequence process from order to delivery to the customer. Organizing around outcomes, use Principles #2, 3, and 6 to limit handoffs and empower two-employee teams to fulfill customer needs.
Combine specialized tasks into single job outcomes	9. Combine previously specialized tasks into new jobs designed around single-dispatch delivery to the customer. Redefine dispatch-controlled jobs to perform all the general steps involved with delivery. Use logical task combinations to prevent multiple dispatches of people assigned to achieve the same outcome.

in the telecommunications industry, GTE Telephone Operations believed it identified such an Achilles heel. Consequently, the organization set out to redesign eight strategic business processes. For GTE, these were core processes such as providing and maintaining network services and systems, billing and collections, and customer contact. In a massive reengineering project, GTE undertook to redesign these core processes before they became crippling weaknesses in the future. By proactively reengineering, GTE senior management hoped to ensure the organization's competitiveness throughout the 1990s. "Incremental improvement wasn't keeping pace with changes in our market. We needed quantum leaps," reported David Allen, Assistant Vice President for Process Reengineering at GTE Telephone Operations. "Because of the rapidly advancing competitive pressures, we not only accepted the challenge of reengineering, but decided to simultaneously reengineer every major process in the company."

Table 9.6. The Impact of Information Technology on Process Innovation. (*Source: Davenport, Thomas H.,* Process Innovations: Reengineering Work through Information Technology, *Boston: Harvard Business School Press, 1993, p. 51.*)

Technology impact	Explanation of impact on work system
Automational	Eliminates human labor from a process
Informational	Captures process information for understanding
Sequential	Changes process sequence, or enables parallelism
Tracking	Closely monitors process status and objects
Analytical	Improves analysis of information and decision making
Geographical	Coordinates processes across distances
Integrative	Coordinates between tasks and processes
Intellectual	Captures and distributes intellectual assets
Disintermediating	Eliminates intermediaries from a process

A sampling of business processes that are the grounds on which many industries compete include—but are not limited to:

- New product and service development.
- Production of products and services.
- Delivery of products and services.
- Billing, receipt, and processing of customer payments.
- Management of customer relationships.
- Marketing and sales.
- Development of innovation and technological capability.

When a strategically important process cannot produce the future performance levels that will be required to ensure marketplace health and success, then the risks of process reengineering seem well justified. The alternative—maintaining the status quo or implementing only continuous improvements—may prove fatal in fast moving markets. The mis-steps and downfalls of former market leaders such as Sears Roebuck, Wang, IBM, or General Motors provide grisly evidence of what can happen to even the greatest companies if they fall behind the pace of change.

The Seven-Step Reengineering Process

Process reengineering can be viewed as a simple seven-step improvement process. Like other improvement processes, the sequence is simple but methodical:

Step 1. Identify the value-added, strategic processes from a customer's perspective.

Step 2. Map and measure the existing process to develop improvement opportunities.

Step 3. Act on improvement opportunities that are easy to implement and are of immediate benefit. (These are the so-called "low hanging fruit.")

Step 4. Benchmark for best practices to develop solutions, new approaches, new process designs, and innovative alternatives to the existing system.

Step 5. Adapt breakthrough approaches to fit your organization, culture, and capabilities.

Step 6. Pilot and test the recommended process redesign.

Step 7. Implement the reengineered process(es) and continuously improve.

These steps occur sequentially, although they may at times overlap chronologically and parallel process. Benchmarking has potential applications in Step 1, Step 3, Step 4, and Step 7 of the reengineering process. At times the comparisons may be of simple performance benchmarks that are useful in establishing gaps and evaluating long-term capability; at other junctures, the benchmarking challenge is to determine the best business practices and the most innovative approaches that can be applied in a complex process or work system. Here the best practice focus may be on process design, information systems, technology strategy, organizational structure, employee skill levels, training, or other managerial systems. (*Figure 9.3* illustrates this seven-step reengineering process as a simple flowchart, and Table 9.7 describes the role of benchmarking at various steps in the reengineering process.)

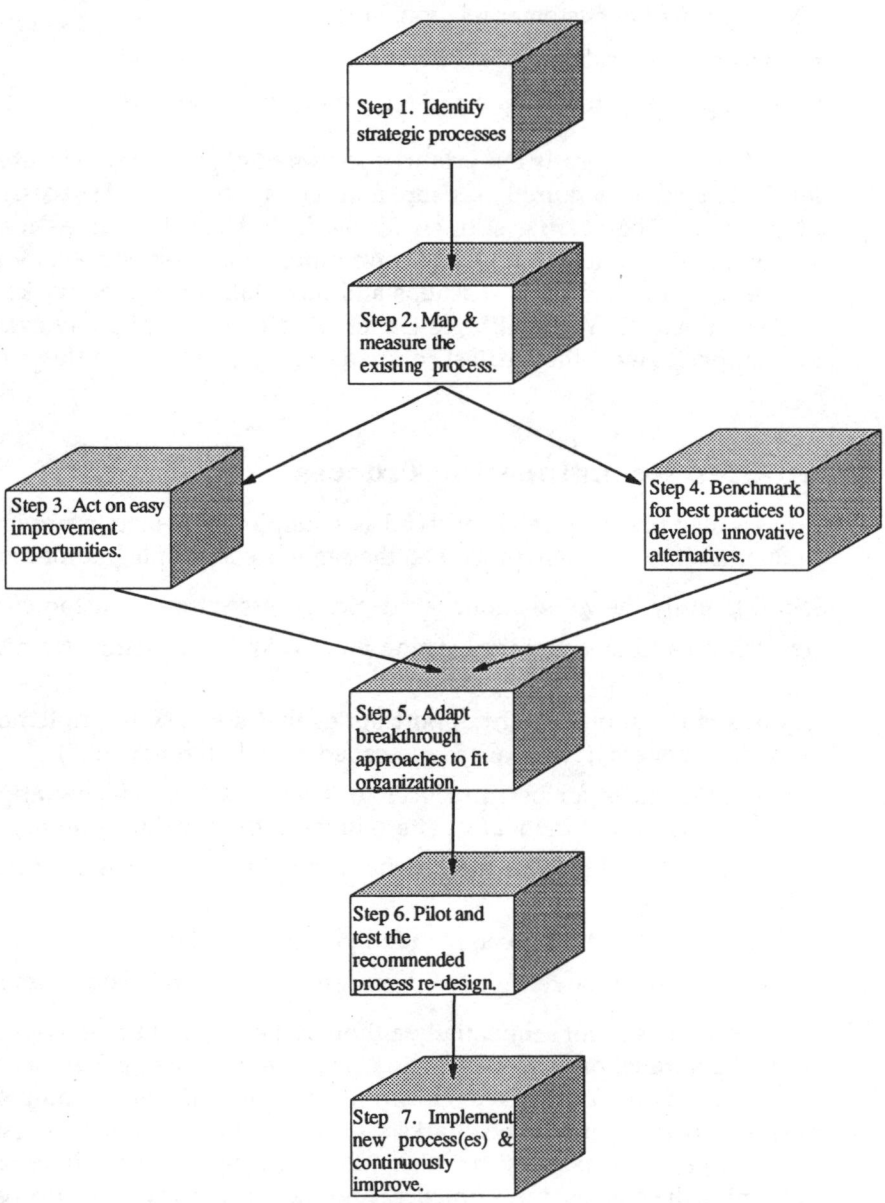

Figure 9.3. The seven-step reengineering process.

Table 9.7. Integrating Benchmarking and Reengineering

Seven-step reengineering process	Tools applied
Step 1. Identify the value-added, strategic processes from a customer's perspective.	Performance benchmark analysis (cost, quality, cycle time, etc.). Customer satisfaction benchmark analysis. Value analysis.
Step 2. Map and measure the existing process to develop improvement opportunities.	Flowcharting and process management tools. Performance measurement tools.
Step 3. Act on improvement opportunities that are easy to implement and are of immediate benefit.	Informal benchmarking for short-term solutions. Implementation planning tools.
Step 4. Benchmark for best practices to develop solutions, new approaches, new process designs and innovative alternatives to the existing system.	Best practice benchmarking among processes and performance systems.
Step 5. Adapt breakthrough approaches to fit your organization, culture, and capabilities.	Process redesign tools, implementation planning tools.
Step 6. Pilot and test the recommended process redesign.	Training, and pilot test techniques. Apply lessons learned from past successful pilots.
Step 7. Implement the reengineered process(es) and continuously improve.	Train employees. Implementation techniques. Use benchmarking to maintain continuous improvement process.

Lessons Learned

A growing body of hard-scrabble learning emerges from the successes and failures of recent reengineering efforts. Five lessons that are especially relevant to reengineering and benchmarking stand out:

- Study the best practices of organizations outside your industry. New approaches and innovative alternatives often lie outside one's own SIC code. Benchmarking partners from other fields frequently trigger important insights that inspire the redesign process.

- Study the most highly-effective work-force competencies underlying an organization's best operating practices. Recruitment and selection, education and training, employee skill sets, cultural elements, compensation, recognition, and performance evaluation systems are often critical success factors enabling the implementation of the most highly-effective operating practices.

- Coordinate interviews and site visits among all teams contacting external benchmarking partners. Such coordination is especially important when two or more reengineering teams identify the same best practice company to benchmark. Lack of coordination can damage an organization's credibility and discourage potential partners.

- Coordinate benchmarking contact with suppliers through the organization's vendor management group.

- Coordinate benchmarking contact with customers through sales account management staff, especially for national or major accounts.

Summary

Business process reengineering is one of the most radical and risky approaches to performance improvement. But the dramatic results delivered by successful reengineering—major magnitude performance gains on the scale of 30- to 90-percent improvements—justifies the risks for many organizations confronted by tumultuous change in their competitive markets.

Before launching into a major reengineering effort—and reengineering by definition entails major effort—managers should answer two pivotal questions: *Is the process strategically important? Does the process currently fail to deliver competitive performance?* If the answer to both these questions is "yes," then reengineering is likely to be the best performance approach to pursue. (Table 9.8 contrasts key dimensions of two primary performance improvement approaches: continuous improvement and process reengineering.)

No organization can live on a permanent diet composed only of reengineering. Managers should think of themselves overseeing an integrated performance improvement portfolio that includes all four improvement approaches—process reengineering, organization restructuring, managed reform, and continuous improvement. Best practice benchmarking is the managerial skill or tool that acts as a catalyst for all four performance-improvement approaches. Best practice benchmarking is a cornerstone concept of all performance improvement.

References

1. Stanton, Steven, Hammer, Michael, and Power, Bradford, "From Resistance to Results: Mastering the Organizational Issues of Reengineering, " *Insights Quarterly*, Fall 1992, p. 8.

2. Allen, David P. and Nafius, Robert, "Dreaming and Doing: Reengineering GTE Telephone Operations." *Planning Review*, March-April 1993, p. 31.

Table 9.8. Contrasting Continuous Improvement and Process Reengineering

Defining characteristics	Continuous improvement	Process reengineering
Starting point	Existing process	Clean slate
Change frequency	Continuous	One-time
Typical scope	Narrow, within process	Broad, cross-functional
Improvement change	Incremental with occasional breakthrough	Radical with targeted breakthrough
Time required	Short	Long
Measures of time	Months	Year(s)
Participation	Bottom-up	Top-down
Integration	Frontline employee involvement	High-level teams
Communications flow	Lateral	Top-down, bottom-up
Risks	Low to moderate	Moderate to high
Primary enablers	✓ Systematic process ✓ Process owner concept ✓ Anticipating future specs ✓ Continuous improvement of existing system	✓ Senior management support ✓ Process owner concept ✓ Vision of future state ✓ Clean slate, "no sacred cows"
Tools	▪ Problem-solving techniques ▪ Process management Techniques ▪ Employee involvement ▪ Suggestion mechanism ▪ Statistical process control ▪ Benchmarking ▪ Empowerment of employees ▪ Management of change	▪ Data collection techniques ▪ Process mapping techniques ▪ Teamwork ▪ Breakthrough thinking techniques ▪ Information technology ▪ Benchmarking ▪ Project management ▪ Management of culture

10

Benchmarking and Time-Based Competition

We have found that the concept of time compression is a powerful force, especially in traditional white-collar functions. Talk to an engineer or an accountant about productivity and they may consider this as a compromise to quality. Then ask them how much time it takes to do things and suddenly the lights go on. Focusing on reducing process time is an effective—and enjoyable—approach to improving quality and productivity.

JAMES H. KEYES
President and CEO, Johnson Controls[1]

Time-Based Competition

Some organizations call it *speed-to-market*, others speak about *fast cycle times* or *fast response*. In still other enterprises, *time-based management* or *time compression* are the operative terms. No matter what the phraseology, a growing number of organizations are focusing their performance-improvement efforts on the competitive dimension of time. To reduce the time it takes to perform operations, organizations are being challenged to simplify their processes, eliminate waste, remove redundancy, and drive down costs. Cycle time reduction or time-based competition is a powerful improvement focus because it correlates so directly with important performance indicators, such as cost, quality, market share, and customer satisfaction. Organizations such as Motorola and Texas Instruments have set aggressive corporate-wide goals to reduce cycle times by 50-percent increments. As they undertake these competitive strategies focused on developing ever-faster process speed, they rediscover the well-worn wisdom: "Time is money."

Time-based competition makes business process speed a primary focus of improvement and establishes cycle time as a key measure of performance. It can play a leading role in any of the four improvement approaches described in Chap. 9. Certainly, cycle time reduction is almost always emphasized in process reengineering and continuous improvement efforts in organizations where being first to market is important—and few industries exist any longer where being first to market is not important. At corporations such as Johnson Controls, General Electric, and Allied Signal, fast response or total operating speed has emerged as a core value of the organization. When Lawrence Bossidy engineered a dramatic turn-around of Allied-Signal in the early 1990s, he galvanized the corporation by drafting a statement of corporate vision, values, and goals that embraced speed.[2]

Bossidy became a devotee of speed as a powerful management concept at General Electric, where he worked shoulder-to-shoulder with CEO Jack Welch. Welch is another senior executive speed fanatic who often rhapsodizes about the importance of cycle-time management. One of GE's seven corporate values urges employees to "understand speed as a competitive advantage and see the total organizational benefits that can be derived from a focus on speed."

The Benefits of Speed

Time-based competition has emerged as a central concept articulated in the Malcolm Baldrige National Quality Award. In the Baldrige criteria, "fast response" is identified as a "core value" or foundation concept of high-performing, customer-focused organizations. "Success in competitive markets increasingly demands ever-shorter cycles for new or improved product and service introduction," note the Baldrige guidelines for quality excellence. "Also, faster and more flexible response to customers is now a more critical requirement. Major improvement in response time often requires simplification of work organizations and work processes. To accomplish such improvement, the time performance of work processes should be measured. There are other important benefits derived from this focus: Response time improvements often drive simultaneous improvements in organization, quality, and productivity. Hence it is beneficial to consider response time, quality, and productivity objectives together.[3]

The benefits of fast response capabilities are borne out in a growing body of business research. Especially noteworthy are recent findings from a McKinsey and Company study which revealed that a new product six months late to market loses 33 percent of its potential profit over its product life cycle. McKinsey's study is a penetrating wake-up call to service providers and product manufacturers alike: Reduce the time it takes to bring a new product or service to market and the organization significantly increases profits. (Table 10.1 summarizes the negative profit impact of being late to market.)

Table 10.1. The Cost of Being Late to Market. *(Source: McKinsey and Company.[4])*

If your new product is late to market by:	Your gross profit potential over the life of the product is reduced by:	A one-month improvement in time-to-market improves the product's lifetime profit by:
6 months	33%	11.9%
5 months	25%	9.3%
4 months	18%	7.3%
3 months	12%	5.7%
2 months	7%	4.3%
1 month	3%	3.1%

No one has spoken more eloquently about the benefits of process speed than General Electric CEO Welch. "There is something about speed that transcends its obvious business benefits of greater cash flow, greater profitability, higher share due to greater customer responsiveness and more capacity from cycle time reductions," opined Welch in GE's 1991 annual report. "Speed exhilarates and energizes. Whether it be fast cars, fast boats, downhill skiing, or a business process, speed injects fun and excitement into an otherwise routine activity. This is particularly true in business," says Welch, "where speed tends to propel ideas and drive processes right through functional barriers, sweeping bureaucrats and their impediments aside in the rush to get to the marketplace. Speed helps force a company 'outside of itself' and prevents the inward focus that institutions tend to develop as they get bigger."[5]

When companies adopt operating speed as a central focus for improvement and measure of performance, managers look anywhere and everywhere for ideas that can help them wring days, weeks, months—even years—out of complex business processes. No wonder Welch has leaned heavily on *speed* and *best practices* as corporate management themes to break down the bureaucratic tendencies that afflict nearly all large, successful organizations. Both concepts lead to expanded operating views that revile not-invented-here, provincial attitudes; moreover, process benchmarking is a primary enabler of speed, for best practices often lead to faster, simpler, less error-prone operations.

Measuring Speed Through Cycle Time

Cycle time is the primary measure of speed in business. It can be defined as the interval between the start and the finish of a process, which can be described in turn as the sequence of events and activities that occur in order to accomplish a piece of work. From a customer's vantage, total cycle time is the time that elapses between expressing a need and having that need satisfactorily fulfilled. During that period, many events and activities may occur.

Virtually any enterprise can measure and manage the cycle time of its most critical processes. In the postal service, for instance, cycle time might focus on the interval between the post office's receipt of mail and the delivery to the addressee;

in a hospital, it might describe the time from patient check-in until the patient returns home. In a mortgage bank, it might describe the time from customer application for a mortgage loan through the actual closing. Cycle time can measure the entire interval from creation through delivery of a product or service, or it can measure the duration of a subprocess, such as the time it takes to check into a hotel or to design a new product, which are components of larger experiences or processes.

In teaching internal improvement teams about cycle-time reduction techniques, Chevron Research and Technology Company represents work as a cycle; in turn, the primary work cycle is composed of subprocesses, each of which has its own cycle time contributing to the total elapsed cycle time. (*Figure 10.1* is the Chevron cycle-time model in which the circumference of the circle—the time from start to finish—represents the cycle time.) This concept—that every work process is comprised of many smaller processes or subprocess—is often referred to as *nesting*. It is important to recognize that within any core process nest many smaller sequences or subprocesses. Consequently, the first step in understanding a process is to identify its many subprocesses and to map them. A detailed process map also allows an improvement team to reach consensus on how it will measure cycle time. Frequently confusion arises when different constituencies measure cycle times in different ways, each including or excluding different subprocesses. After a team agrees on the full cycle to be measured, it can identify the theoretical optimal cycle time. For instance, the theoretical cycle time for underwriting a mortgage loan may be 20 minutes, the actual time it takes to conduct all activities in the underwriting process. However, long waits, bottlenecks, incomplete application forms and other operating inefficiencies may cause the elapsed underwriting time to reach weeks. With a detailed process map, an improvement team can systematically analyze the

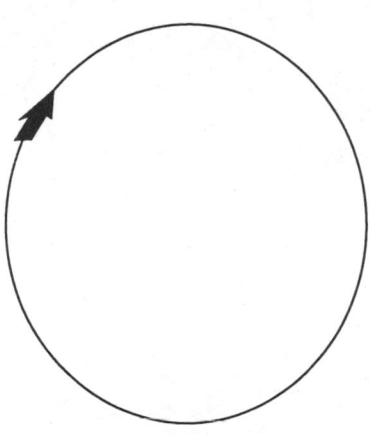

A work process is a cycle.
The circumference of the circle
-- the time from start to finish --
is the cycle time.

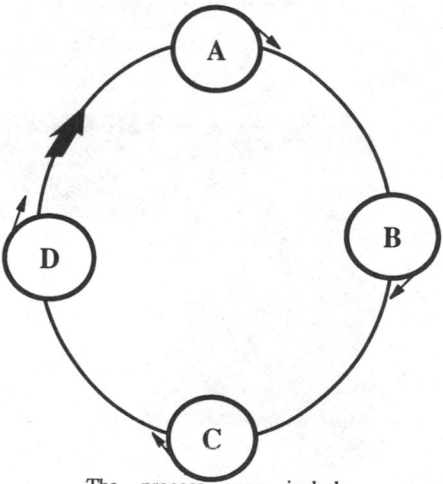

The process may include
sub-processes, each with its
own cycle time.

Figure 10.1. Work is a process or cycle measured in terms of cycle time. *(Source: Chevron Metrics Handbook.)*

obstacles that slow down the cycle; then they can develop solutions to improve processing speed. Frequently, teams reduce the overall cycle time by systematically identifying and implementing improvements in the subprocesses.

Here is where cycle-time analysis and benchmarking go hand in hand. Consider a case in point: A Chevron team set out to improve a complex oil analysis process. The team first examined process quality and found that its process quality was equal to that of competitors. However, when the team next studied cost and time factors, it was surprised to discover that certain competitors achieved superior performance. How? They spent less time in early planning phases, which had a disproportionately beneficial impact on total cost. When Chevron benchmarked how the competitors achieved this time advantage, they discovered the best practice companies approached planning very differently than did Chevron. Without the benefit of this benchmarking, which was spurred by the cycle-time analysis, the Chevron improvement team would never have discovered this new approach to planning. "This is an example where cycle time and benchmarking really create a synergistic effect. They play off each other," observes Chevron Senior Quality Consultant Derek Ransley. "Cycle-time analysis alone would not have provided the solution. Cycle-time analysis is assisted and made more powerful by benchmarking; cycle-time analysis instructs you where to benchmark."

In the 1990s, fast cycle time is essential to business success in many—if not most—industries and nations. The Japanese especially have chosen cycle-time reduction as a competitive strategy. In an international quality survey studying the practices of more than 400 companies worldwide, 84 percent of the responding Japanese businesses indicated that they usually or always employ cycle-time analysis as a primary tool to compress time in business processes. Nearly half as many businesses in the other countries surveyed reported that frequency of use.[6] (*Figures 10.2 and 10.3* summarize the cycle-time analysis findings of the International Quality Study[SM].)

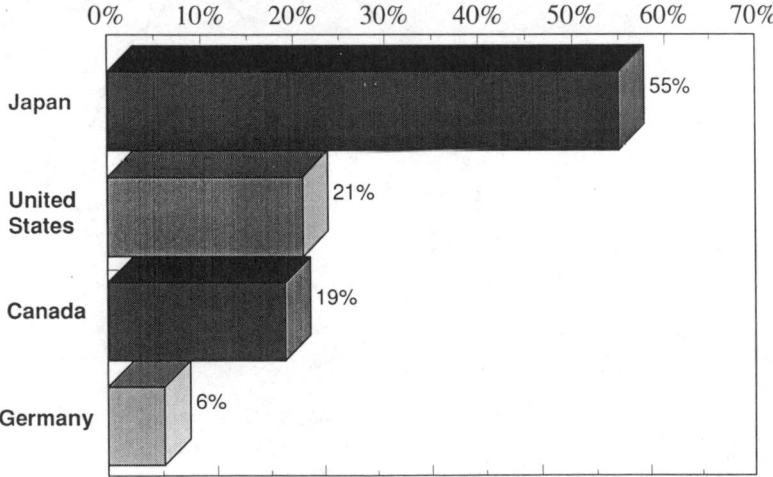

Figure 10.2. Percentage of businesses indicating they always or almost always use cycle time analysis. (*Source: International Quality Study, Ernst & Young and the American Quality Foundation.*)

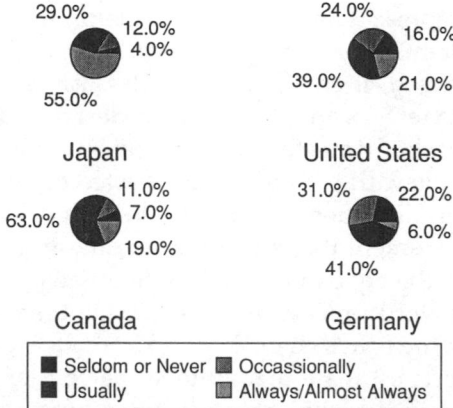

29.0% 12.0% 4.0% 55.0%

Japan

24.0% 16.0% 21.0% 39.0%

United States

11.0% 7.0% 19.0% 63.0%

Canada

31.0% 22.0% 6.0% 41.0%

Germany

■ Seldom or Never ■ Occassionally
■ Usually ■ Always/Almost Always

Figure 10.3. Percentage of businesses indicating the frequency with which they use cycle time analysis. *(Source: The International Quality Study, Ernst & Young and The American Quality Foundation.)*

Cycle-time reduction, therefore, is still fertile ground for improvement for many organizations. Those corporations that have undertaken broad-reaching time compression campaigns find that improvement gains of 50 to 80 percent are within grasp. How is it that such major magnitude improvements are possible? Experience shows that 80 to 90 percent of process cycle time is spent waiting. Many cycle-time reduction and reengineering efforts therefore devise ways to improve work processes by reducing or eliminating waiting. When cycle-time-reduction vigilantes apply process benchmarking here, they find the results can be especially effective. Waiting is symptomatic of processes that have begun to meander from their stated end. Waiting occurs when service or product creation proceeds more rapidly than the efforts of those managing or operating the process. To prevent the process from speeding out of control, operators therefore press on the brakes. In organizations, this braking effect takes the form of waiting for managerial decisions, supplies, repairs, information, and the like. If all these forms of waiting could be removed, the process would proceed from start to finish much more rapidly. Unfortunately, the process operators, who created the many different wait stations in the original process, don't always know how to eliminate these buffers that prevent the process from operating at faster speeds. Benchmarking other enterprises and industries that more effectively manage similar processes often provide the process operators with Copernican insights. These insights provide the impetus to reshape or reform the process in a way that eliminates waiting. Hospitals, medical centers, and health clinics, for instance, have long been stymied by how to manage patient admissions and releases more quickly and efficiently. Several have had great success in reducing or eliminating long admission waits by adapting the practices of world-class hotels, such as Marriott and Ritz-Carlton, that have perfected much faster check-in and check-out systems. Hospitals and luxury hotels may be a world apart in most people's minds; nevertheless they share many generic processes in common.

Rework and redundancy represent another major cycle-time reduction opportunity area in most work systems. For example, every time a letter is retyped, reprinted, reedited, or readdressed, it entails rework. Rework is often systemic; that's to say the process embeds rework and waste into its design. Consider the real case of one of the nation's largest law firms that operated an around-the-clock word processing department. The graveyard-shift manager insisted on different formatting standards for certain legal documents than were required by the day-shift managers. The result: Every document that crossed from one shift to another had to be reformatted according to the style mandated by the manager on duty. This rework caused frustration for word processors, longer cycle times for the firm's lawyers, and increased cost for the firm's clients.

Process management experts estimate that 25 to 40 percent of work effort is typically redo effort. Once again, benchmarking inspires process operators by providing them with vivid examples of highly-efficient work systems that have minimized or driven rework from the performance system. That's precisely how an improvement team at Firstar Bank of Madison, Wisconsin got the idea to wipe out rework in the organization's memo distribution process. The team placed a laser printer in its mail room. Now, when a memo or report goes out from a department, it prints out directly in the mail room, eliminating the need to prepare an internal mail envelope, an internal mail pickup, and a mail sort, just to transmit the memo or report to the mail room for distribution. Global commands on the bank's computer network now also automatically print out in the mail room the appropriate mail distribution list and the employees' mailbox numbers. This cuts time, cost, and redundant effort from the document distribution process and makes delivery much easier for the mail room. In the future, the organization may evolve to a full e-mail system, making hard copy memo and report distribution completely unnecessary. (Table 10.2 identifies some of the most egregious examples of waiting and rework.)

Other Cycle-Time Reduction Methods

Just like reengineering and structured benchmarking techniques, cycle-time analysis is rooted in process mapping and process management. The master keys to effective cycle-time analysis are a process map and time-based performance measures. Working from a detailed flowchart, cycle-time managers can systematically examine the work system with an eye to improving process speed. Some of the telltale signs of cycle-time improvement opportunities are frequent management review, decision loops, bottlenecks, missing steps, process rework, poorly sequenced steps, and steps that do not add value for the customer. Though many of these telltale signs would appear to be cost-based or quality considerations, they invariably affect cycle times. Chevron's Ransley aptly observes: "Cycle time is inseparably linked to cost and quality. All three must be considered at the same time."

From the work of Chevron, Johnson Controls and other companies that have successfully undertaken extensive cycle-time reduction initiatives, a growing body of wisdom has accrued. Table 10.3 summarizes some of this operating wisdom in a

Table 10.2. Where to Begin When Reducing Cycle Times

Where waiting occurs	Where rework occurs
Waiting for management decisions	Retype
Waiting for maintenance	Redesign
Waiting for supplies	Respecify
Waiting for equipment availability	Reorder
Waiting for information	Rework
Waiting for signatures	Repurchase
Waiting for instructions	Reread
Waiting for materials	Redocument
Waiting for next shift	Restate
Waiting for personnel	Rebuild
Waiting for parts	Retool
Waiting for installation	Reprogram
Waiting for priority	Retest

cycle-time reduction checklist that also shows how and where benchmarking empowers the cycle-time analysis and improvement process.

Cycle-Time Reduction Success Stories

Not long ago, cycle-time reduction was a concept found only in the lexicon of manufacturing engineers. Increasingly, it is a rallying cry for CEOs, CFOs, and executives from service and manufacturing industries. The hall of fame of cycle-time success stories is rapidly expanding. Consider a few of the following entries:

- IBM Rochester, a 1990 Malcolm Baldrige Award winner in the manufacturing category, includes cycle-time reduction as one of its six critical success factors for quality improvement. Since 1984, IBM Rochester has reduced product development time for new mid-range computers by more than half, while they have cut their manufacturing cycle time by 60 percent.

- The Cadillac Motor Company, a 1990 Baldrige winner, has reduced cycle time in die-changing from 12 hours to 3.5 minutes through a team of plant workers and common sense—all without any capital investment! The team color-coded the pieces of the die for easy transfer; previously color-coding technical experts were required. They also laid some no-slip floor covering, which prevented the equipment from sliding as the die was being changed.

Table 10.3. Checklist for Reducing Cycle Times. *(Sources: Stan Schulz of Johnson Controls and Derek Ransley of Chevron Research and Technology Company.)*

Opportunity area	Cycle-time analysis questions	Benchmarking opportunities
Process flow	✓ Are bottlenecks slowing processing speed? ✓ Are there redundancies in the current process? ✓ Can some sequential activities parallel process? ✓ What steps can be removed that do not contribute to the end result or do not create value for the customer? ✓ Can over-complicated steps be simplified to increase speed? ✓ Can alternative technology reduce the number of process steps? ✓ Can individual steps or subprocesses be redesigned to improve speed?	Benchmark similar processes of excellent companies to identify simpler, faster, less redundant, more error-free practices.
Resource scheduling	✓ Is information available in advance so that resources (people, material, data) can be available at the time of need?	Benchmark resource planning, scheduling, delivery, and management practices for similar processes.
Decision making/ communications	✓ Are too many approvals required? ✓ Are lines of communication long? ✓ What cross-functional issues may interfere with implementing process improvements? ✓ Do poor lines of communication and coordination create problems and delays?	Benchmark decision making, communication, and empowerment approaches for similar processes.
Capacity/ flexibility	✓ Are different skills or equipment required to accelerate the process? ✓ Are critical resources well utilized and available on short notice? ✓ Do high levels of specialization cause frequent handoffs?	Benchmark cross-training, pay-for knowledge, planning, scheduling, and high-performance teams to seek best practice approaches in these areas.
Organization	✓ Does organizational structure cause frequent task handoffs that slow the process? ✓ Is the layout of the work environment (factory, office, etc.) causing delays and suboptimal processing times? ✓ Does the organization promote teaming within and across functional boundaries? ✓ Is poor job design hobbling the process?	Benchmark plant/office layouts, organizational and team structures, and job designs.

Table 10.3. Checklist for Reducing Cycle Times. *(Sources: Stan Schulz of Johnson Controls and Derek Ransley of Chevron Research and Technology Company.)* (Continued)

Opportunity area	Cycle-time analysis questions	Benchmarking opportunities
Consistency	✓ Are appropriate process controls in place so that outputs are properly controlled, thereby avoiding reprocessing and other unnecessary delays?	Benchmark process control approaches in similar processes.
Supply chain management	✓ Are materials from suppliers causing unnecessary rework or delays? ✓ Does the lack of standardization among suppliers cause delays and rework? ✓ Can suppliers and customers be better tuned into the process to avoid duplication of effort and redundant inventories?	Benchmark various aspects of supplier management, such as communication, standardization, scheduling, training, and so on.
Product and service design	✓ Are there unnecessary design variations among products and services that complicate manufacturing, logistics, and customer service?	Benchmark new product development processes, design-for-manufacturability systems, and so on.

- Focusing on time-compression as a key corporate-wide goal, 1992 Baldrige-winner Texas Instruments Defense Systems & Electronics Group reduced production cycle time by 21 percent in 1992 with a 56-percent reduction in stock-to-production time.

- Using process management and time-compression techniques, 1988 Baldrige-winner Motorola has reduced its corporate auditing process over a three-year period from an average of seven weeks to five days from the time a field auditor first pens a report to the time the final version is delivered to management.

- Globe Metallurgical, also a 1988-Baldrige winner, replies to customer inquiries, no matter where in the world, within 24 hours. Response time used to range from several days to over a week.

- Xerox, a 1989 Baldrige winner, cut the new-product design-to-market cycle by 50 percent through its quality practices.

- Westinghouse Nuclear Fuels, another 1988 Baldrige winner, reduced manufacturing cycle time by 40 percent through its quality efforts.

The pantheon of fast companies is not occupied only by Baldrige winners, although time-compression and fast response are clearly key excellence indicators exhibited by virtually all winners of the national quality award. Johnson Controls, a Milwaukee-based conglomerate, provides a sterling example of how time-compression can provide a central focus for an entire corporation's improvement efforts. This global market leader in control systems, automotive seating systems, facility services, batteries, and plastic packaging, catalogued the following time-compression achievements in a single year:

- *Inventory management and delivery.* The Colmbes, France Customer Service department in the Facility Services and Controls unit redesigned its inventory system to make sure critical items are available the day they're ordered. Previously, standard delivery was six weeks. The one-day delivery program covers 80 percent of customers' orders. And the unit hasn't had to increase warehouse inventory to do it.

- *Taking cycle time out of billing.* During a year, Johnson Controls performs more than a thousand tasks at Bangor Naval Submarine Base in Silverdale, Washington; all need to be billed separately—from fixing a faucet to operating a crane to engineering work. By revising billing practices, the Johnson Controls Facilities Services Bangor team cut the time it takes to bill for a completed service in half, substantially reducing unbilled receivables. The new system also helps the Navy track its costs on a much more timely basis.

- *Recruitment and selection.* Another Facilities Services team streamlined the recruiting and hiring process so it can fill positions faster. From the requisition form which helps to better define the skills needed for the job, through the recruiting, interviewing, and hiring stages, the team improved the process. Positions are now filled in an average of 50 days versus 74 days.

- *New product development.* New electronic controls are getting to market faster, thanks to improved coordination between design and manufacturing teams. While engineers are still designing a product, manufacturing begins to prepare for production. By involving manufacturing earlier in the process, development time has been cut more than 40 percent on new products, such as the metasys temperature sensor.

- *Developing new pattern designs faster.* With new computer software, Automotive Division teams are developing patterns for seat covers faster. Even before a frame and foam model is made, teams can evaluate the seat cover design and fabric and accurately estimate costs. Then, when the foam model is ready, only minor adjustments are needed. Pattern development time has been cut by 30 percent.

- *Completing facility moves more quickly.* When a Johnson Control customer moved its auto production facility, Johnson Controls followed the customer. In a span of less than four days, Johnson Controls moved a seat assembly line from Orangeville, Canada to Taylor, Michigan, installed seven trailers full of equipment, including a 25-foot-long oven, and resumed production. Normally, a move could take four weeks.

- *Quick change artists.* Changing production between 16- and 20-oz. soft drink bottles used to mean shutting down a blow-molding machine for eight to 12 hours twice a week. The multiple molds had to be removed so that spacers could be added or taken out. A new spacer system developed by the Plastics division maintenance team at a Columbia, South Carolina, plant has cut the change-over time to only four hours. That means at least 12 more hours each week that the machines can be working, and the team has reduced wear caused by taking the machines apart.

- *Accelerating new battery development.* The time it takes to develop a new battery has been dramatically shortened in the lab. Computer simulations replicate battery operating conditions, and experiments are designed so that maximum information can be gained from the fewest experiments. In one week, for example, Johnson Controls Battery division tested three key performance factors using 13 designs. To actually design, build, and test 13 batteries used to take up to a year.

- *Manufacturing cycle time speeding product flow.* A 40-percent reduction in the battery manufacturing cycle—from receipt of raw material to delivery to a warehouse—was achieved over a two-year period. The improvement is a direct result of a new production planning system, just-in-time material deliveries, and more reliable manufacturing processes. The bottom-line impact: substantial reductions in product costs and inventory investment while maintaining nearly 100-percent on-time delivery performance.[7]

Benchmarking and Cycle-Time Reduction

"Faster, cheaper, better." This is the holy trinity of customers in the 1990s. It is fitting that *faster* is the first concept in this trinity. Time is money; faster drives cheaper. Simplified processes, which are by definition faster, inevitably deliver lower cost. In most circumstances, faster is also better. Processes that compress time by simplifying sequences, by purging rework and redundancy, by error-proofing tasks, they also produce higher-quality results than their slower predecessor processes. "Clearly, quality, cycle time, and cost are inextricably linked," Motorola's Mort E. Topfer, a Senior Vice President and General Manager for Semiconductor Products Sector, has noted. "As fewer defects are produced, less time is devoted to rework, so both cycle time and manufacturing costs shrink."

Quality has sometimes been described as the "race without a finish line." However, there is a clearly marked finish line from the customer's vantage. The company that can routinely deliver products and services to market faster than its competitors wins. "The bottom line is that when customers decide they want to buy, the first supplier who can fill that need with a product or service will flourish," observes Christopher Meyers, an authority on time-based competition.[8] Customers increasingly demand high quality, a good price, and timeliness all at once. Consequently, the goal is quick response, and the only variants are "fast" and "faster."

As a managerial tool that helps accelerate organizational learning, benchmarking acts as a primary catalyst in the cycle-time reduction process. Once cycle-time measurements are put in place and processes are mapped, benchmarks may be developed and improvement opportunities identified. Successful time-compression efforts, therefore, are neither mysterious nor complex. Cycle-time reduction is achieved through the systematic analysis of complex processes with a clear focus on improving process speed. Those practices that actually compress time and improve speed in one company are almost always eminent targets for innovative adaptation in another organization.

Cycle-Time Solutions

The study of fast companies turns up various approaches that systematically compress speed. In slower organizations, where cycle time has never been systematically analyzed and reduced, these best practices have often been neglected. Consider a few examples of generic practices that tend to reduce cycle time:

- *Establish dynamic information flows.* The instantaneous exchange of critical information is often achieved through Electronic Data Interchange (EDI) and other computer-based communications systems.

- *Reduce bottlenecks.* The *express lane* concept is frequently used to reduce process bottlenecks. Any time a first-in, first-out system is applied at high-volume processing points, easy-to-process items slow down as they wait in queue behind more difficult-to-process items. Directing easy-to-process items through an express lane keeps traffic moving, just like in the supermarket.

- *Reduce the number of inspection sites.* When organizations learn to improve process designs so that errors are prevented, repeated inspections become time-wasting and redundant activities.

- *Reduce decision and approval points.* Every time a process requires management approval or requires a supervisory decision, the process slows and waiting frequently occurs. These nonvalue-added waiting stations can be removed by better training employees, by empowering line workers with decision rights and by designing processes that don't require regular approval and decision points.

- *Simplify the decision-making process.*

- *Standardize information, reporting forms and procedures to reduce variation.* Processes without standards tend to run slowly and out of control. A customer order center, for instance, that receives orders in 100 or 1000 different formats runs more slowly than a center with a single standardized order format. Consequently, a first-line of standardization often involves the creation of well-designed forms that facilitate several process steps and establish information standards. A well-designed form or checklist can be a simple but powerful approach to process automation. Many cycle-time reduction efforts therefore improve the design of forms and formats.

- *Avoid frequent process handoffs.* The use of technology and powerful databases can often help simplify processes by enabling one person to accomplish what previously different departments or specialists individually handled.

- *Improve plant or office layouts that separate work groups or disrupt processes.*

- *Group lots or work with similar requirements.*

- *Organize or "pre-kit" tools, paperwork, gages, and the like, to facilitate rapid workflow.*

- *Consolidate purchases.* Frequent material and supply purchases by work groups, departments, and operating teams take time and resources. Consolidating purchases can achieve cost and time savings.

- *Eliminate unnecessary variation in design.* Frequent changeovers in products or services generate additional work and process steps, all of which slow cycle time.

- *Automate repetitive handling operations.* Repetitive tasks in both service and manufacturing industries can be automated. Automation may range from the use of robots in an automobile plant to the creation of form files and macro computer commands in a law firm. By reducing repetitive handling tasks, process speed increases.

These time-compression tactics may seem fundamental but they work. Why? Because they address the basics of cycle-time reduction: process simplification, error-proofing, rework elimination, waiting reduction, bottleneck removal, and redundancy elimination.

As time-based management has become an integral part of many organizations' competitive strategies, cycle time has emerged as a primary benchmark or operating statistic studied by companies. Cycle-time gaps between competitive organizations can be viewed as a red flag that one organization operates more slowly and probably at higher cost. Savvy process managers carefully scrutinize their organization's cycle-time gaps, recognizing that each gap signals an important process-improvement opportunity. By studying how faster companies achieve their superior cycle times, process managers can adapt those approaches and drive similar improvements in their own organizations.

Benchmarking therefore acts as a time-compression catalyst. It supercharges the cycle-time reduction process in at least three ways. First, cycle-time benchmarks or comparative performance measures provide an early warning system that alerts an organization if its processes are running slowly or suboptimally relative to other competitors and process leaders. Second, benchmarking then helps identify where the lard has become marbled into an organization's processes by providing examples of faster, leaner, and more effective operating systems. Lastly, benchmarking helps generate time-compression solutions by identifying the best practices and approaches of faster companies.

In the 1990s, speed is a distinguishing characteristic of market leaders. Consequently, cycle-time reduction is a primary goal of organizations that wish to remain competitive. Benchmarking, in turn, is a primary means to achieve that end of fast response.

References

1. Keyes, James, 1992 Johnson Controls Annual Report, letter to Shareholders, pp. 2-3.
2. Stewart, Thomas A., "Allied-Signal's Turnaround Blitz," *Fortune*, November 30, 1992, p. 73.

3. The 1994 Malcolm Baldrige National Quality Award Criteria, United States Department of Commerce, Gaithersburg, Maryland, p. 3.

4. Vesey, Joseph T., "The New Competitors: They Think in Terms of 'Speed-to-Market'," Academy of Management Executive, Volume 5, No. 2, 1991, p. 25.

5. Welch, John F., Jr., Letter to Share Owners, General Electric Annual Report, 1991, pp. 4-5.

6. International Quality Study[SM]—The Definitive Study of the Best International Quality Management Practices, a joint Project of: Ernst & Young and the American Quality Foundation, 1991, pp. 28-29.

7. Johnson Controls, 1992 Annual Report, pp. 6-19. (All examples of Johnson Controls cycle-time reductions are reported in the 1992 annual report to shareholders.)

8. Meyers, Christopher, "Six Steps to Becoming a Fast Cycle Time Competitor," a paper published by the Strategic Alignment Group, 1990.

11

Benchmarking and Change Management

It must be considered that there is nothing more difficult to carry out, nor more doubtful of success, nor more dangerous to handle, than to initiate a new order of things. For the reformer has enemies in all those who profit by the old order, and only lukewarm defenders in those who would profit by the new order ... This arises partly from the incredulity of mankind who do not truly believe in anything new until they have an actual experience of it.
MACHIAVELLI, Fifteenth Century

Although sometimes difficult to accept, change is inevitable. Organizations do not have the luxury of choosing whether they are for or against change. Change is the most certain of operating realities. "The question," contends Barry Stein, an authority on organizational change, "is whether you want to deal with change systematically to arrive at some place good or whether you want to take what comes willy-nilly."[1] If organizations decide to approach the management of change in a systematic way, benchmarking is arguably the single most powerful tool within their grasp.

As an instrument for managing change in a positive fashion, benchmarking:

- Creates motivation for change.
- Provides a vision of what an organization will look like after the change.

- Instructs employees what to change.*

- Provides data, evidence, and success stories for inspiring change.†

- Raises awareness of competitor position and headway that stimulates innovative change.‡

- Reduces the cycle time required to achieve change.**

- Identifies best practices for how to manage change.

- Creates a baseline or yardstick by which to evaluate the impact of earlier changes.§

Motivating Change Through Benchmarking

"There are two kinds of leaders—those who anticipate change, and those who wait for a crisis and fail," says Dan Ciampa, President of Rath & Strong, consultants specializing in organizational change.[2] Benchmarking is a favored tool of this first sort of leader. By applying benchmarking as an on-going business practice, they create an early warning system alerting the organization when change is necessary.

Benchmarking is perhaps one of the best sentinels management can install along the watchtowers of the organization. It provides employees and managers the tool, the rationale and the process to embrace change as constant and inevitable. Today's world is larger than any one person's or company's views of it. In other words, no one organization can ever hope to corner the market on effective operating practices and good ideas. To be a leader in today's marketplace, one must by necessity look outward—as well as inward, for constant improvement.

Simply put, the on-going study of best practices can help an organization avoid being ambushed by unexpected change. By systematically studying others and comparing one's own operations and performance with the best and most effective practices of highly-innovative and successful companies, an organization can evaluate when change is a necessity for market leadership—or for survival. The on-going search for best practices quickly draws you outside the confines of your own culture. Best practices benchmarking is therefore a pragmatic approach to managing change. Benchmarking provides the navigational system that helps employees chart a managed course toward proven and effective operating practices and strategies. Marlow Industries, a small-business winner of the 1991 Baldrige Award, for instance, has publicized the fact that after benchmarking the supplier management area, the company recognized the need for improvement and borrowed its supplier certification system from Rockwell International.

One of the most difficult tasks of management is to convince the employees of successful organizations that change may be necessary. Craig Weatherup, president of a highly profitable Pepsi-Cola operating division, told the following "burning

*See Chaps. 4, 5, 9, and 10.
† See Chap. 6.
‡ See Chap. 8.
**See Chap. 10.
§ See Chaps. 3 and 4.

platform" story to dramatize the need for change within his already successful organization: "It seems that a few years ago a North Sea oil rig caught fire. One worker, trained not to jump from the 150-foot high rig into the icy sea but to wait for help no matter how bad things got, leaped anyway. He survived. Asked afterward why he stepped off the edge, the worker said he looked behind him and saw an approaching wall of fire and looked down and saw the sea: 'I chose probable death over certain death.'"[3]

Benchmarking can be a blunt instrument to demonstrate the need for change, even in successful organizations. IBM Rochester, for instance, went outside its business unit for help in developing its highly-successful AS/400 computer during an era of great strength for IBM in the mid-1980s. IBM Rochester solicited user information from 250 customers, adopted Milliken's teamwork approach, Toyota's short-cycle product development system, and Motorola's Six Sigma defect-prevention program.[4]

Benchmarking has also inspired dramatic change at Caterpillar, a world leader in the design, manufacturing, and marketing of earth-moving equipment. Caterpillar heard the bugle cry for change after benchmarking seven companies from the aerospace, computer, electronic, industrial equipment, and automotive industries. This ambitious best practice study convinced the giant manufacturer to streamline the process by which Cat customers replace or repair electrical and electronic components. Guided by many superior operating practices identified during its benchmarking activities, Caterpillar reduced the cycle time from customer problem reporting to parts retrieval from an average of 42 days to one day. Next, Caterpillar reduced second-phase cycle time for parts retrieval and delivery from an average of 54 days to five days. In the critical third phase, when work teams analyzed and determined the root causes of component failures, Caterpillar identified four best practices. Leveraging these practices, Caterpillar then reduced cycle times that ranged from 9 to 138 days to its current 24-hour cycle-time target. All told, Caterpillar reduced the average repair of electrical components from 120 days to seven days, a staggering 94-percent improvement.

Interestingly, Caterpillar's most effective improvements were often simple changes and innovative adaptations. For instance, dedicated analysts now investigate electronic and electrical component problems; the company also created a database to facilitate improved failure analysis, and Cat now uses existing technology to improve its communication systems between dealers and the Caterpillar warranty group. Caterpillar's management was inspired in large part by the organization's benchmarking studies, which served as a catalyst for change. "Driven by the imperative of better satisfying customers, the best practice findings of benchmarking projects have become an inspiration for positive change at Caterpillar," says Tom Kendall, Caterpillar's benchmarking manager.

Benchmarking and Visualizing Change

"If you don't know where you are going," observes the management adage, "any road will take you there." That path to change is meandering and slow, however. Organizations that compete in highly-competitive industries and markets find the long and winding road to improvement takes too long to traverse and often diverges from the organization's true destination. Benchmarking provides a clear vision of the endpoint of positive change. Such a vision acts like a beacon on the horizon; it helps managers chart a course that is more direct and less perilous through the rough waters of change. Best practice profiles provide a vivid picture of what an organization will look like after it accomplishes a proven regimen of change. The president of a mid-size manufacturer of grinding plates for the pulp and paper industry recalls the "two-year debate" he unsuccessfully conducted with the company's plant manager in an effort to persuade the other executive to convert to a new type of manufacturing process. During the third year, the two executives visited the facility of a secondary competitor that had arranged its manufacturing processes just as the president proposed. "Seeing is believing," says this senior executive. Ninety days after the benchmarking visit, the resistant plant manager—on his own initiative— had redesigned and revamped his manufacturing process.

This transformation of employees from change resisters into change champions is a recurring phenomenon among benchmarking teams. Ralph Martinez, the director of quality for GTE's Mobile Communications operating company, marvels at the transformation he witnessed in a vice president during a site visit to an American Express credit card Customer Service center. At first this busy executive didn't want to participate in the benchmarking visit. "This VP said he could not understand what good it would do for him to attend," recalls Martinez. "After he saw the operation, he was the greatest advocate or champion for these changes in our organization." This was especially important because the benchmarking visit surfaced the opportunity for many small but important changes and system improvements. Together these changes could produce dramatic synergistic improvements; yet none of them alone was likely to spur much change. Consequently, the need for a champion was especially important.

Process mapping techniques illustrate graphically—and often dramatically— the benefits that accrue when transforming a process from its current state to a best practice state. Every process has three versions: what the organization thinks it is; what it actually is, and what it should be. Best practice benchmarking sketches the "future state" portrait of what the process should be. Figure 11.1 illustrates this concept that all unimproved processes have three versions. Benchmarking contributes to the creation of the "should be" or best practice state. Using process maps and flowcharts, benchmarking helps people become less resistant to change because they suddenly see the path to the future state. Ford Motor Company, for example, reengineered its accounts payable process in the 1980s. Applying traditional process rationalization techniques and new technology solutions, Ford projected a 20-percent reduction in head count from 500 to 400 people. Then Ford benchmarked Mazda, which had pioneered a paperless or invoiceless accounts payable system with its suppliers. After studying Mazda's much superior approach, Ford was able to visualize a paperless or invoiceless accounts payable system. Inspired by the benchmarking, Ford implemented the process changes to an invoiceless system and reduced the number of employees required to run the payable system by 75 percent. The changes also helped Ford create simpler material controls, provide more accurate financial information, and produce a more efficient payable system.

What you think it is . . .   What it actually is . . .   What you would like
                                                          it to be . . .

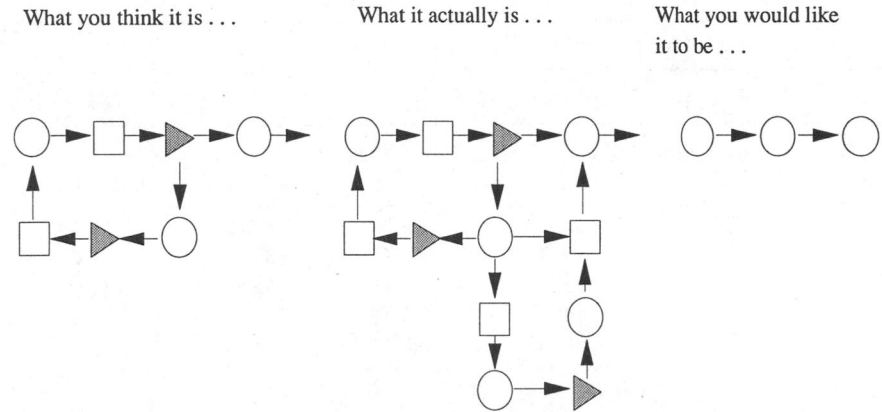

Figure 11.1. Any process has at least three versions.

Deciding What to Change

By comparing baseline benchmarks with those of leading competitors or best practice companies, an organization can evaluate its position relative to the rest of the market. In this context, benchmarking helps assess the need to change—and also suggests what to change. Consider a study of sales-force effectiveness in the pharmaceutical industry. In 1991, a U.S.-based health care products manufacturer found that the sales forces of its three major competitors were significantly more productive than its own; moreover, the company's own sales force productivity was declining. What lay behind this slide?

Research and analysis showed the company's product line was equal to, if not better than, its three rivals. But the study also revealed serious problems with its sales force, including poor knowledge of the product line, poor accessibility and responsiveness of salespeople, poor continuity of sales contact, slow price quotation turnaround time, and poor overall service quality. As a result of its benchmarking studies, the company put the following seven actions into place in early 1991: a 50-hour sales expectations/effectiveness course for all new hires, an ongoing sales effectiveness training program, a new sales compensation structure that emphasized new product revenue streams—increasing the power of salespeople to make pricing decisions, a well-defined career ladder with promotion guidelines, a redesigned recruiting process to attract the best people, and redefined sales territories that made more sense to the sales force. By early 1993, management posed questions about the success of the 1991 changes and the potential for keeping the sales force competitive through the strategic plan ending in 1998. Benchmarking created the impetus to ask these questions; it provided the evidence to evaluate the answers, and it created a needed sense of urgency to retarget goals through each year of the strategic plan. Table 11.1 summarizes the key benchmarks that signaled the organization's management that change was necessary.

Table 11.1. Sample Benchmarking Table for Sale Effectiveness

	Pharmaceutical Industry Sales Force Productivity						
	Sales per salesperson (millions)					Gap versus	
Year	The host company	Competitor A	Competitor B	Competitor C	Competitor average	Average of A, B, and C	Best competitor
1988	$2.05	$3.10	$3.55	$3.70	$3.45	−$1.40	−$1.65
1989	$1.75	$3.25	$3.70	$3.60	$3.52	−$1.77	−$1.95
1990	$1.70	$3.40	$3.95	$3.60	$3.65	−$1.95	−$2.25
1991	$2.15	$3.50	$4.05	$3.70	$3.75	−$1.60	−$1.90
1992	$2.75	$3.55	$4.10	$3.70	$3.78	−$1.03	−$1.35

Benchmarking revealed that the seven actions taken in 1991 resulted in considerable improvement. Nevertheless, significant continued improvement of sales force productivity was crucial to competitiveness. The company's benchmark gaps were significant, and competitor B was offsetting the company's advances with its own continued improvement. The host company conducted two additional benchmarking studies, one on best sales training practices and the other on best sales compensation practices. In the end, the company implemented a proven approach to its sales training system by creatively adapting both competitor B's sales training program and the training program of a leader outside the health care industry. The organization adopted these eight best practices:

- Thirteen weeks of classroom training for every salesperson to occur within 15 months of hire: five weeks devoted to product line, five weeks devoted to sales techniques, and three weeks devoted to customer relationship management.

- Integration of five case studies into classroom learning. The company had commissioned these case studies, on successful and unsuccessful sales contacts, from two leading business schools.

- Extensive video-taped role-playing exercises on effective sales techniques with follow-up critique to reinforce learning.

- Use of award-winning sales people as classroom instructors.

- Regular CEO visits to the classroom to articulate the value, importance, and expectations of the sales force for the success of the company.

- Teaching of salespeople to benchmark competitor's prices, quality, services, and products in order to enhance their presentations to customers.

- Assigning senior sales mentors to directly supervise the field exposure of new salespeople during their first year.

- Debriefings to critique every sales contact during the first six-month probationary period of all new salespeople.

The Two Frontiers of Change

Benchmarking helps organizations find their way through two frontiers of change: First, it provides navigational maps or blueprints describing the most effective practices and approaches for structuring and accomplishing work. Second, benchmarking provides insight and information about how to accomplish cultural change. These two frontiers circumscribe successful change efforts. Consequently, the work systems or processes and the culture must be considered together. The most perfectly designed work processes still produce suboptimal results if the people and the culture reject them. Such cultural rejection might be likened to the way a sick person's immune system rejects a healthy organ transplanted into the sick body. In turn, flawed work systems can hobble healthy corporate cultures and the most motivated employees.

The metaphor of a fast-running river ecosystem is helpful for envisioning these two types of change. The work process or system is easily visible, just as certain types of life and activity above the water line are visible from the river's banks. Plant life, fish life, and geologic activity are all abundant below the water line, but these activities are not as visible from the shore. The work processes are essentially above-the-water line activities and organizational-culture is below-the-water line activities. Best practices benchmarking enables change within the process and the culture, both above and below the water line.

At the above-the-water level of *how work is done*, benchmarking provides valuable design and operating information that can help employees see ways to improve important work systems such as the sales, order-entry, product delivery, billing or customer service process. For instance, a company may be able to streamline the way accounts are paid by introducing PC workstations and a relational database of purchase orders, vendor invoices, and payment records. (*Figure 11.2* depicts the two levels of managing change—above and below the water line.)

Cultural transformations occur below the water line. They embrace softer, less easily charted types of change that link to people's feelings and values, their work-related fears and phobias, their acceptance or resistance to new experiences. Change management at this level blasts away at the bedrock of employee behavior. Not surprisingly, people often shrink from change; they assume defensive postures when confronted with unfamiliar or new experiences. Front-line worker resistance is frequently reinforced by traditional "keep-'em-in-the-dark," "what-they-don't-know-won't-hurt-them" management behaviors. Unfortunately, many organizations have neglected to focus diligently on the cultural level of change management.

Benchmarking reveals that successful cultural change efforts embrace a people-oriented—as well as process-oriented—approach to organizational transformation. Successful change managers help employees to understand the vision or endpoint of change; they communicate the rationale for pursuing that vision; they develop a sense of ownership on the frontline for implementing the change, and they provide reward and recognition for those who make the change work. Transforming organizational culture is much like rearranging the furniture in a familiar room and then turning off the lights. The natural human reaction is to resist the change. Managers and employees tend to view change in polar extremes: it represents a mortal threat or a splendid opportunity. For those who see change as a threat, their reaction is

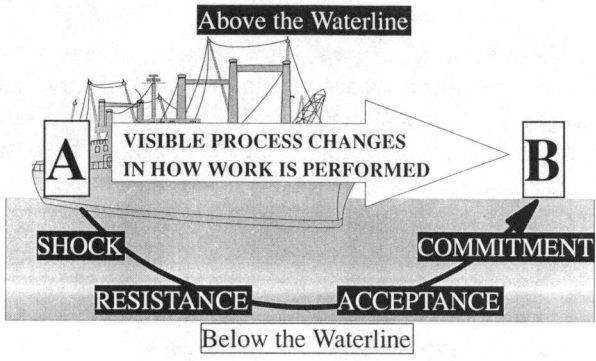

"Below the Waterline" people (emotional) side of change

Figure 11.2.

usually "flight or fight." If they flee, both the employee and the organization lose. If they fight, good ideas are eviscerated when management or employees resort to strong-arm tactics to support their views. "Flight or fight" reactions are debilitating to corporate culture and to business operations.

Among organizations that preside successfully over cultural change, open communications and managerial honesty emerge as best practices. "When you're making revolutionary changes, any adult knows you don't send Hallmark cards," observes Bob Nafius, Director of Strategic Communications at GTE Telephone Operations, which has undergone its own transformation from a once sluggish telephone utility to a company aggressively competing in the world telecommunications market. Adds GTE Public Affairs Vice President Bruce Redditt: "In the past, our communications strategy was 'don't shake up the troops.' Now we feel it's better to talk openly about threats, both competitive threats and threats to jobs." Open and honest communications about difficult changes such as work force reductions, new skill requirements, or job redesign may not inspire cheerfulness among managers and employees, but they enable people to better understand operating circumstances and realities. Understanding the reasons for change is a critical first-step to mastering it. Redditt recalls the GTE strategy: "We brought them into the process as much as possible. We gave them lots of input at meetings, listened to their concerns, and believe me, there were many heated conversations. But we kept emphasizing where the business would be in five years without these changes, where the revenue streams needed to be, and we gradually won support." Ultimately, benchmarking provided a tool to help managers and employees see that change was necessary (it was overtaking the world telecommunications industry) and benchmarking also helped reengineering teams, strategic planners, mid-level managers and frontline employees to craft better solutions than they could have generated alone.

Mastering Change

Cultural or transformational change is not a switch that toggles between off and on. It is a cycle or process that unfolds over time. Envision it similar to the process by which a master photographer develops a picture in the dark room: With all the proper elements in place, the picture slowly appears on the photographic paper, beginning as a ghost-like image. Moment by moment, the picture grows in sharpness and clarity. The photographer—just like a change manager—takes great care not to underexpose or overexpose the image in the development process.

John Scherer, a consultant specializing in change management, conceives of managed change occurring in a seven-step process or cycle that is orderly, if not always neat. Table 11.2 depicts this change process and describes the enabling role benchmarking plays in the cycle.[5]

Transforming corporate culture can be a daunting task. It requires enormous effort and senior leadership support. As an instrument for directing and managing change, benchmarking helps the leadership demonstrate to employees the need, urgency, benefits, and path to change. GTE Telephone Operations, for example, used customer satisfaction research and cost-related benchmarking data to initially launch its *Power of One* campaign, a broad-reaching corporate change process.

Table 11.2. The Change-Management Process. *(Adapted from John Scherer, "Changing the Game.")*

7 steps to managing change	The enabling role of benchmarking
1. Identify what is working and not working in the old culture.	Benchmarking provides an objective way to determine what is and is not working when referenced against the competitive marketplace.
2. Describe the changes needed for the new culture to succeed.	Benchmarking others' best practices helps managers envision what changes will most help usher in the new culture.
3. Model the new culture from the top.	Benchmarking identifies role model behaviors for leading and supporting the cultural change.
4. Train people for the new culture.	Benchmarking can help identify what types of skills and behaviors are necessary to support the change and it also provides insight into how to deliver this training.
5. Reward people for taking on the new culture.	Benchmarking suggests which reward and recognition systems are most effective in supporting the new culture.
6. Structure for the new culture.	Benchmarking helps identify organizational structures that best support the new culture.
7. Assess and maintain the new culture.	Benchmarking provides an on-going means of monitoring the market adequacy of the new culture.

Power of One emphasized the need to reengineer business processes and concentrate on providing value to four stakeholder groups: customers, employees, shareholders, and communities. GTE's Corporate Communications then used the benchmarking data as objective market evidence to establish credibility for senior management's mandate for change. Best practice information, generated by GTE employee benchmarking teams that studied 84 high-performing organizations, further strengthened employee understanding and acceptance of the required changes. Success stories and empirical evidence brought home from other companies by benchmarking teams inspired employee confidence that the company could successfully transform itself.

During this process, employees began to understand that corporate systems and business processes that generated record earnings in 1992 would not lead to market leadership in the future. Management's mandate for change was not a Chicken-Little cry that the sky was falling. Benchmarking provided credibility to the claims; it helped employees understand the reasons why change was necessary and then it enabled the employees to map the course that would lead to the desired change.

"The best way to predict the future is to create it," Peter Drucker, one of the most esteemed observers of modern management theory, has observed. At GTE and other organizations undertaking major change efforts, the senior leadership—informed by benchmarking—initiates the change. But it is the staff and line employees—enabled by benchmarking—who draft and execute the blueprints creating each organization's future.

References

1. "Managing Strategic Change," *Insights Quarterly*, Fall 1992, p. 28.
2. Dumaine, Brian, "TimesAre Good? Create a Crisis," *Fortune*, June 28, 1993, pp. 123-130.
3. Ibid.
4. *The Wall Street Journal*, October 11, 1990, p. 1B.
5. Scherer, John, "Changing the Game," a white paper prepared for GTE's Process Reengineering Implementation Teams, July 1992, pp. 1-4.

12
International Benchmarking

The Japanese Masters

The scene is familiar: A beleaguered organization sends a team abroad to investigate the practices of its competitors. To facilitate the effort, the leadership adopts a policy stating in part that "knowledge shall be sought throughout the world …" They hire outside consultants. As processes and institutions change, the organization's competitive position improves. Does this tableau describe Xerox in 1979 when it launched its first formal competitive benchmarking effort? Does it describe some other struggling American corporation scrambling to reverse its deteriorating market position? No, in fact it represents none of these. A Xerox benchmarking manager did not lead this team; Ito Hirobumi, a little-known bureaucrat, did. The organization is not American; it is the Japanese government. The government did not issue this policy in a memorandum—but in the "Five Articles Oath" of 1868.[1]

In Japan, no word for benchmarking exists, and few businesspeople refer to the practice.[2] The Japanese have captured the essence of striving to be the "best of the best" in the single word: *dantotsu*. Dantotsu implies a degree of awareness about the environment, particularly of others who also strive to be the best of the best. Even though the Japanese do not call the practice "benchmarking," they have become benchmarking masters through generations of practice. "Of all the countries in the world, Japan is by far the most advanced at benchmarking," contends Kenneth A. Bruder of Kaiser Associates, a consulting firm that has worked extensively in the international arena.[3]

The Meiji government of the late nineteenth century, for example, remodeled its anachronistic financial, legal, and educational institutions after American and European models. It sent students like Ito Hirobumi abroad to study. They returned to disseminate the information they learned through important careers in politics and academia. These efforts extended beyond the government. In the late 1800s the house of Mitsui sent a seven-person delegation to the United States to study modern business practices. This effort enabled Mitsui to eventually become one of the largest and most powerful business conglomerates in Japan—in other words, the best of the best.[4]

The general practice of benchmarking, however, never really gained momentum until after World War II. Trying to rebuild their war-torn economy and shattered businesses, the Japanese sent teams to the United States to bring back knowledge. One of these early benchmarkers was Taiichi Ohno, the creator of the Toyota lean production manufacturing system. On a trip to the United States in 1956, Ohno visited competing automakers. Yet, his greatest inspiration during his U.S. tour came from supermarkets.[5]

Observing how supermarket customers take what they need only in the amounts required and the subsequent replacement of goods by employees in time for the next round of shoppers, Ohno reasoned this same practice could effectively increase the efficiency of the manufacturing process. Building on the ideas of Kiichiro Toyoda, the first president of Toyota, and on the fruits of this and other study visits, Ohno created what is known today as the just-in-time manufacturing process.[6]

The Japanese have a long and noble tradition of foreign field study and process examination. Almost obsessively, they collect and consume information. They devour newspapers and foreign books in record numbers. Government agencies and corporations refuse to entertain new ventures without first acquiring, examining, and carefully analyzing mountains of information.[7] The Japanese value information as much as any precious natural resource. Observes one authority on Japanese culture: "For the Japanese, the statement that knowledge is power is not just a pious truism: it is a basic operating principle."[8]

Business Culture

This preoccupation with information gathering has made its way into Japan's business culture, where competitor intelligence is almost an obsession. Consider Mitsui, one of Japan's first corporations to embrace benchmarking techniques; the Mitsui corporate credo is "Information is the lifeblood of the company." Mitsui executives exchange some 80,000 messages each day on the corporation's extensive satellite-enhanced information network. Many of these messages contain intelligence on competitors.[9] This preoccupation with competitor actions, capabilities, tactics, and strategies has helped Japanese businesses achieve many of the competitive advantages they enjoy today.

In other countries, corporations place less emphasis on competitive comparisons. A recent report prepared by Ernst & Young and the American Quality Foundation is particularly illuminating. In a survey of over 500 companies from Canada, Germany, Japan, and the United States, Japanese businesses emerged as the clear leaders in competitive analysis. Some 92 percent of Japanese companies said competitor comparisons are either a major or primary factor in their strategic-planning process. This contrasts with a relatively healthy 82 percent in the United States and a more lackluster 67 percent in Canada and 61 percent in Germany.[10]

Japanese companies continually monitor their competition to stay one step ahead. Recent commentators reported, for example, finding a Japanese executive in a large library with 10 years' worth of annual reports from American companies. By reading the annual letters from the CEO, he said, he was hoping to get a direct impression of how his competitors think.[11]

David Kearns, the former CEO of Xerox, relates a similar tale. During a plane flight on a trip to Japan in the early 1980s, Kearns was catching up on several weeks of reading and came across an article in a leading business magazine about General Motors' capital investments of several billion dollars in new robotics and technology. GM predicted that this investment would help it achieve significant productivity gains and drive other improvements. However, the magazine writer was skeptical that the investment would pay off. In fact, he concluded his story with a cautionary note, saying only time would tell whether these massive investments would prove to be foolish or far-sighted.

When Kearns reached Japan, he visited Toyota. He noted that the article he had just read was posted. The Toyota executives had highlighted the relevant numbers and operating improvements that GM predicted. Unlike the skeptical business writer, Toyota presumed that GM would make every one of its targets right on schedule. However, Toyota had already reset its schedule and operating goals so that it would be ahead of GM by the time GM reached those targets. The Japanese set their sights unbelievably high and pay close attention to their competitors. Whether one calls it benchmarking or competitive intelligence, the Japanese view careful scrutiny of competitors as a ground rule of effective competition.[12]

The Japanese are not satisfied with the mere collection of information about their competitors. They endeavor to absorb and adapt that information to their environment. Observes Chuck Schallhorn, Director of Business Research at Kodak Imaging Group: "It is ingrained in them to constantly compare themselves to other people. They always take ideas to put a Japanese twist on them. They are not satisfied with simple benchmarking, but must improve on it."[13]

Sometimes the Japanese can get over-zealous in their thirst for knowledge. In 1982, for example, Hitachi tried to pay $525,000 to IBM employees for some of Big Blue's trade secrets. In an elaborate sting operation, however, the FBI foiled the plan.[14] Though this effort was halted in its tracks, it is impossible to know if similar cases of extreme borrowing (i.e., industrial espionage) have gone unchecked.

The philosophy of borrowing runs deep in Japanese culture. Traditionally they have been adapters more notably than innovators. A Japanese manager explains this phenomenon: "'You [in the West] chop down the trees, and we [in Japan] build the houses.' In other words, you do the hard work of discovery, and we exploit those discoveries to create new markets. It is instructive to remember that Sony was one of the first companies to commercialize the transistor and the charge-coupled device, technologies pioneered by AT&T's Bell Laboratories."[15] The Japanese certainly do not suffer from the "not-invented-here" syndrome that plagues many American corporations.

Though the Japanese have been sharply criticized by American businesspeople for their prolific borrowing, the difference is essentially cultural. "It is a truism," observes Bernice Cramer, president of Paos Boston, the American arm of a Japanese consulting firm, "that Americans are more open to foreign people and not to foreign ideas. The Japanese on the other hand are open to foreign ideas and not foreign people."[16] Borrowing is prevalent in Japan, adds John Kochevar, an independent consultant with years of experience dealing with the Japanese. The Japanese are a communal society—what's on paper is open and shareable.[17] Acquiring the information is only step one. The Japanese succeed at borrowing because of their willingness to learn. They view interaction as an opportunity. Kodak's Schallhorn, who sits on a professional standards committee, notes that the Japanese companies always accept invitations to participate in the activities of the committee. He observes that U.S. companies do not show the same interest. The Japanese view this participation as another opportunity for them to learn.

Perennial Students

Even the "best of the best" hope to learn more. "The Japanese," says Kaiser Associate's Bruder, "constantly want to know what is going on." Toyota, he notes, constantly benchmarks against GM, Ford, and Chrysler, even though many regard Toyota as the industry leader.[18] As the story of the Japanese executive reading American shareholder letters demonstrates, this willingness to learn pervades all levels of Japanese organizations. "They are just more interested in studying than [Americans] are," says Cramer. "[This is true] even among executives."[19]

Partnerships with American corporations have been a particularly fertile learning ground for the Japanese. Even the condescending approach taken by many of their American partners does not deter them. An executive from an electronics company put it this way: "Our Western partners approach with the attitude of teachers. We are quite happy with this, because we have the attitude of students."[20] This approach pays off. A recent study found that in every case in which a Japanese company emerged from an alliance stronger than its Western partner, the Japanese company had made a greater effort to learn.[21] This results from *dantotsu*—the striving to be the best of the best—and the Japanese approach to continuous learning. Where Western firms see an opportunity to avoid investments and risks, the Japanese see opportunities to improve their skills and, subsequently, their competitive position. They are not satisfied unless they emerge from collaborative efforts with new abilities or technology. One executive from a Japanese company in such a partnership explained: "Collaboration is second best. But I will feel worse if after four years [with this American company] we do not know how to do what our partner knows how to do. We must digest their skills."[22]

The Keiretsu

The Japanese share information almost as much as they consume it, particularly among native companies. Japan is a networked society. Myriad trade associations, government sponsored conferences, technological societies, and, of course, the quasi-integrated structure of the *keiretsu* facilitate information exchange. Historically and culturally, the Japanese expect interdependence and the harmony that sharing brings. For example, the company that first discovers or imports a new idea becomes the recognized market leader—provided it shares the information. This occurred when Ito Yokado brought convenience stores to Japan. Imitators immediately capitalized on the information and experience he made available. None, however, will ever threaten the dominant position he enjoys.[23]

This willingness to share applies equally to *intra*corporate relationships. Because of life-time employment, a kind of corporate community develops within Japanese businesses. Such an environment more closely ties an employee's interests to the company's. One commentator accurately described this environment as "... a living organism with which [the worker] identifies to a degree surpassed only by his identification with his family and his country."[24] As a result, employees readily acquire and distribute information throughout the corporation. There are few fiefdoms bottlenecking information exchange as in many American companies. While facilitating the exchange of information within and into the company, employee loyalty also helps prevent the outflow of sensitive information. "We don't feel any need to reveal what we know," says one Japanese manager. "It is not an issue of pride for us. We're glad to sit and listen."[25]

Historical and cultural factors are not the only reasons for Japan's benchmarking adroitness. A number of institutions have also cropped up that facilitate the learning process. Perhaps the most important of these is the *keiretsu*. These company groups compose a greater community of firms headed by a lead bank. They are designed "to promote mutual interest and to deal with a crisis collectively."[26]

But, the *keiretsu* are neither just a crutch in times of trouble nor merely glorified partnerships. These dynamic alliances allow firms to interact informally on all levels of management as well as among front-line hourly employees. They facilitate sharing and information exchange by providing ready-made introductions. Member executives exchange information at regular meetings. In addition to the usual financial services, the lead banks provide advice and guidance. More importantly, they overcome the constraints of life-time employment by sharing employees. As Bernice Cramer notes, information resides with the employees— "files belong to the individuals, not to the companies." Consequently, employee-sharing among firms allows the dissemination of this otherwise unavailable information. Invaluable sources of information, the *keiretsu* are veritable "benchmarking havens."[27]

The Japanese Government

The Japanese government is also in the benchmarking business. In sharp contrast to the United States where information exchange among competitors invokes the ire of the government, officials in Japan encourage and even facilitate the sharing of research and other breakthroughs between competing firms. It promotes study visits abroad by providing discounts to businesspeople who travel. The Ministry of International Trade and Industry (MITI) and other related government agencies study cross-company practices and teach about them. For certain strategically important markets and technologies it also funds industry consortia—an important avenue for learning best practices.[28]

The Japanese government also frequently assumes an active role. Through standards committees, for example, the government endeavors to set up or guide particular industries. By suggesting or requiring certain processes, these committees are virtual clearinghouses for benchmarking or best practices information. Whatever the means (or the ends), the adversarial government-business relationship so often witnessed in the United States is not nearly as prevalent in Japan. Japan is indeed a networked society where the government is an important link in the circuit.

Industry Associations

Industry associations are also important avenues for benchmarking information. Similar in some respects to the *keiretsu*, these organizations enable companies with similar interests to get together to exchange information. Though industry associations are prevalent throughout the world, Japanese trade groups exploit their resources to a much greater extent.

The technological societies that pervade Japan are also warehouses of information. The Union of Japanese Scientists and Engineers (JUSE) is perhaps the most famous of these societies. As the curator of Japan's famous award for quality, the Deming Prize, JUSE plays an important role in the benchmarking efforts of many Japanese companies. This is done foremost through the Deming Prize itself. Applicants for the prize must demonstrate, among other things, how they determine their quality control practices and more importantly, how they collect information on quality. This suggests that companies are expected, if not required, to seek out and determine the best quality control practices. Furthermore, because of their extensive exposure to the quality practices of many different companies, JUSE inspectors carry an implicit benchmarking system in their heads. These inspectors provide all applicants with a report on their quality efforts and during their on-site inspections give advice or guidance as a matter of course.[29] Moreover, the exhaustive internal examination required to win the prize enables the applicant to more effectively implement the inspectors' suggestions. Whether or not a company wins the Deming Prize, the information they may derive on their own and others' practices makes the challenge worth attempting.

In addition to these more formalized institutions, a strong network of university alumni also get together to exchange information about their companies.[30] More prevalent in Japan than in the United States, these informal dialogues cut across industry lines. Untrammeled by competitive concerns, they provide valuable insights into corporate practices. Furthermore, this network provides important connections within other companies by furnishing the introductions required to truly get an inside look at another company. Whether to facilitate benchmarking itself or as a means to that end, Japanese businesspeople have added incentive to keep in touch with their old university colleagues.

Customer-Supplier Sharing

Corporate relationships—as distinguished from the *keiretsu*—often provide significant benchmarking opportunities. This is especially true between a large customer dealing with a smaller supplier. In these relationships, the smaller suppliers get benchmarking data through the larger customer, but this is not corporate altruism. It is probably closer to enlightened self-interest. The large company/consumer has an important stake in the education of its suppliers. In this way, the customer ensures better materials and the suppliers acquire valuable information about how to improve their processes and their products.

Information transfers from the customer-supplier relationship may also take a more circuitous route. Firms often "open the kimono" to potential customers in an attempt to lure them to their product. In so doing they often reveal important information about their processes and technology. Observes Chuck Schallhorn: "Suppliers may show the merits of their organizations to a potential customer, from which the [target] company may learn a fair amount."[31] This information could then be used within the target company itself. Or, in more extreme cases, it is passed on to other prospective—and probably cheaper—suppliers.

Even though the Japanese language lacks a specific word for benchmarking, these social, cultural, and institutional factors make the Japanese among the best natural benchmarkers in the world. Unlike American corporations that emphasize performance benchmarking (product, feature, price comparisons) and process benchmarking, the Japanese excel at performance and strategic benchmarking. Their deep-rooted preoccupation with competitor intelligence continuously tunes them into performance benchmarks. Moreover, they also concentrate broadly on what strategies and general practices make other organizations successful. When performing strategic benchmarking, the Japanese do not limit their focus to a specific process.

The American preference for process benchmarking may be partly explained by the American manager's burden to produce short-term profits. Improvement of a core process can quickly produce a beneficial impact on the bottom line. In contrast, the Japanese have long been focused on process control and improvement. Their process management practices are often among the most sophisticated in the world. The Japanese, therefore, focus on questions such as, "What companies—without regard to industry—are world-class performers and what drives their success?" Such "strategic" inquiries significantly impact long-term competitive position and direction but have relatively little expected impact on short-term financial results. In their book, *Kaisha: The Japanese Corporation*, James Abegglen and George Stalk observe: "...the risk of falling behind a competitor is regarded by most [Japanese corporations] as a far greater risk than the risk that profits will be depressed. To fall behind one's competitor can mean that profits will never materialize. To minimize the risk of falling behind, the [Japanese corporations] are preoccupied with the activities of their competitors to a degree that is unusual by Western standards." This long-term focus and obsession with market share has made the strategic aspects of benchmarking an indispensable tool to the Japanese.

Strategic Benchmarking

Strategic benchmarking has been effective for the Japanese because they are well equipped to use it. Japanese managers carefully study, understand, control, and manage their operational processes. Consequently, they seem well able to project the benefits of other strategic approaches to their own systems and processes. Indeed, most top Japanese executives have hands-on experience with their corporation's core processes. Robert C. Christopher contends in *The Japanese Mind* that "…nearly all Japan's current crop of top executives are quick to tell you that they started out on the assembly line or in some equally humble job."

Because the Japanese know more about their own processes, they need only identify those companies which they find excellent in some respect and visit them. Upon looking at a nine-step benchmarking process beginning with (1) identify benchmarking area, (2) understand your own process and (3) select benchmarking partners, Paos Boston's Bernice Cramer noted that her Japanese clients always begin with step three; they want first and foremost to study excellent companies. This coheres with the Japanese traditional benchmarking method: business travel and the study visit. These visits yield considerable insights for many Japanese executives because they already have a precise understanding of what information they require and how they can most effectively implement it.

Benchmarking in Japan, then, is generally a less structured affair than process benchmarking in the United States. The "study visit" itself implies a certain informality. As consultant John Kochevar suggests, many Japanese businesspeople make study visits on the pretense of fostering good interpersonal relations. But behind the guise of an affable, casual visit the Japanese hope to gain exposure to new business practices. Study visit reporting is equally informal. Kochevar observes that many Japanese businesspeople can't use a keyboard for typing up reports. Consequently, reports are generally thin; "they get down to the bare bones," says Kochevar. Information therefore frequently makes its way through corporations by way of seminar-style study groups. The members of the study group then turn the information over to engineers for implementation. Bernice Cramer's experience with her Japanese clients further demonstrates this style of information distribution. Her firm regularly sends its clients formal competitive assessment reports. But, she says, the "real" information exchange occurs at the monthly meetings. Japanese corporations need to distribute information directly to individuals.

Obstacles to Japanese Benchmarking

Despite the Japanese penchant for external learning, inhibitors to benchmarking also exist. One such obstacle is the cultural converse of *dantotsu*. As Schallhorn observes, some Japanese corporations are too conscious of rank. The second-tier companies willingly follow the leaders instead of trying to adopt their best practices, innovate on them, and then surpass their competitors. Professional baseball in Japan illustrates this phenomenon, contends Schallhorn. In Japan the Tokyo Giants are the perennial favorite. While watching television there, Schallhorn recalls an interview conducted early in the season with the manager of a competing team. In a statement of almost bitter irony to American baseball fans, the manager remarked, "This year we hope to be number two."[32] This attitude applies equally to many corporations in Japan. Their willingness to accept second-best, like the opponents of the Tokyo Giants, stifles their ability to adapt creatively.

"*Hubris*," adds John Kochevar, is another barrier to benchmarking in Japan. Some companies, enamored of their competitive positions, have apparently forsaken the efforts at continuous learning that helped them achieve their dominant position. Arrogance has led to the blindsiding of many successful companies—and those in Japan have no special immunity. The reversal of fortunes of companies in the microchip industry are evidence of this. Nevertheless, some Japanese companies, just like their earlier American counterparts, believe that they possess superior processes and technology and cannot learn from anyone else. This attitude does not last for long. As Schallhorn notes wryly, "the arrogance level was much higher two years ago" before the Japanese economy was buffeted by recession.

Benchmarking in Europe

Though the Japanese and the Americans are the most advanced in their benchmarking applications, other areas of the world are following their course. This is particularly true in Europe. "The benchmarking concept is growing [there], "says Bob Camp of Xerox. "[It is] paralleling its growth in the United States five or six years ago."[33] Having observed foreign affiliates and divisions of U.S. companies effectively apply benchmarking, the Europeans are gradually adapting benchmarking to their own culture and language.

Traditionally, the Europeans have not had to benchmark. High tariffs and other protective efforts kept companies relatively isolated. But, notes Kaiser Associates' Kenneth Bruder, the barriers are coming down, and "they are horrified of the Americans and Japanese entering their markets." European companies, therefore, have been forced to evaluate their competitive positions. Benchmarking provides them with the answers. Bruder contends that three primary situations currently are mobilizing benchmarking abroad:

1. Foreign competition threatens a company's traditionally strong domestic market; the goal is to use benchmarking to preempt market intrusion and maintain market share;

2. A company desires to enter a new foreign market or new industry segment; the goal is to apply benchmarking so that the company can quickly, effectively, and profitably introduce products to the new foreign markets.

3. A company is locked in tough competition with several foreign and domestic competitors; the goal is to employ benchmarking to help understand one's strengths versus the competition and to generate ideas to surpass the competition.[34]

Unlike Japan, however, European culture does not encourage sharing. "Sometimes the Europeans feel that two or three interviews down the road the information might get to a competitor," observes one manager of a multinational corporation based in Switzerland. As a result, benchmarking has usually been conducted for them, typically by consultants or third parties to ensure confidentiality. "They have been extremely receptive," observes Bruder. "Benchmarking gives them knowledge of how to compete globally." The practice is gradually being accepted as a powerful managerial tool in an increasingly competitive environment.

The Europeans have also demonstrated real initiative in some areas. Take the European Foundation for Quality Management (EFQM), for example. Established by the presidents of Europe's most powerful companies, the EFQM endeavors to establish a European-wide structure to promote quality. Benchmarking is integral to these efforts. "The EFQM is not looking to the U.S. or Japan to provide a model," says Brendan Coyne, editor of *Quality Today*, a British-based professional journal. "Instead it is basing itself upon the world's best practice."[35] For member companies eager to begin benchmarking, the EFQM is a storehouse of information on many premier quality practices.

Benchmarking in Latin America

In Latin America benchmarking appears further out on the horizon. With the North American Free Trade Agreement, Latin American companies—particularly those in Mexico—are waking up to the tools of quality management. "Mexican companies are just beginning … to understand quality tools," says Javier Romero Rio of the TQM Group Mexico. "Benchmarking would be a bit ambitious right now."[36] Ramon Arias of Inter-American Management Consultants confirms this assessment: "Companies are just beginning their quality journey now, just starting to define their strategic quality programs," he says. "In a year, at best, the advanced companies will be doing local best practices searches."[37] For the present, benchmarking in Latin America is limited mostly to United States and Japanese affiliates operating there. But, as Latin American corporations embrace quality management techniques, it is certain that benchmarking will soon follow. Already, countries such as Mexico, Argentina, Brazil, Chile, and Columbia have launched National Quality Awards, often modeled directly after the benchmarking-rich Malcolm Baldrige National Quality Award. Just as the Baldrige award has helped ignite interest in benchmarking in the United States, the Latin American national quality awards can be expected to do the same.

Conclusion

The Greek philosopher Socrates bequeathed to the businesspeople of the world two wise and relevant observations: "The only good is knowledge and the only evil is ignorance," he said, and added, "Know thyself." The Japanese people and their companies, more than those of most other countries, have taken these ideas to heart. Their relentless quest to understand their own operating processes and the best practices of other world-class performers is unparalleled. As a result, the Japanese have become one of the premier benchmarkers in the world, and subsequently, one of the world's most formidable economic powers. Similar seeds of interest in benchmarking have taken root in Europe and many more are also planted in Latin America. In these regions, the harvest lies ahead.

References

1. Reischauer, Edwin O. and Craig, Albert M., *Japan: Tradition and Transformation*, Houghton Mifflin Company, Boston, Mass., 1989, pp. 135, 140, 147-51, 161-62.

2. Meyer, Richard, "Preserving the Wa," *Fortune*, September 17, 1991, p. 52.

3. Bruder, Kenneth A., "International Benchmarking," *Competitive Intelligence Review*, Summer 1992, p. 5.

4. Reischauer and Craig, Ibid, pp. 140, 151, 152-53, 161, 164.

5. Main, Jeremy, "How to Steal the Best Ideas Around," *Fortune*, October 19, 1992, p. 102.

6. Lu, David J., trans., *Kanban: Just-in-Time at Toyota*, Productivity Press, Norwalk, Conn., 1989, pp. 65-66.

7. Christopher, Robert C., *The Japanese Mind*, Fawcett Columbine, New York, N.Y., 1983, pp. 195-96, 207.

8. Ibid, p. 207.

9. Dumaine, Brian, "Corporate Spies Snoop to Conqueror, " *Fortune*, November 7, 1988, p. 68.

10. "International Quality Study, Top Line Findings," Ernst & Young and American Quality Foundation, New York, N.Y., pp. 6, 24-25.

11. Prusak, Laurence and Matarazzo, James, "Information Management and Japanese Success," Ernst & Young Center for Information Technology and Strategy, Boston, Mass., 1992, p. 5.

12. David Kearns speaking at Peat Marwick Management Committee Meeting, New York, N.Y., November 20, 1991 (Chris Bogan was in attendance at the meeting).

13. Best Practices Benchmarking & Consulting, telephone interview with Chuck Schallhorn, April 6, 1993.

14. Dumaine, Ibid, p. 72 and Meyer, Ibid, p. 54.

15. Hamel, Gary and Prahalad, C.K., "Strategy as Stretch and Leverage," *Harvard Business Review*, March-April 1993, p. 80.

16. Best Practices Benchmarking & Consulting, telephone interview with Bernice Cramer, April 1, 1993.

17. Best Practices Benchmarking & Consulting, telephone interview with John Kochevar, March 30, 1993.

18. Best Practices Benchmarking & Consulting, telephone interview with Kenneth A. Bruder, Jr., April 23, 1993.

19. Cramer, Ibid.

20. Hamel, Gary, L. Doz, Yves, and Prahalad, C.K., "Collaborate with Your Competitors—and Win," *Harvard Business Review*, January-February 1989, p. 138.

21. Ibid, p. 134.

22. Ibid, p. 134.

23. Meyer, Ibid, p. 52.

24. Christopher, Ibid, p. 245.

25. Hamel, Doz, and Prahalad, Ibid, p. 138.

26. Yoshimori, Masaru, "Sources of Japanese Competitiveness, Part I," *Management Japan*, Spring 1992, p. 20.

27. Meyer, Ibid, p. 52.

28. Leibfried, Kathleen, McNair, H.J., and C.J., *Benchmarking*, Harper Collins Publishers, Inc., New York, N.Y., 1992, p. 119.

29. "The Deming Prize Guide For Oversea [sic] Companies," Union of Japanese Scientists and Engineers (JUSE), Tokyo, Japan, 1986, pp. 7, 8, 16, 21.

30. Schallhorn, Ibid.

31. Schallhorn, Ibid.

32. Ibid.

33. Best Practices Benchmarking & Consulting, telephone interview with Robert Camp, April 9, 1993.

34. Bruder, Kenneth A., "International Benchmarking," *Competitive Intelligence Review*, Summer 1992, p. 2.

35. Coyne, Brendan, "1992 Fortress Europe," *Quality*, June 1990, pp. 17-18.

36. Best Practices Benchmarking & Consulting, telephone interview with Javier Romero Rio, April 13 and 14, 1993.

37. Best Practices Benchmarking & Consulting, telephone interview with Ramon Arias, February 1, 1993.

13

Benchmarking in the Public Sector

Desert Express

One of the biggest successes of the Persian Gulf War was the division of the Air Force which shipped spare parts between the Wright-Patterson Air Force Base in Columbus, Ohio and the Persian Gulf. The Air Force Logistics Command credited its rapid and reliable parts delivery to benchmarking the methods used by Federal Express.[1] The Air Force is a forerunner in public sector benchmarking. Among state, local, and federal government, benchmarking exchanges are still rare, and rarer still is highly-formal or structured best-in-class benchmarking. Nonetheless, informal benchmarking and an active interest in best practices are growing rapidly throughout the public sector worldwide. Great opportunities exist to improve government performance using benchmarking, and an advance guard of public sector entrepreneurs is beginning to employ benchmarking in its host agencies and governmental departments.

Seattle's AIDS Program

One such innovator is the city of Seattle. The Seattle-King County Department of Public Health benchmarked San Francisco before designing its AIDS care program. For years, San Francisco has been viewed as a model for cost-effective and humane treatment of AIDS patients. The lifetime hospital cost for a person with AIDS is only $41,500 in San Francisco, compared to $60,000 to $75,000 nationwide.[2] Further, research and interviews conducted by the Seattle-King County Department of Public Health indicated widespread use of and satisfaction with the program. Just as importantly, San Francisco's AIDS treatment approach was an effective model for Seattle because the cities have similar populations and minister to large numbers of persons living with AIDS—mostly gay white men. A policy which is effective in a city where intravenous drug users comprise a large portion of the AIDS patient population, such as New York, may be less likely to succeed in Seattle because of dissimilar demographics. Patricia McInturff of Seattle's Department of Public Health says, "Looking at a program serving a different population won't work. We couldn't look at Miami. We looked at peer cities that serve a population similar to Seattle's."[3]

Seattle studied San Francisco's program through interviews and meetings, then adapted it to meet Seattle's needs. "You have to realize that there isn't one answer to a question, but a series of answers. People and cities are different," says McInturff. "We stole everything we could from San Francisco. But we also had the luxury of watching their mistakes."[4] One such mistake was the uncontrolled decentralization of case managers. Originally, case managers who worked for the service agencies guided patients through health and social services. As the AIDS crisis grew, however, so too did the number of volunteers. Eventually, some AIDS patients had as many as three or four case managers. "You had a decentralized system where each agency had a case manager. It became increasingly confusing to patients which case manager was managing the case," says Paul Jellinek of the Robert Wood Johnson Foundation, a national philanthropic organization dedicated to health care. "It got to the point where you needed a case manager to manage the case managers."[5]

Seattle avoided many of these problems largely through lessons learned from its benchmarking partner. The result: Seattle has been successful keeping expensive hospital care well below the national average, effectively controlling the most expensive part of AIDS treatment, while still providing humane care for its citizens suffering from AIDS.

Formal Benchmarking in the Public Sector

The Internal Revenue Service initiated one of the first examples of formal, structured benchmarking in the public sector. In an effort to improve performance and efficiency, the IRS decided to import and adapt the "best practices of the best." It began by providing its employees with benchmarking training from a private consultant. A speaker discussed benchmarking with top IRS executives. Then, instead of applying benchmarking immediately to major strategic issues, the leadership chartered several demonstration projects to illustrate the potential power of benchmarking. Managers inspired by the seminars, with the help of process owners and support people, decided which processes to benchmark. Four demonstration projects were settled on: software measurement, picking and packing in form distribution centers, personnel recruitment and retention, and assistance at walk-in taxpayer sites. The project teams partnered with regional headquarter staff members to ensure that implementation would be feasible.

The IRS team performed literature searches and used professional society contacts to find benchmarking partners, most of whom were in the private sector. They searched for successful processes that were applicable to their own mission and procedures. The IRS then embarked on a 10-step benchmarking process, using the same methodology employed by the Xerox Corporation. "Benchmarking methodologies are all very similar—they vary mostly in the amount of detail outlined in the steps," says Joel Parfitt of the IRS Quality Office. "There are some basic steps that you have to do."[6] IRS processes were examined across departments from both a value and a customer's viewpoint.

While the four benchmarking teams were still in the process of blazing new ground for the IRS, Parfitt was already planning for the future. He intends to help redesign teams use benchmarking as they document, study, and reengineer the major strategic areas of the IRS.

Public Sector Keiretsu

Despite the experience of the IRS, most public sector benchmarking does not imitate the formal and structured benchmarking processes of the private sector. Governmental benchmarking is usually informal, often consisting only of conversations about other practices. This approach to benchmarking is strikingly similar to that of the Japanese *keiretsu*, tightly-knit communities or operating networks of companies formed around a central bank. In Japan, *keiretsu* alliances provide a strong network and a fertile setting for the sharing of information and best practices. *Keiretsu* companies sometimes share employees, furthering the transfer of valuable knowledge and operating practices. This informal information transfer represents the most common form of public sector benchmarking.

Ryan Evans, the Director of Budget and Research in Dallas, Texas, is a firm believer in the value of this type of informal network. He regularly benchmarks 12 "sister cities" which are of similar size and demographics to Dallas to find out if they run their programs more efficiently or cost-effectively than Dallas does. If they do, Evans learns how they operate. "I make a dozen calls a week to other cities," observes Evans. "If you find out another city of similar size has a parks and recreation department that's operating $10 million less expensively, you're going to ask why."[7] Evans annually compares Dallas' largest departments with those of the other cities to improve them. What's more, when it comes time to develop new programs, such as Dallas' recycling initiative, he looks to other municipalities' programs for ideas. "Benchmarking is done in the public sector for the same reasons it is in the private sector," Evans says. "We cannot 're-create the wheel' all the time. We must learn from what others are doing."[8]

Obstacles to Public Sector Benchmarking

With such obvious and valuable applications, why isn't public sector benchmarking more prevalent? A wide array of reasons explains the public sector's lag in the field of benchmarking:

- *Financial restrictions.* Public agencies tend to push benchmarking aside in favor of more pressing concerns. Meryl Libbey of the Innovations in State & Local Government Awards Program at Harvard's John F. Kennedy School of Government says, "The public sector has more immediate human interests to invest in. Should an agency spend money on poverty or quality training?"[9] Ironically, as many businesses have learned, benchmarking can return many times on the investment, saving money to use for those pressing problems and even helping to devise more successful solutions to them. Examining an established and successful agency can save the benchmarker time and money.

- *Public sector skepticism.* Convinced that renovation and adaptation will be too burdensome, some public servants are unwilling to take advantage of the opportunity benchmarking represents. Their reasoning: Different municipalities have different priorities and complications. Because of the large number of variables found in the making of public policy, imported ideas and plans need so much adaptation and renovation that they lose value. For these skeptics, the concept of innovative adaptation through benchmarking isn't compelling.

- *Lack of emphasis on research and development.* Research and development is often sacrificed because of scarcity of funds and the absence of a single supportive centralized network.[10] Research and development needs to be valued by public sector management before a process improvement project, such as benchmarking, can be undertaken.

- *Lack of a clear mission.* Meryl Libbey notes that the public sector has difficulty initiating benchmarking projects because there is no "values consensus." While many private sector companies have clear and articulated goals and mission statements to guide their efforts, public sector agencies often suffer from an inability to define an all-encompassing purpose or objective. Rarely is there a consensus on the goals of a government agency; indeed, changing social values and ideas mean that an agency's mission is constantly in flux. Without clear organizational goals, it becomes difficult to identify functions and "core competencies" on which to focus. An agency performs a large number of functions, and it is usually difficult to prioritize them. Moreover, without clear goals it is often difficult to set clear standards to determine how well the agency is doing its jobs. In turn, this situation makes it difficult to decide which processes to improve.

- *Cultural barriers.* The "not-invented-here" syndrome still thrives in the public sector. W.C. Enmon, special advisor to the governor of Texas on TQM, observes that imported ideas are routinely denigrated. "If it's not invented here, it's no good" is a common view, he notes.[11] Furthermore, some public sector purists argue that private sector best practices have no relevance in the public sector.

- *Fear of failure.* Public sector employees are often risk-averse. They know that if they make a mistake and waste public money, they may face recriminations in the media. Consequently, they are very cautious before embarking on new initiatives. They do not want to tarnish their image before the public.[12]

- *Lack of constant pressure to improve.* Few incentives exist to spur public sector bureaucrats to pursue continuous improvement and change with the zeal evidenced in the best private sector organizations. There is no public counterpart to the bottom line, which is a constant reminder of the importance of performance.

- *Quality initiatives are still in the development stage.* The public sector has just begun its quality journey, and for the most part is not yet at a stage where benchmarking would be productive. Texas' W.C. Enmon observes: "There is a place to rightfully introduce benchmarking. Government agencies need to hold off until the prerequisite quality and data analysis processes are in place."[13]

- *Lack of objective standards.* Some public sector managers find it difficult to determine with whom to compare themselves. Public sector organizations are not market-driven, and they frequently lack comprehensive performance measurement systems that create common standards concerning what and how to compare systems.[14] Long lag times between public policies and the results they produce also make it more difficult to determine what is a successful policy. For instance, what short-term performance measures should one use to evaluate national environmental or education policies? Tangible results may be years off.[15] Finally, since reasonable people often disagree on the objectives of public policy, program evaluation becomes even more complicated. Public sector managers lament the lack of objective evaluation criteria.

The Growing Popularity of Government Benchmarking

Despite these obstacles, a growing number of government and public sector organizations are overcoming these hurdles. Even without formal performance measurement systems in place, they are studying other programs and exploring others' most effective practices. These successful early forays suggest that public sector benchmarking can be effective—and the tide of interest is rising.

Consider the Reno police department, one such public sector champion that has found valuable operating lessons outside the government. Police officers in Reno, Nevada, were facing a significant increase in traffic accidents in the mid-1980s. Their response was a crack-down on speed limit violators. With an arsenal of 21 new radar guns, Reno police doubled the number of traffic citations. But accident rates held, and angry Reno citizens rejected two referendums which would have increased police funding. Then the Reno police adapted a page from the playbook of Harrah's Casino Hotels. Harrah's has one of the most extensive customer-service databases in its industry; each year it surveys thousands of customers and carefully tracks the results on a monthly basis. Using Harrah's customer requirements and needs setting practices as a model, the Reno police department listened to its customers. It surveyed thousands of Reno citizens. Police now target sites with large numbers of accidents instead of writing tickets in random locations, and electronic signs hooked up to radar guns warn motorists if they are exceeding the speed limit. Nine of 10 Reno citizens approve of the department, up from 4 of 10 in 1988, and accidents are down 20 percent despite a decline in ticketing.[16]

Other public sector benchmarking successes pop up on all points of the compass. After launching a service quality initiative two years ago, the British Columbia Provisional Government has interviewed with agencies across the world to learn about their processes. Among others, it has benchmarked British Columbia Telephone, Federal Express, Xerox, and the Australian government. The British Columbia Ministry of Transportation and Highways has studied BC Tel's practices in the areas of union relations, industrial relations, management incentive compensation plans, and employee participation programs.[17]

Neil Sullivan, a former policy chief for the city of Boston, made a de facto benchmarking trip to Pittsburgh in 1988 with then-mayor of Boston, Ray Flynn. The goal: to discover how Pittsburgh schools were attracting middle-class parents. Sullivan remarked in the *Boston Globe* in 1993, "We were more than impressed; we were influenced."[18] He added that the cooperation he witnessed between school officials and labor leaders in Pittsburgh will be a model for the Boston school reform package he is currently developing at the Boston Private Industry Council. In the same article, the *Globe* staff writers acknowledged the benefits of studying exemplary practices when they wrote, "In this Boston mayoral election year, when the city is struggling with foundering schools, fear of crime, and high inner-city unemployment, civic leaders might look [at Pittsburgh] for a cure to at least some of the Hub's urban ills."

As suggested by the *Globe's* remarks, growing support for public sector benchmarking exists in the media. *Financial World* magazine annually devotes a cover story to "benchmark cities" which employ best practices other municipalities ought to emulate. In one such cover story, *FW* staff writers observed: "Our goal is to help government managers emulate their private-sector counterparts in the practice of benchmarking. Each of the cities we have singled out has accomplished something in a way that other cities can replicate … Benchmarking is clearly the most sensible route for cities interested in progress"[19]

The development of criteria to evaluate the quality of state and local government management is a forceful step forward, helping to determine public sector best practices. *Financial World* also rates the 30 largest cities and profiles their most successful public sector programs. Other publications, such as the *Best Places Rated Almanac* and *Money* magazine have, also, devised systems to evaluate the quality of U.S. cities and their governing bodies. This early stage of public sector performance measurement—a prerequisite of effective benchmarking—is where corporations were 10 to 15 years ago.

Public Sector Associations

Organizations are also springing up to facilitate public sector benchmarking. One example is the National Governor's Association, which facilitates sharing among states and has developed a resource list of key contacts in state governments. The Innovations in State & Local Government Awards Program rewards innovation in public service—including all policy areas for state and local government. Winners are often benchmarks for other municipalities. Other organizations include the National Association of Police Chiefs, the Total Quality Performance Consortium in Los Angeles, and the Federal Quality Institute, all of which encourage the development of total quality management (TQM) practices in government, including the sharing of effective practices.

Other state governments and agencies are planting the first seeds of benchmarking efforts. The New York State Governor's Office of Employee Relations has developed a total quality initiative involving 11 state agencies. Best practices from other states and private companies are being collected and two reports have been published: The first examines and compares total quality management initiatives in state government and the second report documents management development in the 50 states. The Governor's Office then intends to conduct a more broad-reaching best practices search. It plans to look to both the public and private sectors.[20] Oregon has created a program called "Oregon Benchmarks" to improve its strategic planning and performance measurement. The program is designed to encourage state agencies to "tie dollars spent to results obtained" by making it easier to compare performance among agencies and offices. Other states regularly use it as a model.[21] Just like Texas Instruments, an active benchmarker, the state of Texas' quality initiative specifically values benchmarking as a critical enabler of performance improvement. When Texas' licensing bureau needed to be streamlined, it copied Missouri. The United States Bureau of the Census searched for best practices concerning large-scale mailings and organizational management structures (matrix management and functional management) in the private sector.[22] The Social Security Administration looked to Met Life for effective practices.[23]

Conclusion

All these and other public sector organizations have begun to tap the potential of benchmarking as a power tool to encourage innovation and continuous improvement in the government and nonprofit sector. The twin concepts of benchmarking and best practices will likely grow in importance as more of the public sector agencies and offices follow the lead of the Air Force, Seattle, British Columbia, the IRS, Dallas, New York, Texas, and the many others who are actively benchmarking or developing benchmarking programs. Perhaps Robert Herhold of McDonnell Douglas summed it up best: "Benchmarking is neither a science nor an art, it is a statement of your culture." The culture of many public sector organizations is changing, shifting from a comparatively closed insular view to one that embraces the value of others' experiences and innovative adaptation.

References

1. Best Practices Benchmarking & Consulting, Inc. "Public Sector Benchmarking: A Research Study." Interview with Carolyn Burstein, Federal Quality Institute, April 1993.

2. Barrett, Katherine, and Greene, Richard, "Focus on the Best." *Financial World*, March 2, 1993, pp. 36-37.

3. Best Practices Benchmarking & Consulting, Inc. "Public Sector Benchmarking: A Research Study." Interview with Patricia McInturff, Seattle-King County Department of Public Health, April 1993.

4. Barrett, Katherine, and Greene, Richard, "Focus on the Best." *Financial World*, March 2, 1993, pp. 36-37.

5. Ibid.

6. Best Practices Benchmarking & Consulting, Inc. "Public Sector Benchmarking: A Research Study." Interview with Joel Parfitt, IRS Quality Office, April 1993.

7. Barrett and Greene.

8. Best Practices Benchmarking & Consulting, Inc. "Public Sector Benchmarking: A Research Study." Interview with Ryan Evans, Director for Budget and Research, Dallas, Tex., April 1993.

9. Best Practices Benchmarking & Consulting, Inc. "Public Sector Benchmarking: A Research Study." Interview with Meryl Libbey, Innovations in State & Local Government Awards Program, April 1993.

10. Ibid.

11. Best Practices Benchmarking & Consulting, Inc. "Public Sector Benchmarking: A Research Study." Interview with W.C. Enmon, Special Advisor to the Governor on TQM, April 1993.

12. Best Practices Benchmarking & Consulting, Inc. "Public Sector Benchmarking: A Research Study." Interview with Jay Chatzkel, the National Academy of Public Administrators, April 1993.

13. Best Practices Benchmarking & Consulting, Inc. "Public Sector Benchmarking: A Research Study." Interview with W.C. Enmon, Special Advisor to the Governor of Texas for TQM, April 1993.

14. Best Practices Benchmarking & Consulting, Inc. "Public Sector Benchmarking: A Research Study." Interview with Marty Russell, Total Quality Performance Consortium, April 1993.

15. Best Practices Benchmarking & Consulting, Inc. "Public Sector Benchmarking: A Research Study." Interview with Jeff Bradt, Minnesota Pollution Control Agency, April 1993.

16. Boroughs, Don. "Bureaucracy Busters," *U.S. News & World Report*, November 30, 1992, p. 52.

17. Best Practices Benchmarking & Consulting, Inc. "Public Sector Benchmarking: A Research Study." Interview with Ray Mau, British Columbia Provincial Government, April 1993.

18. *Boston Globe.* "Pittsburgh, Once Steel King, Reigns As Livable City," August 25, 1993, pp. 1, 18.

19. Barrett and Greene, p. 36.

20. Best Practices Benchmarking & Consulting, Inc., "Public Sector Benchmarking: A Research Study." Interview with Catherine Gerard, New York State Governor's Office of Employee Relations, April 1993. [Also see "Total Quality Management Initiatives in State Government," (August 1992) and "State of the States: Management Development" (August 1993).]

21. Barrett and Greene, p. 58.

22. Best Practices Benchmarking & Consulting, Inc. "Public Sector Benchmarking: A Research Study." Interview with Ken Riccini, the United States Bureau of the Census, April 1993.

23. Best Practices Benchmarking & Consulting, Inc. "Public Sector Benchmarking: A Research Study." Interview with Carolyn Burstein, the Federal Quality Institute, April 1993.

14

Managing Best Practice Knowledge

With great anticipation, a Fortune 100 company performed a comprehensive benchmarking study on best practices in financial management and capital investment. The subjects were relevant to all corporate divisions and operating units. Consequently, the study findings were eagerly awaited by the senior executives who commissioned the project. These executives understood that the company could potentially save tens to hundreds of millions of dollars if it could substantially improve its decision-making process for major capital investments.

The benchmarking project proceeded well through its early phases. Team members developed useful insights that promised significant improvements in the company's financial and capital investment processes. But then, despite the benchmarking team's successful investigation and findings, the project went from sizzle to fizzle. The benchmarking research phase was successful and produced excellent information to help the company improve. However, implementation proved to be nettlesome. The benchmarking team could not find an effective method for sharing its breakthrough information.

Finally, team members settled on using traditional memos and management presentations to communicate project results. Unfortunately, both methods provided disappointing: the best practice recommendations were ignored or soon forgotten by the divisions that could most benefit from them. For this company and many others, *effectively sharing best practices* proved a herculean challenge much more difficult than researching and collecting the benchmarking information.

Once a company finds best practice information, it still must successfully share and deploy that knowledge. Best practice information that is never implemented resembles a national currency racked by hyperinflation: It may look and feel great, but it has little value! Those companies that effectively deploy best practice information reap the compounding dividends of true intellectual leverage. What then are the best ways to share best practice knowledge? The answer requires both hardware and software solutions.

Companies that successfully leverage best practice information manage both the *hardware* or physical systems that enable information exchange and the *software* or cultural elements that affect employees' willingness to share and learn from others. Hardware systems range broadly but may include databases, electronic mail, videos, newsletters, articles, newspapers, memos, bulletin boards, competitor reporting forms, demonstration rooms, and meetings. The success of any of these options depends largely on the quality of the enabling software systems.

The software solution resides in the managerial systems which develop a culture that supports and encourages best practice information sharing. A public library in which no one comes to read or borrow books is a low-returning community asset. A best practice knowledge-management system is much the same. No matter how expensive or impressive the technology for disseminating information, the system is of little value to the organization if employees do not regularly use it.

Developing Effective Software Solutions

Companies develop the soft systems that support a best practices culture by focusing on four primary activity areas: education, communications, reward and recognition, and employee development. A good starting place in any organization is with education and training. Employee education can be a powerful tool to directly and indirectly encourage best practice sharing as an effective operating strategy. By bringing outside speakers, ideas, concepts, tools, and practices into the organization, education can indirectly nurture a culture that encourages best practice sharing. Moreover, some education may be directly targeted at benchmarking, best practices exchange, and the use of the technical systems such as databases that enable employees to share benchmarking information with each other. This is the case at companies such as General Electric, Hewlett-Packard, Motorola, and Xerox. In addition, every manager might ask himself or herself these basic questions:

- In what ways can your unit invest in employee education that brings outside ideas and speakers before your employees?

- In what ways can you ensure that all employees are trained and enabled to use existing systems that foster idea exchange and best practice sharing?

The benefits of benchmarking can also be communicated in other ways. To encourage best practice information gathering and innovative adaptation, some organizations make idea exchange and competitor assessment a regular part of meetings. They lay out work areas to encourage frequent employee interactions and foster information transfer; they design libraries and bulletin boards to spotlight other industries' innovations and effective operating procedures. Such is the case at companies as diverse in size as multibillion-dollar giant Wal-Mart and $100 million Manco, Inc. Among these and other organizations, knowledge of innovative operating practices is more precious than gold. Consequently, in business meetings managers and line employees routinely study and discuss the successful practices of other business units, other companies, and other industries. It's open hunting season on good ideas 365 days a year. At one meeting, manufacturing employees may discuss the application in their business of a time-based service guarantee like that employed by Dominos Pizza; at another session, the topic may be on adapting 3M's approach to managing and measuring innovation. Even when another company's ideas are not actually adopted or adapted, the routine examination of other industries' innovative practices stretches managers to continuously challenge and reconceive their organization's own business practices.

Reward and recognition systems are instrumental in further encouraging innovative adaptation. Many managers reinforce acts of creative adaptation with a pat on the back, a word of praise, and other forms of informal recognition. A growing number of companies is experimenting with more formal reward programs that recognize creative borrowing through gain-sharing and other awards. Finally, at the vanguard of innovation are companies who are embedding benchmarking and creative adaptation skills in performance reviews, where they become important factors influencing promotion, incentive pay, job definition, and employee development plans.

Best practice knowledge systems must also be well managed to be successful. Usually this means a designated person or group must shoulder the responsibility for maintaining the system. Hewlett-Packard's Bill Boller proposes use of a "reference librarian" to manage best practice database systems. He believes that people are more likely to use such systems if an intermediary does the research for them. His experience is that many people need to be "spoon-fed" because busy work schedules discourage them from actively exploring and using such a database on their own. A reference or research librarian saves users time by entering relevant information and articles into the database and by researching and retrieving subject requests.[1]

Some companies may choose not to use a designated reference librarian; nevertheless, some individual or group must be responsible for ensuring that the system runs properly, that information is accessible, and that employees are encouraged and enabled (usually through training) to effectively use the system. Moreover, an organized approach to managing best practices knowledge facilitates learning, eliminates redundant work efforts and repetitive research, formalizes communication channels, and creates a corporate-wide network of information contacts and subject experts. Figure 14.1 summarizes the most important lessons learned by organizations that foster effective best practice information sharing.

Developing Effective Hardware Solutions

After developing a climate or culture that supports best practice information sharing, a company can turn its attention to hardware issues and begin to develop a knowledge-management system appropriate for the culture. Unfortunately, cookie-cutter design approaches won't work. There is no single best approach to knowledge management and best practice. Field experience suggests each organization must tailor a system that fits its employees' work habits and cultural preferences. In a study of the best practice knowledge management systems of 11

Education:
1. Focus on importance of best practices.
2. Encourage bringing in outside ideas.
3. Bring in outside speakers.
4. Post information.

Meetings:
1. Regularly share best practices info.
2. Regularly relate success stories.
3. Layout work area to encourage info exchange.

Reward & Recognition:
1. Evaluate best practice info sharing & innovative imitation in performance review.
2. Reward great borrowed ideas.
3. Link best practice info sharing to compensation.

Responsibility:
1. Include best practices info collecting/sharing as part of job descriptions.
2. Designate best practice champions.
3. Use all systems to broadcast & communicate.

Figure 14.1. Developing the culture.

leading multinational corporations, the database emerged as the most popular method of communication; e-mail ran a close second and produced some fervent advocates.[2] Both systems provide means for communicating best practices information; however, passionate debate brews among users over which is the superior system. Users of each advocate the strengths of their system and the problems of the other. "You can't convince me that I need a database" contends Warren Jeffries of Xerox Corporation. He denigrates the benefits of databases, observing that they create persistent maintenance problems in order to keep information current. Jeffries favors the grassroots flexibility of e-mail. Why? Because in Xerox many more people are accustomed to using e-mail to informally request or seek information than they are accustomed to probing corporate intelligence or benchmarking databases.[3] However, Jim Madigan of Eastman Kodak counters this view with equal passion. He enthusiastically endorses the benefits of benchmarking databases and describes the capabilities of Kodak's worldwide IBM mainframe database system, which Madigan asserts is both "user friendly" and "well-used."[4]

Best Practice Databases

The best practice database presents a fast, centralized method for storing and retrieving information companywide. Consequently, databases are especially popular as a vehicle for managing best practice information. However, more than a few companies have encountered system management snags which have prevented them from effectively leveraging accumulated benchmarking knowledge. Some of the more frequent problems cited in fully leveraging best practice databases include:

- Managers and users fail to update the database frequently enough. This causes information to age and discourages frequent use.

- Inflexible database structures discourage frequent use by employees.
- The time and expense of data entry and maintenance become prohibitive.[5]
- In the absence of a data screening process, superfluous and redundant information lard the database.

IBM Rochester

Currently no single database format has emerged as a national standard. For instance, some databases have restricted access, such as IBM Rochester's "Big Blue" system. IBM Rochester limits access to a relatively few selected people because the database houses confidential information including competitor intelligence and benchmarking reports. Based on a mainframe computer, the "Big Blue" database accommodates benchmarking information in various formats, such as text, graphics, tables, and matrices. Users can search the database by keyword and by a dictionary or index of terms. In a large corporation like IBM, where many teams, work groups, and business units are actively conducting benchmarking efforts, a central database can be especially valuable, for it helps the organization leverage existing work and avoid redundant research. At IBM, benchmarking coordinators for each organization have the access code for the "Big Blue" system, while a team of top experts decide what information goes into the database.[6]

Updating and maintaining such a large system can prove challenging. Indeed, lost information is a generic problem plaguing many best practice databases. Sometimes benchmark teams neglect to update the corporate database. Other times, benchmark participants delay for a month or more before recording site visit information in a corporate database. The quality of information rapidly deteriorates with such long delays.

Large corporate best practice databases may need to be updated on a weekly basis. Consider a large company organized around six process teams, each of which conducts three benchmarking studies per year; if each study produces field visits to three companies, the organization will have 54 benchmark report entries to make each year. This number can grow much larger if employees enter the results of informal benchmarking studies, conference proceedings, key article summaries, competitive assessments, interview notes, and other types of information that can benefit the organization.

Eastman Kodak

At Eastman Kodak, benchmarking efforts are supported through an unrestricted database that's operated through an IBM mainframe computer and that's available to employees throughout Kodak's worldwide operations. The database is menu-driven with six categories that offer Kodak's best practices, articles, benchmarking opportunities, research synopses, abstracts, results, and references to people who house important benchmarking reports. Users can perform keyword searches to retrieve titles and dates of benchmarking studies, or they can enter information and add studies by filling out a report. About 25 times each day, Kodak benchmarkers access the system. Kodak has built such high usage by training employees how to use the database, making frequent presentations to promote the database and by encouraging its use in internal newsletters. Moreover, Kodak's benchmarking database was built on an existing system; consequently, employees were already familiar with how to use the database and such familiarity has spurred frequent use of the benchmarking functions.[7]

AT&T

AT&T employs a database system that can be described as a "sophisticated electronic rolodex." Known as the AAA system, this database contains one-page personnel profiles that can be used to direct benchmarkers to people and information sources that may help them in their benchmarking efforts. These profiles include information about each person's knowledge of companies, products, regions, and languages. Each individual supplies information about himself or herself. All levels of the company can use and be a part of the database. Usually information remains in the organizations rather than the database, as the database directs users to the appropriate expert. Special features of the system include searches that can be saved and repeated weekly which provide the user with updated information. AT&T employees use the system about 200 times a month and receive information through e-mail. Three people work on the system full-time: a database administrator, a technical advisor, and a customer service manager.[8]

Motorola

Centralized databases are not right for every company. Motorola, for instance, abandoned a centralized competitor intelligence database that was designed "to monitor not just the competition, but also the entire business, political, and economic environment for the company's worldwide interests." Unfortunately, no one was charged with keeping information current; consequently Motorola employees came to believe information was not updated and few people used the system. Despite the large expense to initiate the system, Motorola eventually abandoned it.[9]

Chastened by this experience, Motorola decided to use a much smaller and less ambitious database for managing benchmarking information. Its current system tracks benchmarking reports by title and description, providing the report number and geographic location where hard-copy files can be found. Motorola then supports this limited database by regularly informing upper managers through memos of important benchmarking and competitor intelligence information and by distributing key articles and reports to executives.[10]

Electronic Mail

Electronic messages transmitted to specific mailboxes across computer networks, usually called just *e-mail*, presents another option for communicating best practice information. E-mail provides a method to rapidly transfer information directly to a target group of employees who have mailboxes on the e-mail system. Unlike best practice databases, which actually house subject information under different topical headings, e-mail is a general-purpose broadcast system. Information requests go out to all users on the e-mail system or to some subset of users. If the benchmarker does not know where an internal expert resides, he or she blindly transmits an inquiry and hopes that the appropriate person actually sees the inquiry.

Xerox

Xerox supports both e-mail and database systems; however, most people choose to use e-mail rather than the database, notes Xerox Customer Services Benchmarking Manager Warren Jeffries. Jeffries contends that people want an easy-to-use system and that the database, with a 50-page user manual, is too complex. Like other organization's databases, Xerox's benchmarking database contains some superfluous information and is encumbered by so-called "ancient history" or information that has not been frequently updated. "The well is three quarters empty," observes Jeffries, who admits that he only uses the database once a year.[11]

Xerox therefore communicates information to employees directly at their workstations which are connected through a worldwide Xerox e-mail network. People from all levels of the company use the system to broadcast information needs; in turn, they receive information by hard-copy transmittals or e-mail. Mail inquiries must therefore be addressed to a specific person or organization since no e-mail equivalent exists to database interest-groups which cluster information in a central place, like an electronic bulletin board or library. To facilitate e-mail dispatches to the appropriate employees, special-interest distribution lists are often used. In this way, one inquiry or distribution can be simultaneously directed to many people. However, on specialized benchmarking projects, a Xerox employee must still work hard to identify the appropriate contacts who might provide insight or assistance based on their past work. If a benchmarker doesn't know who to direct an e-mail inquiry to, responses can be poor—even if the organization has previously performed benchmarking studies in the interest area. "Networking is the lifeblood of benchmarking, both internally and externally," observes Jeffries. If a Xerox employee leaves the company, best practice information or expertise he or she possesses also departs because there is no central archival system such as a best practice database. Jeffries contends this is not a major problem in a company such as Xerox, where benchmarking is viewed as a basic business skill that should be conducted on an ongoing basis. Jeffries argues that information or knowledge lost through employee turnover is of little value because the information ages quickly. The broad-reaching networking features of e-mail, he contends, far outweigh the archival benefits of a centralized database.[12]

ALCOA

ALCOA also uses an e-mail system for all forms of communication, including competitor intelligence and benchmarking information. All ALCOA employees can access the system and once a week the company transmits a compilation of competitor intelligence and benchmarking summaries to a broad-reaching distribution list. ALCOA business units develop this information, which can be transmitted in many formats, including text files, spread sheets, charts, and graphs. ALCOA managers scan externally created documents into its computer system and each week transmit these documents along with internally generated benchmarking summaries. A card catalogue system to identify employees with specialized knowledge and expertise has also been developed but not yet fully deployed. Like other organizations using e-mail for benchmarking communications, ALCOA benchmarkers locate internal experts and information sources by trial and error.[13]

Choosing Between Database and e-Mail Systems

The experience of "power benchmarkers" such as Xerox, Motorola, AT&T, and ALCOA demonstrate the importance of tailoring a benchmarking knowledge management system to the individual organization's culture. Clearly, one size will not fit all corporations. Tables 14.1 and 14.2 summarize the advantages and disadvantages of each system for best practice knowledge management.

Other Knowledge-Management Vehicles

For enterprises that lack the technology, budget, or need for organizationwide databases or e-mail systems, a potpourri of other communication options exist for leveraging best practice and competitor intelligence information. Moreover, all these communication vehicles can also be used to supplement existing database or e-mail systems.

Table 14.1. Managing Best Practice Knowledge Through Database Systems

Databases	
Advantages	Disadvantages
• Fast retrieval of information	• Must be updated frequently to be effective
• Centralized location for storing information	• May include superfluous information if not screened
• Companywide access	• Time/expense of data entry
• Easy to find information/sources due to search functions	• Expense of system, software, upkeep

Table 14.2. Managing Best Practice Knowledge Through e-Mail Systems

e-Mail	
Advantages	Disadvantages
• Fast method of contacting people	• Must have access to right user group to find desired contact
• Simple and well-used system for all communication purposes	• Does not have centralized location for storing information
• Companywide access	
• Wide distribution of information	• All employees do not actively use the system

Videotapes.   Videos present an innovative approach to studying work processes in action. Unfortunately, most best practice videotapes capture internal processes since many corporations restrict visitors from filming their core work processes. Hewlett-Packard, for instance, has developed a videotape library of internal best practices,[14] while DuPont has used the medium for distributing benchmarking study presentations. Due to the high cost of producing professional-quality videos and due to the limited frequency with which they can be easily updated, videotapes are best regarded as a supplemental best practice communication form.

Internal Publications.   Digital Equipment Corp., McDonnell Douglas, and Hewlett-Packard all employ newsletters, articles, and internal newspapers to communicate best practice information. An inexpensive and ongoing method for reaching large audiences, best practice publications keep readers informed about innovation, continuous improvements, and breakthrough findings. However, lead times and circulation cycles can be long and the list of topics can be eclectic.[15]

Memos and Miscellany.   Memos, bulletin boards, competitor reporting forms, demonstration rooms, reverse engineering labs, and benchmarking meetings represent other approaches for sharing best practice and competitor information within an organization. Many may already exist in some form in your own organization.

The best system is one that fits the size and needs of the individual company. A small company with limited budget might not need a formal e-mail or database system. A best practice newsletter or benchmarking meetings can work well when effectively administered. At larger, multisite organizations, an ideal system would likely combine the network advantages of a computer database and an e-mail system. The first generation system would be a database that archives information and provides users with best practice abstracts and contact names. Users would also be able to make contact with internal experts and knowledge sources by e-mail, phone, or fax. A more fully evolved, versatile system would integrate database and e-mail systems, as do new groupware applications such as Lotus Notes. These advanced systems would also permit benchmarkers to transfer information and best practice knowledge directly among themselves. To assure its success, the system management would actively promote the benefits of the best practice database and keep information up-to-date, while preserving important historical data to enable ongoing trend analysis. A research librarian would be an added benefit, but the system would be simple enough for people to use on their own. Lastly, the organization's leadership would recognize the organizational benefits of innovative adaptation and would therefore actively support a culture that leverages best practice knowledge.

Guidelines for Knowledge-Management

The following guidelines emerge from the experience of organizations that actively manage their best practice knowledge:

1. Design a knowledge-management system based on user needs, requirements, and usage habits.
2. Choose a system that is consistent with your organization's culture and user patterns.
3. Set a realistic budget to develop and maintain the system. Then adhere to the budget.
4. Design a system that adds value to information.
5. Work with existing systems. Do not duplicate existing resources.
6. Organize information simply. Complexity discourages usage.
7. Screen information that the company archives.
8. Frequently update information in the system.
9. Encourage and train people to use the system.
10. Designate an individual to manage and maintain the system.
11. Seek top-management support for the system and ensure a high-level champion actively promotes and encourages use of the system.[16]

References

1. Boller, Bill. December 1992 interview.
2. Best Practices Benchmarking and Consulting, Inc. "Best-Practice Knowledge Management Systems: A Study of the Knowledge Management Practices of 11 Leading American Corporations," 1992.
3. Jeffries, Warren. December 1992 interviews.
4. Madigan, Jim. December 1992 interview.
5. Fuld, Leonard M. *Monitoring the Competition.*

6. Eyrich, Hank. December 1992 interview.

7. Madigan, Jim. December 1992 interview.

8. Stark, Marty. December 1992 interview.

9. Fuld, Leonard M. *Monitoring the Competition.*

10. Chitwood, Lera. December 1992 and January 1993 interviews.

11. Jeffries, Warren. December 1992 interviews.

12. Ibid.

13. Jackson, Jeff. December 1992 and January 1993 interviews.

14. Boller, Bill. December 1992 interview.

15. Boller, Bill, (Hewlett-Packard), Anderson, Peg, (Digital), and Anderson, Barbara, (McDonnell Douglas), December interviews.

16. Fuld, Leonard M. *Monitoring the Competition.*

15

Benchmarking and the Twenty-First-Century Organization

Top Ten Excuses For Not Benchmarking

10. *Our entire travel budget consists of one pair of running shoes to be shared by the entire department.*

9. *We could probably find the money to travel, but we can't find any plausible benchmarking partners in Honolulu.*

8. *We are too busy working hard to learn how to work smart.*

7. *We already have one of the best processes in the world or maybe that was in the country ... or was it in the state ... would you believe the best process on the third floor?*

6. *I don't know how to start. Besides at least 5 books on benchmarking, multiple benchmarking conferences, organizations such as IBC, SPI, and CCI, there's not much material available.*

5. *We're waiting for benchmarking slogans to appear on posters and coffee cups.*

4. *We are too busy with ISO 90210 which is the new quality standard that is required if you intend to advertise your products on TV.*

3. *Benchmarking is just a fad. (Note: this excuse worked well in 1990, 1991, and 1992)*

2. *Our process is not fully documented ... This also has a long history of successful use.*

1. *Benchmarking was "NOT INVENTED HERE."*

ERIC KENNEDY and DAN ROURKE
IBM Corporate Benchmarking Steering Committee Members

The Organization of the Future

Imagine the ideal organization of the twenty-first century. It is a Wall Street darling with a history of superlative financial performance. Its operating capabilities and characteristics are also sterling. Envision it as a Malcolm Baldrige National Quality Award winner and perhaps a Deming Prize winner, too. The twenty-first-century organization worships at the altar of customer satisfaction, ultimately defining its own success through the eyes of its customers. This ideal organization takes great care in hiring, developing, training, and retaining its employees; it involves them intimately in the affairs of the organization and increasingly empowers them to act on the customers' and the company's behalf. The twenty-first-century organization uses technology as a tool of liberation, not as a way to rationalize jobs. It is lean and habitually focused on productivity, cost management, and real growth.

The high-performing organization of the twenty-first century instills or embeds the relentless impulse for continuous improvement into all its processes and systems. Consequently, it is also a fast company, making speed, quick response, and time to market the bases for competitive advantage. Fast learning is therefore a defining characteristic and prerequisite for the twenty-first-century enterprise that operates at high velocities without crashing. Moreover, best practice benchmarking is the catalyst that enables the organization to learn and to improve quickly. Like a powerful genetic code driving continuous improvement, best practice benchmarking is deeply ingrained in every process. As a fundamental business skill, it enables the organization to continuously test its capabilities to uncover improvement opportunities, to spur adoption of the best practices and to press relentlessly toward ever greater performance. Unlike organizations that unwittingly foster insular, not-invented-here attitudes, the twenty-first-century organization embraces a "we-can-learn-from-everyone" culture.

A Call to Action

Organizations that reengineer for success in the twenty-first century will therefore embrace benchmarking for best practices. In doing so, they will heed the following operating imperatives:

- Employ benchmarks and best practices as management tools to support strategic thinking and planning.

- Integrate benchmarking into the management of every-day processes.

- Codify your benchmarking language and approach to leverage learning.

- Deploy benchmarking as a fundamental business skill with versatile applications.

- Manage benchmarking exchanges and best practice knowledge.

- Use networking and technology to expedite and optimize benchmarking.

- Train front-line employees, supervisors, and managers to be receptive learners and skillful benchmarkers who actively practice innovative adaptation.

- Formally evaluate and recognize benchmarking as an essential skill.

- Create a corporate culture supportive of innovative adaptation.

1. *Employ benchmarks and best practices as performance indicators and management tools to support strategic thinking and planning.* Benchmarking empowers the strategic thinking and planning process. As critical comparative operating statistics, *benchmarks* represent objective reference points that provide the basis for setting and validating the adequacy of performance goals and targets. *Benchmarking*, in turn, yields best practice information that triggers insights, rapid knowledge acquisition, and operating breakthroughs that lead to superior performance. By integrating the benchmarks or comparative measures with best practice knowledge acquired through the benchmarking outreach process, organizations identify their most important improvement opportunities, set the agenda and priorities for change, and map the course by which to attain superior performance.

Benchmarking adds value to the strategic thinking and planning processes in many ways. Among the most important considerations embraced by the twenty-first-century organization, benchmarking:

- Enables organizations to project competitors' future performance.

- Validates the adequacy of short-term and long-term organizational goals.

- Identifies the critical success factors that can be leveraged to achieve market leadership.

- Stimulates long-term planning to close gaps and make business processes competitive.

- Identifies best practices for using technology.

The first call for action is an exhortation to make benchmarks and benchmarking essential parts of the strategic thinking and planning processes.

2. *Integrate benchmarking into the management of every-day processes.* The twenty-first-century organization will manage a portfolio of short-term and long-term change initiatives and it will integrate benchmarking into all four of the fundamental performance improvement approaches: continuous process improvement, organization restructuring, process reengineering, and managed reform. Every major process—from new product development and sales to billing and after sales service—will be viewed as a candidate for continuous improvement and benchmarking. Process owners and process teams are key to successfully integrating benchmarking and continuous improvement into work processes that sweep like a fast-running river through different departments and across traditional lines marking functional responsibilities. "Processes also need clearly defined owners to be responsible for design and execution and for ensuring that customer needs are met," observes Thomas H. Davenport, an authority on process innovation and reengineering. "The difficulty in defining ownership, of course, is that processes seldom follow existing boundaries of organizational power and authority. Process ownership must be seen as an additional or alternative dimension of the formal organizational structure that, during periods of radical process change, takes precedence over other dimensions of structure. Otherwise, " Davenport says, "process owners will not have the power

or legitimacy needed to implement process designs that violate organizational charts and norms describing, 'the way we do things around here.'"[1]

From among hundreds of processes, the twenty-first-century organization concentrates its daily management efforts on those few core processes that create the greatest value and are essential to the organization's ongoing success. Some process management experts contend there are fewer than 20 major processes within any company, no matter how large or small. IBM, for instance, has identified 18 core processes; Ameritech has spotted 15, Xerox has noted 14, and Dow Chemical has observed 9. At Ameritech, individual process owners have been assigned to all 15 processes. "Each of the 15 process owners has their own benchmarking advisor," adds Orval Brown, director of process architecture and benchmarking at Ameritech. "They also have a separate quality improvement advisor, and they are not the same people." At Ameritech, Brown heads a business process management staff that provides technical support and has a "dotted line" connection to the benchmarking advisors. Just as Ameritech and other organizations have done, successful companies of the future will find it desirable to establish process ownership and to designate individual champions or advisors to help integrate benchmarking into daily performance improvement.

3. *Codify your organization's benchmarking language and approach to leverage learning.* Benchmarking language and methods vary superficially from organization to organization. The differences between one company's nine-step process and another enterprise's five-step process are inconsequential. It matters little which proces model or terminology your organization adopts. What matters greatly, though, is consistency of language and process within the organization to ensure easy communication and information sharing among teams, work units, departments, or divisions. A systematic benchmarking approach, whether in the form of a multistep process model or a simple checklist, guides the benchmarker from planning to data collection to data analysis to best practice identification, adaptation, and implementation.

4. *Deploy benchmarking as a fundamental business skill with versatile applications.* As a catalyst for learning and performance improvement, benchmarking has both informal and formal applications. Informal benchmarking includes simple idea adaptation and learning from past successes so that they can be repeated. More formal benchmarking examines complex processes and work systems. Process benchmarking may target practices that are best-in-company, best-in-market, best-in-industry, or best-in-class (without regard to country or geographic boundaries.) The twenty-first-century organization will apply benchmarking throughout its portfolio of performance-improvement efforts. Consequently, individual employees and improvement teams will employ all types of benchmarking to their improvement efforts; they will be adept at choosing the most appropriate application for each specific improvement effort. Successful benchmarkers of the future will not be wed to the notion that projects must take three, six, nine, or more months. They will blend quick tactical improvements with breakthrough projects that require many months to implement successfully.

5. *Manage benchmarking exchanges and best practice knowledge.* The twenty-first-century organization will be skilled at knowledge management. Consequently, it

will develop systems to coordinate benchmarking information exchanges and manage best practices knowledge dissemination within the organization. Organizations of the future will establish managerial positions to direct best practices development and to act as a central point of contact for inbound and outbound information exchange. The benchmarking champion will help the organization focus and concentrate benchmarking activities. He or she will guide the organization as it integrates benchmarks and best practice information into its strategic planning and continuous-improvement processes. The benchmarking champion will facilitate benchmarking integration into day-to-day business operations. Texas Instruments, whose Defense Systems & Electronics division won a 1992 Baldrige Award, has already set out on this path. In the early 1990s, Texas Instruments named Laura Longmire, a veteran manager, to the post of "Benchmarking Champion for World-wide Operations." The Texas Instruments benchmarking champion is charged with "coordinating, facilitating, training, and implementing benchmarking" throughout the corporation. A small but distinguished legion of other leading companies has recently created similar positions, and the trend seems sure to continue. When 76 large service and manufacturing organizations—including many of the nation's most admired corporations—were surveyed about their perceptions of benchmarking, 79 percent said "companies will have to benchmark to survive."[2] The benchmarking champion will facilitate full deployment of benchmarking to ensure survival in the twenty-first-century corporation.

6. *Use networking and technology to expedite and optimize benchmarking.* Successful organizations of the future will increasingly draw on associations and networks, common interest groups, consortia, databases, and other forms of technology to help them access, share, analyze, and absorb benchmark information. Already such groups have sprung up in telecommunications, yellow pages publishing, newspaper publishing, and various other industries. More general networking groups have also been organized by the Strategic Planning Institute's Council on Benchmarking and the American Productivity and Quality Center's International Benchmarking Clearinghouse. All these organizations and many other industry associations can help focus, streamline, and optimize the process by which an improvement team determines who and what to benchmark and then explores the practices that enable the best to achieve their superior performance.

The technology revolution makes information quickly and easily accessible to large and small organizations. Moreover, computer networks and user friendly database and e-mail applications make the analysis and sharing of benchmarking information easier than ever before. Proliferation of these networks, versatile groupware applications such as Lotus Notes, and easy-to-use public databases will enable the twenty-first-century organization to accelerate benchmarking activities and to facilitate best practice knowledge sharing.

7. *Train front-line employees, supervisors, and managers to be receptive learners and skillful benchmarkers who actively practice innovative adaptation.* To accelerate the pace of learning and improvement through benchmarking, the twenty-first-century organization will create benchmarking skill competency among all employees, supervisors, and managers. Comprehensive training is the key to creating bench-

marking competency throughout the organization. Training will help employees develop basic benchmarking skills that they can apply in their jobs and to understand the role of best practices in continuous performance improvement, strategic planning, reengineering, time-based competition, change management, and innovation.

8. *Develop benchmarking as a skill that is formally evaluated and recognized.* Benchmarking training will be buttressed by managerial systems that support ongoing skill competency. Three interlocking managerial systems will be essential in reinforcing benchmarking's importance in the twenty-first-century organization. First, job descriptions will reflect benchmarking as a basic responsibility of competent performance. Second, individual performance appraisals will regard benchmarking and innovative adaptation as fundamental skills to be reviewed and developed. Third, recognition and compensation systems will view benchmarking and best practice excellence as a cause for praise, promotion, commendation, and incentive pay. Together, these three managerial systems will continuously highlight benchmarking as a required skill for sustained success in the next century.

9. *Create a corporate culture supportive of innovative adaptation.* In many workplaces, openly borrowing and adapting other people's ideas runs against the grain of years of training, education, and corporate culture. In the twenty-first-century organization, the unwritten prohibition against importing other's good ideas will be permanently repealed. The new era will simultaneously celebrate invention of new ideas and creative application of existing excellent ideas.

The twenty-first-century organization will overcome current cultural barriers to innovative adaptation through a systematic approach to shaping and managing corporate culture. No one managerial system will produce a "we-can-learn-from-everyone" culture. Rather, the successful future organization will create many smaller integrated systems that work together to encode the values of creative adaptation and innovative imitation in employee behavior. These encoded values, in turn, create habitual behaviors that are passed on to succeeding generations. Some previously observed but highly-effective managerial approaches to forge a "we-can-learn-from-everyone" culture include:

- Create lending libraries that focus on competitors and other high performers' winning strategies and systems.

- Routinely parade outside speakers and ideas before employees.

- Review the products and practices of other companies and visit their facilities whenever possible.

- Publicize the benefits of borrowing from the best.

- Sponsor regular meetings and discussions where employees and managers exchange ideas and explore best practices.

- Lay out work areas to encourage impromptu meetings, idea sharing, and information exchange.

- Employ best practice information in the problem-solving and continuous improvement process.

- Engage high-level executives directly in benchmarking and innovative adaptation.

- Make competitive information-gathering everyone's responsibility, especially including functions such as Sales, Marketing, and Personnel.

- Regularly identify, study, and celebrate internal success stories and best practices with the goal of repeating them.

- Institutionalize learning by making evaluation-and-improvement cycles a required part of performance reviews; focus these reviews on learning from what went right, as well as what went wrong.

Conclusion

The call to action for the successful twenty-first-century organization is a call to make best practice benchmarking a cornerstone of the organization's approach to conducting business. Best practice benchmarking is much more than just a business tool. For the twenty-first-century organization, it represents a fundamental approach to competing and managing. The best practices strategy complements traditional continuous improvement efforts. The best practices strategy sets out to leverage the learning and experience of the best, rather than merely reforming the practices of the worst, which has been the traditional path of performance improvement. The best practices approach is behaviorally powerful and compelling. It concentrates on performance improvement through organizational learning, effective management of intellectual assets, and leverage of others' proven experience.

In the twenty-first century, the effective use of intellectual capital will increasingly become a prerequisite of marketplace success—perhaps even of survival. Successful organizations will therefore employ best practice benchmarking as a primary catalyst of fast learning; best practice benchmarking will be regarded as a power tool for leveraging internal and external experience. In the future, return on intellectual assets will be equally as important a performance indicator as return on physical assets. In this new era which is dawning, best practice benchmarking will be the long-handled lever of the knowledge-age. It will be a primary instrument helping the twenty-first-century organization achieve the full benefits of intellectual leverage: By creatively adapting the best proven ideas, practices, and approaches of others, an organization can leverage their partners' people, knowledge, resources, and experience with very little capital investment. Innovative adaptation creates a compounding effect on the internal rate of improvement; it produces a high-yield knowledge dividend. Consequently, best practice benchmarking represents the ultimate competitive weapon: a low-investment, organization-wide, renewable resource that produces rapid learning and performance improvement by leveraging others' learning and most effective practices. Unleashed in a global marketplace, this approach to management and continuous improvement promises rich returns. And that is why, for so many organizations throughout the world, the best practices revolution has just begun.

References

1. Davenport, Thomas H., *Process Innovation: Reengineering Work Through Information Technology*, Harvard Business School Press, Boston, 1933, p. 7.

2. Lambertus, Todd, "Surveying Industry's Benchmarking Practices," a report by the American Productivity and Quality Center, 1992, p. 9.

Appendices

Steal This Idea®—The Art of Innovative Adaptation: How Can Your Organization Apply These Ideas?

Customer Service and Satisfaction:

800 Telephone Number Waiting Queues. WordPerfect Corporation and Lotus Development Corporation have borrowed features from drive-time radio disc jockeys to make waiting on their Customer Service 800 telephone lines more interesting. WordPerfect and Lotus both employ live "hold jockeys" who announce the number of callers in your queue and your average waiting time. At Lotus, hold jockeys describe various Lotus products and play music. At WordPerfect and Lotus they seem to favor jazz and soft rock. *What can you do to make waiting less nettlesome and more enjoyable for your customers?*

Express Checkout. KPMG Peat Marwick borrowed the concept of a supermarket's express checkout to implement an express line in its Word Processing pools. The checkout enabled work teams with changes of 1 to 5 items to go through an expedited route, rather than wait at the end of the line to be processed. This change can be of great value for any word processing function or other functions that handle high-volume work orders. The idea can be applied to other service functions, such as copying, graphic production, research, and so on. *Where in your organization could you implement the express checkout lane concept to improve processing speed and customer satisfaction?*

Surveying Customers on the Backs of Their Rebate Checks. Microsoft has developed an ingenious way to raise response rates on its customer surveys. In its product boxes, Microsoft offers an unusual rebate. It innovates away from the traditional rebate, in which a customer must fill out a market survey and send it to the company to receive a rebate check. Microsoft includes a check, made out to cash, for a $10 to $15 rebate. Customers usually answer the brief questionnaire they find beneath the signature endorsement line of the check. After depositing the check, the banks return the canceled check/survey to Microsoft which eliminates its mailing costs. This survey reduces the hassle and wait for customers and consequently improves their response rate. *How could your company use a similar technique for its surveys?*

Survey with Product Shipments. A Midwestern tool manufacturer sends a survey with product shipments to customers. Customers do not have to wait weeks or months after a delivery to comment on the product and service. Customers can immediately complain, praise, and make suggestions about the product and service. *How can you improve customer feedback about your service and products?*

New Product Wish Line. Microsoft Corporation understands the importance of a steady flow of new products. It created a new product ideas telephone number for its users who have ideas concerning new features and applications. "Microsoft provides several avenues to give product feature requests," notes Microsoft's directory of products and services. "The following is a voice number for Microsoft Excel and Word. You can offer suggestions on other products by calling the dedicated support number at Microsoft Product Support Services. We welcome your ideas for future versions. Microsoft Wish - 206-936-WISH." Other companies that want to hear from their customers concerning what new products, services, and features they should provide have implemented a similar line. *How can your company apply a "wish line" to encourage your customers to share their new product ideas with you?*

Demonstrating Value to Your Customers. Staples does not merely advertise low prices; the Massachusetts-based office supply chain demonstrates value at point-of-sale. To reinforce the value it provides its customers, Staples prints out the actual discounts on customer receipts. This feature highlights customer savings and reinforces the company's low price and high-value image in the minds of customers. The receipts itemize all purchases and then automatically calculate savings off retail prices. "At List Prices, Your Purchase Would Cost $216.67," declares a typical Staples' receipt. "At Staples' Low Prices, You Paid $82.46 And Saved $134.21." *Where can you provide discount information to your customers, helping them appreciate the value and savings they receive by purchasing supplies and services from your company?*

Personalized Customer Contact. A Midwestern manufacturer sends pictures of their staff or team to customers to personalize the telephone communication process. Another company personalizes its customer contact by sending a picture of the cell manufacturing team or the production team in the product box. Both methods show that the companies want their customers to get to know them better. Rather than the "this product was inspected by No. 12" message, a photograph personalizes No. 12. *How can you personalize your contact with customers?*

Customer Involvement in Product Design. In the business furniture design industry, manufacturers such as Weaver's Business Interiors create design software. In addition to giving it to design engineers at design companies, they give the "design software" to their customers for little or no cost. The customers use the software to design new products. This provides better, clearer information for the manufacturers. *How can you involve your customers in designing new products?*

Company Environment/Atmosphere:

Atmospheric Music.   A $250 million retailer of electronics and furniture pipes music into the furniture show rooms to enhance the presentation of the furniture. Additionally, the music relaxes customers. *How can your company use sound to enhance its primary products and services?*

Flowers in the Rest Rooms.   The president of a Wisconsin-based independent supermarket chain observed that many of his customers during daytime hours were women with young children. During shopping trips, these moms frequently needed to park their shopping carts and change diapers in the store's rest rooms. When this executive's anchor store remodeled, management expanded the size of the rest rooms. Borrowing the idea from Stu Leonard's Dairy in Connecticut, it placed flowers in the rest rooms to make the environment more pleasant. It was a small touch that won frequent praise from patrons. *What areas in your building can you enhance with flowers?*

Adopt a Common Space.   The CEO of a Midwest hospital borrowed an idea she heard on Minnesota Public Radio and creatively adapted it to improve operations in her hospital. The public radio story reported about an innovative approach to beautifying public green space. To overcome the persistent problem of litter-strewn islands on public streets, Minnesota encouraged citizens to assume "ownership" of common green areas in their neighborhood and to keep them clean. The program had a felicitous effect. In this green-space program, the hospital CEO found an elegant and simple solution for a nagging problem. Common areas in the hospital, such as hallways and patient waiting areas, lay in no particular department's jurisdiction. Consequently, they became littered, wall-marked, and floor-scuffed without any hospital employee initiating immediate cleanup. These areas were also often the ones most visible to patients. By adapting the Minnesota green-space program to the hospital, the CEO solved a persistent problem for her facility and improved conditions for both employees and patients. *Where can you start an "adopt-a-green-space" program in your organization?*

Employee Satisfaction, Reward, and Training:

Reward through Understanding.   Solectron established an innovative approach to combining a suggestion system and a reward program. Solectron enters employees who submit ideas that the company uses into a lottery. The winner of the monthly lottery may take a day off or request any officer or employee in the company to take his or her job for a day. It allows line officers to have senior executives serve in their jobs—at times which have been production line or purchasing type jobs. The exchange improves morale, increases communication, and keeps executives in touch with operating reality. *How could your company achieve similar benefits by allowing line employees to choose executives to serve in their jobs?*

Creating a University in the Lunch Room. The company has "Brown Bag University" (or "luncheon learning") during the winter. Technical specialists make presentations to clerical people on the features of the company's products. This enables the clerical people to develop a better understanding of the company's products, services, and features. *How can you make employee training a part of your everyday activities?*

Educational Appetizers. Observing that Mexican restaurants serve appetizers such as chips and dip to keep customers happy while they wait for the main course, a training team at GTE Telephone Operations is borrowing the idea and applying the concept to training. They will prepare one-hour training "appetizers" for business units. These bite-size courses can be delivered immediately, while the business unit requesting the training schedules a more in-depth session, which can run one or more days but may be scheduled several weeks away. *How can you apply the appetizer concept to your organization to keep customers happy while they wait for delivery of your primary product or service?*

Supervisor Training Program. Companies develop a certification/development program where supervisors earn college credit for the courses. *How can you improve or develop your training programs?*

Work Family Program. IBM and AT&T implemented flextime and dependent care initiatives. These programs promote an environment responsive to evolving demographics. *Would similar benefits be valuable for your employees?*

Training Module. MONY developed a training module methodology and empowered the staff to identify and make improvements on it. MONY applies these established guidelines for its training approach and deployment. *What guidelines can you set for your training?*

New Client Sales Commission. MONY increases the commission for sales to new clients. This increases the incentive for the field underwriter to target new prospects. *How can you increase new client sales?*

Personalized Notes. MONY sends birthday cards, anniversary cards, and promotion congratulation cards to staff members. This gesture demonstrates that all employees are an important part of the company. *How can you personalize relations with your employees?*

Compensation Incentives. MONY gives new field underwriters a conditional $100,000 check upon hire. When underwriters achieve the conditions after a set number of years, they receive the money. The money gives new hires a tangible goal to focus on and increases retention of field underwriters. *What incentive programs can you install to motivate and retain your employees?*

Flexible Benefits Package. Companies allow employees to buy and sell their benefits based on their need. At American Finance, employees can select their benefits on an 800 number. *How can you improve your benefits package?*

Letter Display. The company publicly displays letters of praise. This recognition awards good work and motivates people to work harder. *How can you honor a job well done?*

Casual Dress Day. The company assigns one day of the week to be "casual dress day." This relaxed atmosphere improves employee morale. *How can you improve employee morale by creating a more relaxed environment?*

Internal Communication:

Communication Process. Firstar Bank of Madison placed a laser printer in its mail room. When a memo goes out from a department, it prints out directly in the mail room. Global commands make it automatically print out the appropriate mail list and employees' mailbox numbers. This cuts the time and cost of the memo distribution process and makes it easier for the mail room. *Where could your company improve its memo distribution process?*

Memos. The CEO of a professional service firm remarked that repeatedly putting the distribution list on the top of memos takes up time, space, and some cost. After a few mailings, committee members are seldom interested in seeing the same mailing roster. *How can you improve speed and time by cutting out unnecessary distribution lists?*

Voice Mail and e-Mail. Various companies avoid wasting paper and time by using e-mail and voice mail for memo and communication functions. On e-mail, if a person wants a hard copy, he or she can simply print out the screen. *How can you more effectively use your voice mail or e-mail system?*

Waiters Wearing Pagers. Sisters Hotel Restaurant in Sisters, Oregon, has its waiters and kitchen staff work closely together by using a pager system. The waiters wear pagers on their belts, which work throughout the restaurant. When the kitchen has completed a meal for the waiter, the kitchen pages him. The waiters feel the pager vibrate which indicates that a meal is ready. Sisters borrowed the idea from medical professionals/doctors who wear pagers when they are on call. You can use this idea for any Customer Service reps that need to be in constant contact with their suppliers, customers, or team members. *Can you identify areas to apply this idea in your organization?*

Mobile Headsets for Customer Service Representatives. Weaver's Business Interiors, a Wisconsin-based furniture design company, observed McDonald's drive-through window employees using cordless microphones and headsets. These mobile communication units allowed restaurant employees to remain in constant contact with customers, even when they stepped away from the drive-through window to fill an order. Weaver's imported the idea from the fast food industry into its customer service department. Sales people and customer service representatives no longer miss calls because they stepped away from their desks. *How might mobile communication technology keep your company in better touch with your customers?*

Fax Cover Sheets. At KPMG Peat Marwick, a secretary noted that clients sent faxes with a Post-It note cover which saved paper and telephone transmission time. She proposed that KPMG follow the example of their clients. KPMG followed her advice and saved thousands of dollars. *What small changes can you make that will save you time and money? How can you more frequently implement employee suggestions?*

Process Simplification

Billing Notices. An executive in a manufacturing company remarked that customer checks/payments should come back with the necessary account and billing information. The design of the envelope should ensure that the customer encloses critical billing information, that is, a window envelope that requires the customer to reinsert the receipt voucher inside. *How can you assure that customers include necessary account and billing information?*

Phone-In Orders. An industrial distribution company uses car phone-in orders for easy pickup. Contractors call from their trucks and therefore do not have to come in to the order desk. The company borrowed the idea from takeout food and fast food drive-through windows. *How can you use phone-in orders to provide faster service?*

Situation Boards. After finding a problem with a product or service, a team member notes the time and problem on a dry-erase board. It alerts all team members, management, and quality staff of the system problems. In addition, it highlights problems and facilitates assistance for short-term and long-term problems. The board has a yellow area for hindrance problems and a red spot for stoppage problems. *How could your company use a situation board to note and solve problems?*

Story Boards. Walt Disney uses story boards to help outline complex projects that involve many people working at different times. Other companies have borrowed this quality tool from Disney. One owner described how he reduced the time it took to produce a mail order catalogue for his company from 12 months to four months. *How can you save time and money by outlining and simplifying projects on a story board?*

Form Printing. MONY prints forms only on demand. To save storage space, it does not keep them in stock in the agency. *What can you do to provide more storage space and reduce clutter in your company?*

Cost-Benefit Analysis.   MONY does a formal cost-benefit analysis for all changes to procedures and/or systems. The analysis increases awareness of how the decision affects the bottom line and promotes a better use of resources. *Where can you use a cost-benefit analysis?*

Macros for Professional Letters.   A law firm created macros for professional letters to speed the letter-writing process. *Where can you use macros? How can you speed your letter-writing process?*

Warehouse Layout.   L.L. Bean places the most frequently ordered products at the front of its warehouses. The layout saves time when packaging an order. *Where can you improve the layout of your supplies or products?*

Magnetic I.D. Cards.   Sony makes employee I.D. cards with a magnetic time keeping system. The system is more efficient than the old manual system. *What other uses can you find for magnetic I.D. cards?*

Miscellaneous:

Networking on the Golf Course.   Frank Waldron, VP of Mutual of New York, tells the story of how one of his managers effectively opened a new office in New Orleans—thought to be a very closed area. The manager joined the local country club and then took the golf starter who coordinates tee starts out to dinner. He confided to the tee starter that he was new in town and did not know many people. Since he thought golf would be a good way to meet people, he asked the starter to pair him with influential people who needed a player in a foursome. The starter soon called him to say that the president of Company A needed a fourth for a foursome. The president and his contacts helped the manager open the new office. *How could you adapt this strategy—of seeking assistance of line people—to help you to penetrate a market?*

Power Recruitment Questions.   Frank Waldron, at MONY IFS, developed two power questions to help him in recruiting top salespeople. First, he asks candidates, "Are you lucky?" He wants to determine if candidates have confidence in themselves and in their good luck. Second, he orders them to tell 10 people that they want to become a life insurance salesperson. At the second interview, he asks them whom they told. Again, he wants to judge their confidence and enthusiasm. *How can you improve your interview questions to find the best candidates?*

Paycheck Stuffers.   MONY IFS includes mission statement stuffers with its paychecks. Employees are likely to read the mission statement stuffer when they find it with their paycheck. *How can you effectively emphasize your mission statement?*

Wallet-Sized Card with Vision Statement.   Companies effectively communicate the vision statement on the card. Employees can quickly reference the card, which they can keep in their wallet. *How can you use a wallet-sized card to communicate important information?*

Internal Work Guarantees.   GTE, Marriott, Westinghouse, and Dun & Bradstreet use internal service guarantees to align internal departments more closely. Westing-house used the guarantee to align its R&D and field units. The guarantee forced R&D to determine the field units' budgets and to find when they needed the research. As the work guarantee stated, if R&D did not meet the schedule or budget it would pay for the additional costs. *How can you use internal guarantees to help your departments work closely together?*

The Boss on the Telephone Line.   At some companies, officers and managers answer toll-free lines once a month. This increases their knowledge of customers' needs and breaks down some barriers of a hierarchical organization. *How can you use this method to increase upper-level management's knowledge of customers' needs?*

Line Managers' Visits.   Companies where line managers visit agencies promote a better understanding of the environment and development needs. This helps improve communications, "big picture" knowledge, and customer/supplier knowledge. *What area can you focus on to improve knowledge and communications within your company?*

Barcode Mass Mailings.   Companies use barcodes to sort and route mail. The process reduces handling and increases productivity. *How can you improve your mailing process?*

Zip Code Plus Four Digits.   Companies cut postage costs 15 to 20 percent by using the extended zip code when sending out major mailings. The postal service gives a discount on mass mailings with the four extra digits because it facilitates the sorting and delivering process. *Can you use zip code plus four digits on your mass mailings? How else can you reduce your mailing costs?*

Recruiting Field Underwriters.   Mary Kay Cosmetics recruits field underwriters through its selling process. This focuses recruiting and improves employee reten-tion. *Can you use this form of recruiting in your company?*

Conversion Illustrations.   MONY provides conversion illustrations on term pol-icy anniversaries for review with its policyholder. This results in a yearly review of the client's needs and increased conversions to permanent insurance. *Where can you apply a similar review policy in your company?*

Peer Review Sessions.   MONY provides peer review sessions demonstrating critical processes, product specifications, and program specifications. The sessions promote shared knowledge, integrated processes, and error prevention. *What areas can you improve through use of peer review sessions?*

Client Referrals.   The company partially compensates its new field underwriters based on their client referrals. This frequently results in increased sales. *How can you increase your client referrals?*

Job Rotation.   Employees rotate between departments and work areas. The rotation increases their knowledge and gives them new perspectives on existing processes. *In what ways can you improve employees' knowledge of all existing products?*

Data Collection.   MONY uses application information for its market research data. This eliminates the need for any other method of data collection. *How can you efficiently use your resources and eliminate redundant information?*

"Ten Most Wanted" List.   Paul Kahn, CEO of AT&T Universal Card Services, noted that UCS borrowed the idea of America's 10 Most Wanted Criminals and applied it to quality improvement. UCS created the "10 Most Wanted Quality Improvements." This colorful and dramatic technique focuses the company on the most important quality improvements. When a quality improvement (QI) has been "arrested," UCS brings another QI onto the list. It helps the company to continually "identify, focus, resolve, and then refocus." *Can you apply this "most wanted" approach in quality or other areas of your company?*

Fast Response on Customer Phone Inquiries.   Federal Express, known for its exceptional service, always answers customer phone calls in less than three rings. This service standard ensures that customers feel well attended. Borrowing from this approach, a Midwest manufacturer has set its own customer response on the FedEx standard. It instructs both its customer representatives and its automatic call answering system to respond in three rings or less. The standard communicates a sense of urgency in responding to customers. *Where can you set customer response standards to communicate a sense of urgency in serving your customers?*

Saying Thank You to Customers with Flowers.   Robert Groenevelt, CEO of Real Veal, Inc. which provides food formulas and scientific testing services to cattle ranchers, recalls how his company borrowed a customer service technique from car dealers who send flowers to customers after a purchase. Real Veal adapted the technique and now sends flowers to customers after their first order. Groenevelt observes that this gesture has resulted in a positive reaction from his customers. *What opportunities are there for your organization to impress new customers with similar gestures to affirm their importance to you?*

Thanking Customers in Advance for Returning a Survey.   A Midwest company was impressed by a supplier that sent a note and small token of appreciation for responding to their customer service survey—before the company had even sent the survey. The company's CEO said that he felt obligated to respond when the survey arrived. Consequently, he adapted this market research technique to his own company which had an 18-month survey process. The company sent a small personalized coaster thanking the customers in advance for taking the time to fill out the forthcoming survey. Response rates on the surveys rose from 18 to 43 percent. *How might you improve response rates by adapting this technique in your own customer communications?*

Customer Greeters.   WalMart places "greeters" at the entrances to its stores to welcome customers and to assist them in finding their way around the store. Borrowing this concept of the greeter, Leeson Electric Corporation in Grafton, Wisconsin, greets all prospective customers at the Milwaukee airport who have come to view Leeson's electric motors' manufacturing facilities. Using greeters gave the company an hour before their scheduled meetings to present information to the customers about the company, its programs, and its products. *Where and when could you use greeters to personalize your relationship with your customers?*

Fast and Efficient Customer Processing.   What do a commercial airline and a state college have in common? Both must process information about tens of thousands of people when delivering services. With this similarity in common, Waukesha County Technical College looked to the airline industry for inspiration and ideas on how to quickly and efficiently register its students each semester. Waukesha County Technical College President Dr. Richard T. Anderson said that prior to borrowing processing practices from the airline industry, registering the college's 35,000 students each semester was "a frustrating, inefficient and time consuming process."

Waukesha administrators were impressed that the airlines used technology to help them quickly process ticket orders by phone and in person. Consequently the school adopted and adapted various processing techniques from the airlines. Today, the college registers 40 percent of its students by telephone using keypad prompts and operators connected to databases. Another 30 percent of students are preregistered in their classes, borrowing "one-stop shopping" techniques from super malls, so that these students don't have to go through the traditional on-campus registration process. Registration is greatly simplified for the 30 percent of students who still go through on-campus registration. Using airline-like databases and other information technology to assist processes, Waukesha has automated the registration process. A registrar can determine at registration whether a course is over-subscribed, which saves many time-consuming enrollment corrections. Now a student can register, schedule, and pay for courses at the same time and place. The system has proven so effective that the college employs eight registration clerks to process 35,000 students. Anderson says that they register 35,000 students faster than eight registration clerks processed 3000 students in 1976. *How can your organization employ technology and other techniques used in the airlines industry to more effectively process your customers?*

Keeping Facilities Clean.   Walt Disney Company inspired John Graham, president and CEO of GIC Management, Inc., a real estate development and management company, by maintaining immaculately clean facilities at its theme parks. To improve cleanliness, GIC followed Disney's example. It made picking up litter everyone's responsibility and a point of pride. Graham reports that the improvement was quick and impressive at GIC's multitenant facilities. *Where can your organization improve facilities' maintenance by borrowing from Disney's practices?*

Making it Easy for Customers to Communicate with Your Organization. Thomas J. Henkels, CEO of MRM Services, a Midwest company specializing in laying gas pipe for utilities, was impressed with how Hyatt and other hotels communicated after hours with their customers. He liked their method of using doorhanger messages for privacy, maid service, or breakfast orders. The Wisconsin-based company decided to use doorhangers to communicate with utility customers by presenting them with an 800 number to call if they had any problems. The doorhangers caught customers' attention and enabled them to directly call MRM Services rather than the utility to resolve service problems. *What innovative techniques can you employ to make it easier for your customers to communicate with your organization?*

Using Customer Histories to Personalize Service. Thomas O. Lied, CEO of Lied's Nursery Company, was impressed by a TV repair company that kept comprehensive customer service records on his purchases. Lied applied the concept to his nursery company, which provides commercial and retail customers with landscape products. The customer records database enables Lied's to keep track of customers' past purchases, project future purchases, and demonstrate personal interest in each account. In an industry not well known for personalized service, Lied's has been able to satisfy customers with its personalized service. *How can you create customer service records to help you improve and personalize service to your customers?*

Just-in-Time Professional Services. Samuel D. Eppstein, a managing partner of Eppstein Keller Uhen Architects in Milwaukee, visited Milwaukee Power Tool Company, an advanced manufacturer using just-in-time cell manufacturing techniques in the production of power tools. He was impressed by the speed of scheduling and information processing achieved through Kanban cards, which are used to order new parts for the manufacturing cells. Eppstein Keller Uhen Architects has adopted the use of Kanban cards to its own professional service firm, where it is using them to expedite scheduling and job processing. *What other JIT techniques can be applied to services within your company?*

Target Excellence in the Field. John Dedrick, president of West Bend Mutual Insurance, is always looking for ways to inspire excellence among his property-casualty insurance company's agents. Inspired by J.I. Case and Anheuser Busch, both of which employ Dealer Excellence Programs to set standards and recognize excellence among its dealers, West Bend Mutual applied the concept to its agency structure. Dedrick says that the program has been effective in communicating and recognizing high-performance standards. *How can your organization raise and recognize its standards for its dealers, agents, or salespeople?*

Steal This Idea®
Exercise

Thomas Edison: "Keep on the outlook for novel and interesting ideas that others have used successfully. Your idea has to be original only in its adaptation to the problem you are currently working on." World War II military designers borrowed from the Cubist art of Picasso to create more efficient camouflage patterns for tanks. What ideas can you borrow? ROGER VON OECH

Background

One characteristic of high-performing organizations is a "we-can-learn-from-anyone" attitude. These organizations are able to accelerate their rates of learning and change by borrowing ideas from both internal and external sources. These high-performers then tailor the ideas to suit different situations and needs. They have escaped the "not-invented-here" syndrome. Their culture, reward, and recognition systems support borrowing ideas as much as creating ideas. This exercise is designed to focus on the value of learning from others.

Objectives

- Learn to "borrow shamelessly"
- Apply this concept to your organization's operations
- Develop skills for the spreading of best practices

Course of Action

Step 1. Using the attached *"List Ideas" worksheet*, each individual should generate a personal list of ideas applicable to his or her business setting. List many ideas regardless of the scale of the idea—small ideas are as good as large ones. Consider ideas you have observed inside and outside your own organization. At this point, do *not* screen ideas by merit—include all ideas (for example: use of Post-It notes in place of fax coversheets, use of bar code technology instead of timecards).

Step 2. Rotating among members of your discussion group, present your top ideas. Use the *"Score Ideas" worksheet* to centrally record the ideas, identify the areas where the idea *is* used, and where it *can* be used. In the next column, note where each idea originated.

Step 3. Total the number of ideas generated by members of your discussion group. Total the number of ideas whose source was outside your team. Compute the "Traditional" score and the "Learn From Anyone" score using the formula at the bottom of the page. Discuss what increasing the weight given to ideas from the outside does for motivating a "learn-from-anyone" attitude.

Step 4. Each team will then present to the full groups an executive summary describing:

- The best ideas generated
- The "stealing" possible
- The scoring of the ideas.

Steal This Idea Worksheet—List Ideas

Short Name for Idea	What is the Idea?	Result(s)
1.		
2.		
3.		
4.		
5.		
6.		
7.		

Steal This Idea Worksheet—Score Ideas

✓	Which idea?	Use of the idea (use a check if ideas is already used and an X if it could be used in this area)									Origination of the idea (check off)	
		Field offices	Business technology	Human Resources	Corporate Services	Legal	Marketing & sales	Production & Ops.	Finance & Accounting	Customer service	Within your team	Outside your team
	1.											
	2.											
	3.											
	4.											
	5.											
	6.											
	7.											
	8.											
	9.											
	10.											
											Total #1:	Total #2:

"Traditional" Scoring [Score = (Total #1 × 5) + (Total #2 × 1)]   "Traditional" Score: _____
"Learn From Anyone" Scoring [Score = (Total #1 × 5) + (Total #2 × 5)]   "Learn" Score: _____

If high scores are best, which scoring system encourages learning and continuous improvement?

Developing the Culture: Executive Exercise

Benchmarking for Best Practices
Executive Exercise

Quality is not an act; it is a habit.

ARISTOTLE

No single action or managerial system produces a "we-can-learn-from-everyone" culture. It is the cumulative result of many smaller actions, processes, and systems working together. In organizations where the culture supports borrowing from the best as a worthy operating approach, remarkable things occur. Lots of improvements— small and large, incremental and breakthrough—are quickly imported into the organization. Best practices benchmarking is a catalyst for fast learning. Organizations that systematically seek out and study best practices experience the beneficial effects of "intellectual leverage." They enjoy the compounding effect of good ideas and creativity that spring from a reservoir extending far beyond the boundaries of any one organization.

This executive exercise explores ways that you can make best practices benchmarking part of your business unit's culture. In your groups, discuss as many of the following managerial questions as time allows. You may address the questions in any order you choose.

Involving Employees in the Search for Best Practices

✓ What actions can you take to make competitive information gathering everyone's responsibility, especially including functions such as sales, marketing, and personnel?

✓ What can you do to employ best practice information in the problem-solving and continuous-improvement process?

✓ How can you sponsor or promote idea exchange and best practice explorations during regular meetings and discussions in which you participate?

Do List

Notes

_____

_____

_____

_____

_____

_____

_____

_____

_____

_____

_____

_____

_____

_____

_____

_____

Education and Awareness About Best Practices

✓ In what ways can your unit invest in employee education that brings outside ideas and speakers before your employees?

✓ If you were to create a lending library that focuses on competitors and the winning strategies and systems of other excellent companies, what materials would you place in circulation?

Do List

Notes

_____

_____

_____

_____

_____

_____

_____

_____

_____

_____

_____

_____

_____

_____

_____

The Role of Leadership in Benchmarking

✓ What personal actions might you undertake to visibly support benchmarking as an activity that fosters quality excellence?

✓ What efforts do you personally make to learn about the practices of domestic and international competitors?

✓ How often do you visit or study other business units within your organization with the goal of importing their most effective operating practices and strategies?

✓ What actions could your company take to engage high-level executives directly in benchmarking and innovative adaptation?

✓ If you were to make someone responsible for being a best practices' champion, what role and function would he or she play in your organization?

Do List

Notes

_____

_____

_____

_____

_____

_____

_____

_____

_____

_____

Championing Benchmarking in the Organization

✓ In your organization, what are the most effective ways to publicize the benefits of borrowing from the best?

✓ What approaches have you developed to regularly identify, study, and celebrate internal success stories and best practices with the goal of repeating them?

Do List

Notes

_____

_____

_____

_____

_____

_____

_____

_____

_____

_____

_____

_____

_____

_____

Developing Internal Best Practices

This exercise is intended to identify internal best practices among Sandoz operating companies. Internal best practices are the winning strategies, approaches, processes, and systems that each operating company has developed in its functional areas. By identifying internal best practices, Sandoz can begin developing a library of best practices that can be shared and leveraged by all Sandoz operating companies.

Instructions

Step 1. Briefly review the attached best practices matrices. Focus on the topical areas assigned to your team. Use the process descriptions to help you *identify practices, strategies, techniques, or systems that have proven highly effective in your operating area.*

Describe your "best practices" in the matrix. Frame your comments as useful tips, successful strategies, lessons learned, or proven practices. Be specific! For example:

- Team members are co-located in the same work areas to facilitate rapid and clear communications during new product development projects.

- Our group sets team—not individual—goals to align the interests of all cross-functional team members. Team bonuses are linked to one goal—reducing development time by 50 percent.

Step 2. Share best practices among your group. Learn what others regard as their best practices. Please put your name on your work sheet. All work sheets will be collected and compiled to begin building a best practices library for your reference.

Discussion Questions:

- What are your unit's greatest strengths and best practices?
- What areas represent your greatest improvement opportunities?
- Which practices from other Sandoz companies could be adapted by your unit?
- What actions might you take to continue sharing best practices and winning strategies among Sandoz companies after this meeting?

Step 3. Be prepared to share your team's work and best practices with the entire group. A brief wrap-up will be conducted following the individual group discussions.

Financial and Information Management

Core process area	Subprocess or activity area	Your best practices (Useful tips, successful strategies, and lessons learned)
Manage financial resources	✓ Develop budgets ✓ Conduct financial forecasting ✓ Manage resource allocation ✓ Manage cash flow	
Process finance and accounting transactions	✓ Process accounts payable ✓ Process payroll ✓ Process accounts receivables, credit, and collections ✓ Close the books	
Report information	✓ Provide external financial information ✓ Provide internal financial information	
Manage the tax function	✓ Employ effective technology ✓ Communicate tax issues to management	
Manage physical resources	✓ Manage facilities ✓ Plan fixed asset additions ✓ Manage risk	

Financial and Information Management

Core process area	Your best practices (Useful tips, successful strategies, and lessons learned)
Information capacity planning	
Information strategy planning	
Hardware development and planning	
Software development and planning	
Information technology management	
Technology transfer	
Information management and integration systems	
Information quality evaluation and audit	
Information technology and business planning linkage	

Operations and Manufacturing:
Producing and Delivering Products and Services

Core process area	Subprocess or activity area	Your best practices (Useful tips, successful strategies, and lessons learned)
Plan for and acquire necessary resources or inputs	✓ Acquire capital goods ✓ Hire employees ✓ Obtain materials and supplies ✓ Plan and design products/services	
Convert resources or inputs into products	✓ Design production floor layout ✓ Test products and service ✓ Develop and adjust production process ✓ Use automation and robotics ✓ Schedule production ✓ Move materials and resources ✓ Make product ✓ Package and store the product ✓ Stage the product for delivery	
Make delivery	✓ Arrange product shipment ✓ Deliver products to customers ✓ Install (if specified)	
Manage, produce, and deliver process	✓ Document and monitor order status ✓ Manage inventories ✓ Assure quality ✓ Schedule and perform maintenance ✓ Monitor environmental constraints ✓ Automate inventory control ✓ Plan logistics ✓ Manage purchasing, warehousing, and distribution systems	

Operations and Manufacturing:
Supplier Management

Core process area	Your best practices (Useful tips, successful strategies, and lessons learned)
Supplier relations management	
Vendor certification	
Contract development and negotiation	
Supplier quality assurance	
Supplier evaluation	

Human Resources
Management and Development

Core process area	Subprocess or activity area	Your best practices (Useful tips, successful strategies, and lessons learned)
Recruiting and selection	✓ Personnel selection ✓ Employee orientation	
Managing and developing human resources	✓ Human resource strategy creation ✓ Personnel management ✓ Performance appraisals ✓ Career planning ✓ Union relations management ✓ Diversity management ✓ Succession planning ✓ Employee training and education ✓ Employee well-being and morale assurance ✓ Personnel relocation management	
Involving employees	✓ Employee empowerment strategies ✓ Employee involvement strategies ✓ Employee opinion survey process ✓ Suggestion systems	
Rewarding and recognizing employees	✓ Benefits ✓ Compensation systems ✓ Reward and recognition systems ✓ Pension management ✓ Health care management	

Marketing and Sales

Core process area	Subprocess or activity area	Your best practices (Useful tips, successful strategies, and lessons learned)
Marketing product or service to customer segments	✓ Develop pricing strategy ✓ Develop advertising strategy ✓ Develop market messages to communicate benefits ✓ Estimate advertising resource and capital requirements ✓ Identify specific target customers and their needs ✓ Develop sales forecasts ✓ Sell products or services ✓ Negotiate terms	
Order fulfillment	✓ Process order entries ✓ Manage customer preparation and expectations ✓ Make deliveries and pickups ✓ Install/remove systems	
Maintaining products and services	✓ Manage service calls ✓ Dispatch services ✓ Give after sale service ✓ Provide product maintenance ✓ Provide support information	
Process customer orders	✓ Accept orders from customers ✓ Enter orders into production and delivery process	

Researching and Developing New Products

Core process area	Subprocess or activity area	Your best bractices (Useful tips, successful strategies, and lessons learned)
Performing basic research	✓ Hiring and selecting staff ✓ Designing experiments ✓ Setting scientific targets, goals, etc. ✓ Performing scientific performance reviews ✓ Creating an environment where ideas are challenged in positive ways	
Developing new product/service concept and plans	✓ Translate customer wants and needs into product/service requirements ✓ Plan and deploy quality targets ✓ Plan and deploy cost targets ✓ Develop product life cycle and development timing targets ✓ Develop and integrate leading technology into product/service concepts	

Design, build, and evaluate prototype products or services	✓ Develop product/service specifications ✓ Conduct concurrent engineering ✓ Implement value engineering ✓ Document design specifications ✓ Develop prototypes ✓ Apply for patents	
Refine existing products/services	✓ Develop product/service enhancements ✓ Eliminate quality/reliability problems ✓ Eliminate outdated products/services ✓ Test effectiveness of new or revised products or services	
Prepare for production	✓ Develop and test prototype production process ✓ Design and obtain necessary material and equipment ✓ Install and verify process or methodology	
Manage the product/service development process	✓ Translate customer needs to products/services ✓ Extend existing products intonew markets ✓ Improve and upgrade existing products/services ✓ Transfer technology among groups, teams, and functions ✓ Test effectiveness of new or revised products or services	

Benchmarking Do's and Don'ts[1]

Introduction

As different companies adapted benchmarking to their own needs, a variety of apparently different benchmarking processes with different numbers of "steps" began to emerge. Respondents to an International Benchmarking Clearinghouse survey identified processes of between 4 and 33 steps.[2] Examples include ALCOA's 6-step model,[3] AT&T's 9-step model,[4] and Xerox's 10-step process.[5] In fact, all benchmarking processes are much the same and are built from the same basic model, which was first articulated by Xerox. These "Do's and Don'ts" guidelines include activities common to most benchmarking models. These guidelines are organized by generic steps that appear in most benchmarking models. These generic steps to benchmarking include:

A. Select the process or function to benchmark.

B. Understand the existing process or function.

C. Identify benchmarking partners.

D. Collect data and information.

E. Identify gaps and reasons for them.

F. Develop programs to address findings.

G. Implement changes and monitor results.

Marketing and Sales

Core process area	Subprocess or activity area	Your best practices (Useful tips, successful strategies, and lessons learned)
Market management	✓ Market research and information capture ✓ Market selection ✓ Product or service planning and development ✓ Pricing ✓ Marketing communications ✓ Sales leads qualification ✓ Sales prospecting management ✓ Sales process ✓ Sales territory planning ✓ Proposal process	
Customer satisfaction management	✓ Customer engagement ✓ Determining customer requirements ✓ Customer communications and feedback ✓ Setting customer standards ✓ Customer relationship management ✓ Customer service telephone response ✓ Developing customer loyalty ✓ Complaint handling ✓ Customer satisfaction measurement	
Understand markets and customers	✓ Determine customer needs and wants ✓ Conduct qualitative assessments ✓ Conduct customer interviews ✓ Conduct focus groups ✓ Conduct quantitative assessments ✓ Develop and implement surveys ✓ Predict customer purchasing behavior ✓ Measure customer satisfaction ✓ Monitor satisfaction with products & services ✓ Monitor satisfaction with complaint resolution ✓ Monitor satisfaction with communication ✓ Monitor changes in market or customer expectations ✓ Determine weaknesses of product/service of reaction to competitive offerings ✓ Identify new innovations that are meeting customers' needs ✓ Determine customers' reaction to competitive offerings	

Preliminary Considerations Before Starting Benchmarking

1. Benchmarking tends to be a term that is often used to describe a variety of activities. These, in some cases, are more accurately described as surveys or competitive assessment studies. Although these can be value-added activities, they do need to be distinguished from benchmarking. Surveys and competitive assessment activities can identify gaps, but unlike benchmarking, don't provide information on how to close the gaps or make step change improvements.

2. Benchmarking can be conducted proactively or reactively. There's no reason to wait for a crisis before considering benchmarking.

3. Benchmarking should not be viewed as a stand-alone process. It is best integrated into existing quality improvement, process management, and strategic planning models. The objective is to have it considered as a tool that can be routinely applied when appropriate.

4. Management of benchmarking is important, especially for large companies. There is a need to ensure that projects and partners are not duplicated, learnings are shared, and that legal and ethical practices are followed.

5. Not everyone needs to be a benchmarking expert. The mass training of the whole organization may not prove to be cost-effective. Consequently, AT&T and some other organizations have adopted a model that creates a smaller cadre of in-house experts who provide just-in-time training to employees and facilitate specific projects.[6]

One way to share for mutual benefit is to accumulate lessons learned in the form of two lists: the *Do's* —that is, those things that enhance benchmarking—and the *Don'ts* —activities that can create pitfalls.

A. Selection of the Process to Benchmark

Do's

1. As a general rule, the process or function selected should be one of the most critical to your business strategy. An exception is if this is your first or second project and the team is still developing its skills on how to conduct effective benchmarking projects. In this case, the project may target a less strategic topic; it should still be one of interest, have a good chance of success, be modest in scope, and require less than six months to complete.

2. Projects must be well defined, and generally they should require less than a year to complete the research, analysis, action planning, and preliminary implementation.

3. Management must support the project, provide adequate resources (staff, funding, time, and the like), and be prepared to champion implementation of best practice findings.

4. The organization must be willing to change.

5. The team should be comprised of a workable number (4 to 6) of committed cross-functional stakeholders. Ideally, a benchmarking expert will help facilitate the benchmarking process, unless the team is already experienced in benchmarking.

6. The team must understand who the customer is for the study. The customer's expectations from the study are established by direct interaction.

7. If the organization has already developed an existing Quality Improvement (Process Management) model, this model should be used to identify the customer, expectations, measures, and so on.

8. It is very useful to have an explicit mission statement which documents the project's deliverables, purpose, and metrics.

9. Notice of your team's benchmarking study should be shared inside your organization, particularly with anyone assigned to coordinate the company's benchmarking activities.

10. Benchmarking is more readily accepted if your organization has been exposed to quality concepts and are used to thinking of work as a process. The value of teamwork and group facilitation is common for quality improvement and benchmarking and is essential for success.

11. Identify all those that may be affected by the project and secure their ideas, contributions, and support.

12. Ensure all functions that will implement changes or improvements are represented from the beginning on the benchmark team. (If this beneficiary group is too large to accommodate everyone on the team, they should be involved as secondary participants who provide counsel on project design, focus, action planning, and so on.)

13. Understand the ongoing roles and communication requirements of different constituents throughout the benchmarking project/process.

14. Consider implementation issues early in the planning phase and throughout the project.

Don'ts

1. Don't initially benchmark areas where the organization already performs well.

2. Don't benchmark topics or processes that aren't important.

3. Don't benchmark processes that are so broad in scope, so poorly defined, or so poorly circumscribed that the team cannot agree on its mission and cannot focus its efforts.

4. Don't undertake major strategic benchmarking projects if the team lacks management support, has no well-defined customer, doesn't have stakeholders involved in the project, and witnesses no organizational commitment to change.

5. Don't undertake benchmarking projects with a team that is too large to be effective (10 or more) or too small to be credible (1 to 2.)

6. Don't undertake complex process benchmarking efforts with team members that don't understand the benchmarking process and don't have access to an experienced benchmarking facilitator.

7. Don't confuse a metric or benchmark with a process, that is, don't make the metric or operating statistic the end of the benchmarking investigation.

8. Don't benchmark unless all those affected by likely changes are represented on the benchmarking team or are given opportunity to contribute their ideas and interests to the benchmarking team.

B. Understand the Existing Process

Do's

1. Develop a roadmap of the benchmarking project. (This can be modified over time and is a good communication tool during the study period. It is also a good record afterwards. Include reviews with sponsors, customer inputs, as well as team activities.)

2. The ideal way to prepare a novice benchmarking team is through just-in-time (JIT) delivery of training; JIT education provides benchmarking tools, concepts, and skills to team members when they are ready to apply them during the project.[7] The trainer therefore presents benchmarking modules at each stage of the benchmarking process.

3. Develop an understanding of the critical success factors and the process drivers. (These include work and information flow, practices, structural factors, culture, environment, technology, etc.)

4. The use of deployment flowcharting and Ishikawa cause-and-effect (fishbone) diagrams can be extremely useful. This alone may help uncover easily implemented and valuable improvements.

5. If the benchmarking team is examining a complex process, it often helps to separate it into its major subprocesses and examine each one from a cost, time, and quality perspective. This proves most effective when the team can produce real cost, time, and quality measures. Such performance measures helps the team focus on the most critical parts of the overall process.

6. Identify the most important performance measures. These often include cost, time, quality, productivity, and effectiveness measures.

7. Check in with the project sponsor(s) to ensure that the game plan is acceptable and implementable.

8. Begin the search of the literature to develop information about the process. Librarians add real value.

9. If no performance measurement data have been collected previously, collect data during a short interval, such as one week to one month, to help analysis.

Don'ts

1. Don't overlook the step of understanding the existing process, even if it may be tedious and time-consuming. Those teams that fail to fully understand their own process also fail to understand the critical success factors of the project; consequently, they often fail to focus their information search.

2. Don't overlook the power of flowcharting.

3. Don't try to use too many measures or inappropriate measures.

4. Don't forget to check the team's plans with the internal and external customers and the process sponsor.

5. Don't forget to gather input from customers of the process.

6. Don't overlook the importance of secondary research.

7. Don't create a flowchart that is so detailed it requires many months to complete and includes minutiae.

8. Don't give up if the team discovers it doesn't possess any existing performance measures.

9. Don't believe that the team needs a year's worth of data to proceed. Avoid paralysis through too much analysis.

10. Don't try to overhaul the business unit's entire performance measurement system when flowcharting.

C. Identify Benchmarking Partners

Do's

1. Establish a list of selection criteria for benchmarking partners. Include profitability, stature, relevancy to your process, and ease of collection, but don't be misled by the "Halo Effect": (i.e., a single company, such as a Baldrige winner, can't be best at every process.)

2. Start with literature searches to help identify partners.

3. Ask "Who else in our company would have an informed opinion about best practice companies?"

4. Ask customers, suppliers, consultants, academics, security analysts, Baldrige winners, professional and trade associations for opinions on best practice companies.

5. Check with benchmarking clearinghouses and internal coordinators or consultants.

6. Remember that any company which is significantly better than yours in a given process is a good partner.

7. Always consider companies outside your industry. They are more likely to produce practices that are new, better, or never before considered by your company. Also, information is often more readily shared across industries than among industry competitors.

8. Be creative in your choice of partner(s).

9. Limit the number of partners to four to eight. That is plenty for any one benchmarking study.

Don'ts

1. Don't limit benchmarking partner selection to the small number of Baldrige winners. They may not be best in your area of interest. Moreover, they are inundated with benchmarking requests and they have had to be very selective.

2. Don't limit your search only for *the best* practice company; searching for a single best company is an illusive target and only wastes time.

3. Don't permit data collection convenience to be placed ahead of best practices in partner selection.

4. Don't limit selection to companies in your industry.

5. Don't discount companies that are smaller just because of their size.

D. Collect Data and Information

Do's

1. Use secondary sources to develop background and specific process information about benchmarking partners.

2. Make sure your team understands the benchmarking Code of Conduct and the appropriate protocols. This was developed by the International Benchmarking Clearinghouse and the Strategic Planning Institute's Council on Benchmarking.

3. Do a thorough job of questionnaire development. Test the questionnaire with an internal partner or a friendly external partner before distribution.

4. Limit the questionnaire to a small number of open-ended questions and to no more than two pages of evaluation-by-numbers, multiple choice, or yes/no questions.

5. Seriously consider the pros and cons of asking questions by written response, telephone interview, or face-to-face interviews. The questionnaire becomes the collection vehicle when a written response is requested, when a guide for phone interviews is required, or when an agenda is needed for in-person meetings.

6. Be specific about the time commitment you are requesting from your benchmarking partner and be prepared to send a written questionnaire or a scoping document ahead of the team's visit.

7. Identify the incentive for your benchmarking partner to participate in the study. This usually includes a summary of findings. (Your partner may have additional requirements that include a reciprocal visit or study.)

8. Limit the benchmarking site visit team size to about three people and be specific about each person's role. A team needs one member to ask questions, one to take notes, and one to listen for questions and issues missed by the questioner to watch the clock and the overall process to ensure the team stays on track. Teams are well advised to prepare and practice for the visit with dry runs and rehearsals.

9. Debrief the visiting team as soon as possible following the site visit. The airport, a nearby cafe, or hotel will serve the purpose.

10. Be prepared to respond to spontaneous questions from the benchmarking host; in turn, ask only questions that your team is willing to answer about your own organization.

11. Share nonproprietary flowcharts or printed materials that describe the benchmark area before or at the start of a visit (not at the end of a visit).

12. If in doubt, particularly when benchmarking competitors, seek advice from legal counsel.

13. A third-party consultant is often valuable or necessary for gathering and blinding data from competitors.

14. When developing your questionnaire, ask yourself, "What is the reason for asking this question?"

Don'ts

1. Don't approach a benchmarking partner without being thoroughly prepared. There are those that feel that benchmarking starts by picking up the phone—don't be one of them.

2. Avoid making cold calls to benchmarking contacts. Others in your company may have personal or professional contacts who can facilitate the initial contact. Such introductions greatly increase the likelihood of the team's success.

3. Don't design a questionnaire that is large and overwhelming.

4. Don't assemble too large a group for a benchmarking visit. They will slow the process, overwhelm the host, and hinder information exchange.

5. Don't permit the site visit team's debriefing to be deferred for even a few days following the visit.

6. Don't get side-tracked by interesting or amusing stories that don't add to the team's knowledge.

7. Don't ignore or reject information that is reported in a different format than that used by your organization.

E. Identify Gaps and Reasons for Them

Do's

1. Examine data for consistency. (A phone call to your partner may be sufficient to clarify any inconsistencies.)

2. Set up data so that direct comparisons are readily made. Typically, this takes a tabular form. However, graphical representation such as bar charts, spider diagrams, or pie charts can have greater impact.

3. Look for and verify the cause and effect relationships between the measures and the reasons for gaps.

4. Look for trend data. (Remember—the data is a photograph at one instant in time. The competition will continue to improve.)

5. Aim to develop consensus from the team on key issues.

Don'ts

1. Don't aim for too much precision.

2. Don't process or develop too much information.

3. Don't provide only narrative analysis.

F. Develop Programs to Implement Findings

Do's

1. Agree on key findings and recommendations. Seek informal reviews and obtain feedback from project sponsors and constituents before making recommendations.

2. For each recommendation for change, make sure the owner is identified.

3. Recommend specific tasks and clearly designate which team member has responsibility for accomplishing the task. The implementation team, which is often different from the benchmarking team, has the detailed responsibility for the implementation plan, timetable, and actions.

4. Identify cost and realistic expected benefits for recommended changes.

5. Agree on how results and recommendations are to be communicated and to whom.

6. Incorporate feedback from customers and process owners into the final report.

7. Obtain commitment and support from management for the implementation plan.

8. Make it as easy as possible to implement recommendations by anticipating any hurdles that will have to be overcome.

9. Make sure that any impact on peripheral groups is considered in the implementation plan.

10. Offer to make personal presentations for every group that will be asked to participate in improvements.

Don'ts

1. Don't rely solely on a final report to communicate your recommendations.

2. Don't underestimate the difficulty of persuading others to buy into the team's plan.

3. Don't make long, dull, or unenthusiastic presentations, or long, ponderous reports that provide too much detail.

4. Don't get overly ambitious with your timetable for change.

5. Don't forget to identify who is the owner of the change activity and who has the responsibility for specific activities.

6. Don't overestimate the benefits or underestimate the costs of implementation.

7. Don't overlook the impact the implementation may have on peripheral groups.

8. Don't forget to seek input from customers, management, and process owners.

G. Implement Changes and Monitor Results

Do's

1. Make sure all those involved in the process change thoroughly understand the rationale behind the change and what is expected of them.

2. Make sure that management is supportive of the change and takes interest in the process.

3. Consider rolling out the process change with a pilot program.

4. Set up the monitoring activity and process before initiating the change. Focus on key performance measures.

5. Anticipate resistance to change—and don't take it personally.

6. Find ways to cascade improvements to other parts of the organization.

7. Make sure that others peripheral to your process are not adversely impacted by the change recommendations.

Don'ts

1. Don't forget that without implementation, benchmarking and all the team's efforts to date are worthless.

2. Don't implement without monitoring.

3. Don't overlook the fact that other factors may impact the results.

References

1. Derek Ransley, senior quality consultant at Chevron Research and Technology Company, compiled the Do's and Don'ts list, with contributions from Christopher Bogan, Sam Bookhart, and other benchmarking experts.

2. "Planning, Organizing, and Managing Benchmarking Activities: A User's Guide," American Productivity and Quality Center.

3. Bernowski, Karen, *Quality Progress,* January 1991.

4. "Benchmarking: Focus on World-Class Practices," AT&T, 1992.

5. Camp, Robert C., *Benchmarking: The Search for Best Practices Which Lead to Superior Performance* , Quality Press, Milwaukee, Wisconsin.

6. "Benchmarking: Focus on World-Class Practices," AT&T, 1992.

7. Ransley, Derek L., "Teaching Managers to Benchmark," *Planning Review,* January-February 1993.

Benchmarking and Baldrige

Learning is the essence of continuous improvement—and the central management concept of the Malcolm Baldrige National Quality Award. Vicarious learning, through benchmarking or competitive comparisons, permeates the Baldrige criteria. Benchmarks, benchmarking, and competitive comparisons influence more Baldrige points than any other management concept found within the Baldrige criteria. Powerful management concepts, customer satisfaction, cycle time reduction, process management and control, and employee involvement and empowerment do not influence as many points and Baldrige areas as benchmarking and competitive comparisons. Together benchmarks and benchmarking concepts influence more than half the total points awarded during a Baldrige assessment.

What makes an applicant successful in winning the Baldrige Award? From year to year, the answer varies on specific details, but remains the same in spirit. The Baldrige is dedicated to continuous improvement on behalf of a company's customers. While the number of categories has remained constant—at seven—their criteria have evolved. One important change lies in the increasing importance of benchmarks and benchmarking. No longer can companies set goals based primarily on internal operating targets. They must adopt an external view by comparing themselves to the "best." Based on the results of these benchmarks, the company is motivated to set higher standards.

Congress established the Malcolm Baldrige National Quality Award in 1987 to promote quality awareness, to recognize quality achievements of U.S. companies, and to publicize successful quality strategies. Baldrige judges review applicants in seven categories: Leadership, Information and Analysis, Strategic Quality Planning, Human Resource Development and Management, Management of Process Quality, Quality and Operational Results, and Customer Focus and Satisfaction.

For each of these areas, one may ask: Why "reinvent the wheel?" When a company can adopt or adapt ideas that have already proven to be highly effective, it makes sense to import them. In this respect, benchmarking is a remarkably simple management concept. Yet for many companies the concept of borrowing or adapting another's winning ideas or practices remains foreign. To this end, the Baldrige guidelines have greatly expanded the importance of benchmarking and finding appropriate benchmarks. In 1988, the award included no direct references to benchmarking. In 1989, benchmarking influenced 80 points and then increased 500 percent to 400 points by 1993.*

The points influenced by indirect references to benchmarking varied over the years;† however, the points influenced by direct and indirect references to benchmarking increased from 175 points in 1988 to 550 points in 1993 (a 314 percent rise.)**

Indirect and direct references to benchmarking are seeded throughout the Baldrige application. Benchmarking's role has grown to maturity with the development of the award's criteria. Item 1.1 (Senior Executive Leadership) changed from learning about competitors' quality (1989-1991) to a more detailed requirement of benchmarking competitors (1992-1993). In 1994, it grew into an open-ended requirement to engage in "benchmarking" (without regard to industry) in the course of developing and maintaining an environment for quality excellence. In 1992s Item 1.2 (Management for Quality), the company was asked to describe benchmarks that it had found. This inquiry disappeared in 1994, as the standard of general comparison became higher; the focus of Category 1 sharpened from simply learning about the quality of competitors to learning from and about the "best." Not only must companies find benchmarks as operating statistics, they must also learn from them. No longer can a company's leadership remain satisfied with merely learning about the quality of its competitors. In addition, the leadership can· learn from these competitors. "Shamelessly borrowing" the successful ideas and practices of competitors and other industry leaders saves the benchmarker time, money, and effort.

No category has evolved more dramatically than Category 2, which explores a company's use of information and its analysis of that data for management purposes. In 1988, Category 2.2 (Use of Product of Service Quality Data) required that companies collect data about worldwide competitors' products and services. In 1989 and 1990, Category 2.1 (Scope of Data and Information for "Management by Fact") asks for the scope of data, including benchmarks, in the company's information systems. Category 2 exemplifies benchmarking's increasing importance with its new division solely related to benchmarking. Category 2.2 blossomed into "Competitive Comparisons and Benchmarks" in 1991 and remains as such in all subsequent application criteria. The category requires detail about benchmarking

---

*This and all subsequent point totals of direct references to benchmarking include references to "benchmarking" and "benchmarks."

The points influenced by direct references to benchmarking increased 100 percent from 1989 to 1990 (160 points), by 34 percent from 1990 to 1991(215 points), by 81 percent from 1991 to 1992 (390 points), and by 3 percent from 1992 to 1993 (400 points.)

†Indirect references to benchmarking include comparisons to competitors, industry averages, and industry leaders. If an indirect and a direct reference to benchmarking exist in the same item, the points are reflected as a direct reference. This reflects the growing emphasis placed on benchmarking in the award.

Indirect references decreased by 37 percent from 1988 (175 points) to 1989 (110 points), increased by 23 percent from 1989 to 1990 (135 points), increased by 104 percent from 1990 to 1991(275 points), decreased by 42 percent from 1991 to 1992 (160 points), and decreased by 6 percent from 1992 to 1993 (150 points).

**The overall increases in the points influenced by direct and indirect references to benchmarking include: 9 percent from 1988 (175 points) to 1989 (190 points), 54 percent from 1989 to 1990 (295 points), 66 percent from 1990 to 1991(490 points), 12 percent from 1991 to 1992 (550 points), and 0 percent from 1992 to 1993 (550 points.) Though benchmarking's total influence remained the same from 1992 to 1993, it influences over half the total possible points of the award in both years.

information criteria, scope, use, and evaluation and improvement. In 1994, the concept grows further to note that "benchmarking information and data refer to processes and results that represent superior practices and performance and set 'stretch' targets for comparison. Sources of competitive comparisons and benchmarking information might include: (1) information obtained from other organizations through sharing; (2) information obtained from the open literature; (3) testing and evaluation by the company itself; and (4) testing and evaluation by independent organizations."

The 1993 and 1994 criteria additionally focus on benchmarks in item 2.3 (Analysis and Uses of Company-Level Data). Analysis of benchmark and competitor data becomes a driver for identifying improvement opportunities, for supporting decision making and for setting improvement priorities and targets on a broad-range of operating dimensions such as product/service quality, productivity, resource allocation and nonfinancial and financial performance.

In 1992, 1993, and 1994, benchmarks and benchmarking encouraged "breakthrough approaches" and "stretch" objectives in Category 2.2. The criteria now much more explicitly state that companies must dedicate themselves to learning from others. Also, they must collect comparative operating statistics to evaluate their progress in relation to others. Though the wording of the categories remains similar from 1992 to 1994, the criteria now ask for the scope, sources, and uses of benchmarking information instead of benchmark data.* Why the shift? The Baldrige authors are reflecting trends in the marketplace of management where companies are encouraged to go beyond simplistic study of operating metrics or benchmarks. In addition to gathering and comparing the metrics of competitors and other excellent organizations, companies must also consider the practices the best use to achieve their superior results. Besides requiring companies to discover the "best" and borrow their ideas, the criteria now call for companies to discover novel methods of improvement to surpass competitors. The criteria encourage invention, innovation, and creativity. Benchmarking in the current Baldrige criteria is not static; it is a dynamic process to drive continuous improvement.

The role of benchmarking in strategic goal setting, which is articulated in Category 3, Strategic Quality Planning, has become more focused over the years. In 1989, Item 3.2 (Plans for Quality Leadership) determined the selection basis, sources, scope, types of analysis, and roles in company planning that benchmarks play. Items 3.1 (Strategic Quality Planning Process) and 3.2 (Quality Leadership Indicators in Planning) explored similar areas in 1990. In 1991, benchmarks helped determine companies' plans by assisting them in setting goals. Benchmarking and competitive comparisons also determined two- to-five year projections of improvements for key measures and performance indicators in 1991-1994 applications. As of 1991, benchmarking goals also became more focused because companies set a time projection period. Companies began to envision their future quality and results and to compare them to those of their competitors. The projections identify gaps that motivate companies to apply their ideas for quality improvement and make them work within a defined period. Moreover, in 1994 Item 3.2 is retitled "Quality and Perform-

---

*This also relates to the shift in Items 5.2 and 5.3 from 1992 to 1993 and 1994.

ance Plans" and it requests companies to explain "estimates or assumptions" regarding competitors projected quality and operational performance.

Though most categories show a significant evolution in the importance of benchmarking over the years, Category 4 (Human Resource Development and Management) shows little change. It requires a comparison of the most important measures and indicators of employee well-being and satisfaction to that of industry averages, industry leaders, local/regional leaders and key benchmarks.

In Category 5 (Management of Process Quality), benchmarking has gradually expanded to include more areas. In 1988, Item 5.3 (Design of New or Improved Products and Services) demanded comparisons of quantitative quality to competitors. Benchmark data became a necessary part of process improvements in 1990 and 1991. In 1992-1994, item 5.3 split process management into two items covering "Product and Service Production and Delivery" in a newly constituted Item 5.2 and "Business and Support Service Processes" in 5.3. Companies must relate the role that benchmarking information plays in achieving better quality, cycle time, and overall performance in both process areas.

The importance of benchmarks, competitive comparisons and best practices have grown to permeate all items of Category 6 (Quality and Operational Results). Growing from one item that originally focused just on quality and operating quality comparisons, benchmark information and competitive comparisons are currently required for 6.1 Product and Service Quality Results, 6.2 Company Operational Results, 6.3 Business and Support Service Results and 6.4 Supplier Quality Results. In 1994, the criteria expanded to note that "comparisons and benchmarks for business and support services (6.3b) should emphasize best practice performance, regardless of industry." Baldrige had grown to reflect the operating reality that most support services are generic processes, such as accounting, legal services, information services, public relations, software support, etc.; consequently, benchmarking need not—indeed, should not—be limited to one's own industry.

A principal goal of highly effective managerial systems is to achieve customer satisfaction. Of course, customer satisfaction must be evaluated against others to assess its relative adequacy in competitive markets. Consequently, benchmarking casts its shadow directly over Category 7—Customer Focus and Satisfaction. Over the years (1989-1994), customer satisfaction comparisons expanded from one item to four as their relative importance grew.* Criteria have expanded to include market share comparisons (1990), customer commitment measures (1991), current customer requirements (1991), future customer requirements (1992), consideration of competitors' customers and satisfaction comparisons (1994).

Since the Baldrige quality-assessment guidelines were first introduced in 1988, benchmarking has steadily grown in scope, stature, and scoring influence in the award criteria. Today benchmarking is a foundation concept underpinning all aspects of the National Quality Award's comprehensive total quality assessment and management system. From the Baldrige view, a company cannot achieve world class stature and quality excellence without fully embracing benchmarking as a fundamental management discipline.

*Direct and indirect references to benchmarking and competitive comparisons increased five-fold in the category from 1989 to 1994 (from 30 points to 150 points).

Selected
Bibliography

"A Special Report for Manufacturers: What Are Your Competitors Up to?" *Agency Sales Magazine* 18, No. 5 (May 1988): 33-34.

AT&T Benchmarking Team. "Benchmarking: Focus on World-Class Practices. "August 1992.

Allio, Michael K. "The Argument Against Adopting a 'Process' Mentality." *Planning Preview* 21, No. 1. (January/February 1993): 50-51.

Alster, Norman. "An American Original Beats Back the Copycats." *Electronic Business* 3, No. 19 (October 1987): 52-58.

—."Competitive Benchmarking Paves the Road to Quality." *Human Resources Briefing* 6, No. 4, The Conference Board, May 1990.

Altany, David. "Benchmarkers Unite." *Industry Week* (February 3, 1992): 25.

—. "Share and Share Alike." *Industry Week* (July 15, 1991): 12-17.

—. "Copycats." *Industry Week* (November 5, 1990): 11-18 (reprinted in *Quality Digest*, March 1991, pp. 52-59).

Balm, Gerald J. "Benchmarking, The IBM-Rochester Way." *Commitment Plus*, the newsletter of the Quality & Productivity Management Association, 66, No. 12 (August 1991): 1-4.

—. "Benchmarking: A Practitioner's Guide for Becoming and Staying Best of the Best," QPMA Press, 1992.

Band, William. "Benchmark Your Performance for Continuous Improvement." *Sales & Marketing Management in Canada (Canada)* 31, No. 5, pp.36-38.

Beall, Donald R. and David Kearns. "Quality Improvement Begins at the Top." *World* 20, No. 5 (1986): 18-23.

Beck, Larry. "We Are Successfully Taking on All Competition. " *Modern Materials Handling*, 41, No. 10 (September 1986): 56-59.

Bemowski, Karen, "The Benchmarking Bandwagon: AT&T and ALCOA Share Their Steps to Success." *Quality Progress* (January 1991): 19-24.

"Benchmarking." *Productivity* 12, No. 9 (September 1991): 10-12.

"'Benchmarking' Shows How Company Measures Up to Customer Satisfaction. " *American Management Association*. May 8, 1989.

Biesada, Alexandra. "Benchmarking." *Financial World* (September 17, 1991): 28-47.

—. "Strategic Benchmarking." *Financial World* (September 29, 1992): 30-58. (Includes profiles of companies written by other authors.)

—. "The Second Opinion." *Financial World* 160, No. 25 (December 10, 1991): 88-90.

Bogan, Christopher E. and John Robbins. "Steal This Idea. " *Mortgage Banking* (December 1992): 55-61.

Bogan, Christopher E. and Michael J. English, "Benchmarking: A Wakeup Call For Board Members (and CEOs Too)", The Planning Review, (July/August, 1993): 28-33.

Camp, Robert C. *Benchmarking: The Search for Industry Best Practices That Lead t Superior Performance.* Quality Press: Milwaukee, WI.

Cavinato, Joseph. "How to Benchmark Logistics Operations." *Distribution* 87, No. 8 (August 1988): 93-96.

Chapple, Alan. "Benchmarking Pits Industry Against World's Best in Class." *Engineering Times* 14, No. 4, Published by the National Society of Professional Engineers (April 1992): 1 & 6.

Chvatal, Kris. "Xerox Benchmarking Studies Paying Off: Costs Have Gone Down, Quality Has Improved." *Electronic Buyers' News* (July 2, 1990): 27.

"Clearinghouse Commentary: Benchmarking at TI: Strategy, Deployment, and Current Studies." *American Productivity & Quality Center* 11, No. 11 (May 1992): 5-6.

Cole, Jesse. "Bettering the Best." *Sky* (January 1993): 18-22.

"Benchmarking, The IBM-Rochester Way." *Commitment-Plus* (Newsletter of the Quality and Productivity Management Association), August 1991.

Cook, Brian M., "Terry Rock & Convex Computer: Trying To Be Best in the Business." *Industry Week* (February 18, 1991): 39-46.

Crane, Katie, "Often the Best Nuggets Aren't in the Written Report." *Computerworld* (April 16, 1990): 71-76.

Davies, Paul. "Perspectives: Benchmarking." *Total Quality Management* (December 1990): 309-310.

Davis, Tim R.V. and Michael S. Patrick. "Benchmarking at the SunHealth Alliance." *Planning Preview* 21, No. 1 (January/February 1993): 28-31, 56.

Day, Charles R., Jr. "Benchmarking's First Law: Know Thyself!" *Industry Week* (February 17, 1992): 70.

Demers, Mary L. "Team Xerox Ties the Score." *CDIS Newsletter* 2, No. 6.9, Dataquest, San Jose, CA (June 1987): 18-19.

Denton, D. Keith. "The Service Imperative." *Personnel Journal* 69, No. 3 (March 1990): 66-74.

DeToro, Irving J. "Strategic Planning for Quality at Xerox." *Quality Progress* 20, No. 4 (April 1987): 16-20.

Deutsch, Claudia H. "Businesses emulate the very best." *The Arizona Republic.* Vol. 3, Issue 522, January 27, 1991, p. I⁻4.

Deutsch, Claudia H. "When Imitation Is More Than Flattery."

Doades, Ronald, "Making The Best of Best Practices," Public Utilities Fortnightly, (August 15, 1992):15-18.

Drozdowski, Thomas E. "GTE Uses Benchmarking to Measure Purchasing." *Purchasing* 94, No. 6 (March 1983): 21-24.

Dumaine, Brian. "Corporate Spies Snoop to Conquer. " *Fortune* 118, No. 11 (November 7, 1988):68-76.

Eccles, Robert G. "The Performance Measurement Manifesto." *Harvard Business Review* (January-February 1991): 131-134.

Enslow, Beth. "The Benchmarking Bonanza." *Across the Board* (April 1992): 16-22.

Eyrich, H.G. "Benchmarking to Become Best of Breed," *Manufacturing Systems*, April 1991.

Fifer, Robert M. "Cost Benchmarking in the Value Chain." *Planning Review* 17, No. 3 (May/June 1989): 18-27.

Fifer, Robert M., Timothy R. Furey, Lawrence S. Pryor, and Jeffrey P. Rumburg. *Beating the Competition: A Practical Guide to Benchmarking*, Kaiser Associates, Inc., 1988.

Fitz-enz, Jac. "Benchmarking Best Practices," *Canadian Business Review* 19, No. 4 (Winter 1992): 28-31.

Fitz-enz, Jac. "Value-Added Benchmarking: A Tool for Getting Precisely What You Want," *Employment Relations Today* (Autumn 1992): 259-264.

Flower, Joe. "Benchmarking: Springboard or Buzzword?" *Healthcare Forum Journal* (January/February 1993): 14-16.

—. "The Source." *Healthcare Forum Journal* (January/February 1993): 30-36.

Fray, Earl N. "The Evolution of Performance Measurement." *Industrial Management* 30, No. 5 (September/October 1988): 9-12.

Frederickson, Carl. "Put Your Member Survey to Work." *Credit Union Executive* 30, No. 1 (Spring 1990): 18-23.

Fuld, Leonard M. *Competitor Intelligence: How to Get It—How to Use It*, John Wiley & Sons, 1985.

—. *Monitoring the Competition: Find Out What's Really Going on Over There*, John Wiley & Sons, 1988.

—. "Taking the First Steps on the Path to Benchmarking." *Marketing News* (September 11, 1989): 15, 20.

Furey, Timothy R. "Benchmarking: The Key to Developing Competitive Advantage in Mature Markets." *Planning Review* 15, No. 5 (September/October 1987): 30-32.

Gale, Bradley. "Allocating Capital More Effectively." *Sloan Management Review* 29, No. 1 (Fall 1987): 21-31.

Gardner, Elizabeth. "Putting Guidelines into Practice," *Modern Healthcare* (September 7, 1992).

—. "What Comes First, Data or Guidelines?" *Modern Healthcare* (September 7, 1992): 28-29.

Geber, Beverly, "Benchmarking: Measuring Yourself Against the Best," *Training* (November 1990): 36-44.

Glavin, William F. "Competitive Benchmarking, A Technique Utilized by Xerox Corporation to Revitalize Itself to a Modern Competitive Position." *Review of Business* 6, No. 3 (Winter 1984): 9-12.

Guilmette, Harris and Carlene Reinhart. "Competitive Benchmarking: A New Concept for Training." *Training and Development Journal* 38, No. 2 (February 1984): 70-71.

Hammer, Michael, "Re-engineering Work: Don't Automate, Obliterate." *Harvard Business Review* (July-August 1990): 104-112.

Heidbreder, James E. "Looking for the Light—Not the Heat." *Healthcare Forum Journal* (January/February 1993): 25-29.

Henricks, Mark. "How Do You Measure Up?" *Small Business Reports* (June 1993): 29-39.

Hoover, Charles W. Jr., "Return on Investment in Continuing Education," *Chemtech* (June 1990): 338-341.

Hyatt, Joshua, "Steal This Strategy." *Inc.* (February 1991): 49-57.

IBM Class Module Handout, "Measurements/Benchmarking," 1990.

Jacobsen, Gary and John Hilkirk. *Xerox: American Samurai*. McMillan Publishing, 1986.

Juran, J.M. *Juran on Leadership for Quality: An Executive Handbook*, The Free Press, 1989.

Kaiser Associates Brochure. "Benchmarking: The Concept and Its Application."

—. "Industry Analysis and Competitive Analysis."

—. "Improving Your Competitive Position Through Benchmarking."

—. "Benchmarking R&D, Engineering, and Other Technical Functions."

—. "The Four Critical Elements of Effective Competitive Management."

Kanin-Lovers, Jill. "Total Compensation Analysis: A Broad Perspective." *Journal of Compensation & Benefits* 4, No. 4 (January/February 1989): 221-224.

Kapp, Sue. "Profiles: Gunning for No.1." *Business Marketing* 71, No. 4 (April 1986): 12, 14.

Karch, Kenneth M. "Getting Organizational Buy-In for Benchmarking: Environmental Management at Weyerhaeuser," *National Productivity Review* 12, No. 1 (Winter 1992/3): 13-22.

Kearns, David T. "Xerox: Satisfying Customer Needs with a New Culture." *Forum* (February 1989): 61-63.

Kelsch, John E. "Benchmarking: Shrewd Way to Keep Your Company Ahead of Its Competition." *Boardroom Reports* (December 15, 1982).

Krause, Irv and John Liu. "Benchmarking R&D Productivity." *Planning Preview* 21, No. 1 (January/February 1993): 16-21, 52-53.

Layne, Richard. "The Best, the Biggest, and the Debate." *Information Week* (September 18, 1989); 6-12.

Lewis, Bryon C. and Albert E. Crews. "The Evolution of Benchmarking as a Computer Performance Evaluation Technique." *MIS Quarterly* 9, No. 1 (March 1985): 7-16.

Lorincz, Jim. "Purchasing Research: How do you Measure up?" *Purchasing World* 34, No. 5 (May 1990): 30-33.

Main, Jeremy. "How to Steal the Best Ideas Around." *Fortune.* (October 19, 1992): 102-106.

Marinaccio, Lou. "A White Paper on Benchmarking." KPMG Peat Marwick, pp. 1-7.

Martin, Patricia. "Benchmarking: A Leg up on the Learning Curve." *Manage* (May 1991): 30-33.

Maturi, Richard J. "Benchmarking: Studying, Evaluating Others." *Investors Daily* (April 28, 1989):

—. "Benchmarking: The Search for Quality." *The Financial Manager* (March/April 1990): 26-31.

McCage, William J. "Examining Processes Improves Operations," *Quality Progress* (July 1989): 26-32.

McComas, Maggie, Christopher Knowlton, and Patricia A. Langan. "Cutting Costs Without Killing the Business." *Fortune* 114, No. 8 (October 1986): 70-78.

McGonagle, John J., Jr. "Benchmarking and Competitive Intelligence," *Journal for Quality and Participation* 15, No. 5 (September 1992): 30-35.

McNair, Carol. *Benchmarking: Adding Distinctive Value to Every Aspect of Your Business.* Harper-Busn, May 1992.

McNair, C. J., and Kathleen H. J. Leibfried. "Janssen Pharmaceutica: Focusing Through Competitive Analysis," *National Productivity Review* Vol. 12, No. 1 (Winter 1992/93): 95-110.

Motorola Course Description. "Benchmarking—BMK 160," 1989.

Nevens, T. Michael, Gregory L. Summe, and Bro Uttal. "Commercializing Technology: What the Best Companies Do." *Harvard Business Review* (May-June 1990): 154-163.

O'Dell, Carla. "Building on Received Wisdom." *Healthcare Forum Journal* (January/February 1993); 17-21.

Peterson, Donald E. and John Hillkirk. *A Better Idea: Redefining the Way Americans Work,* Houghton-Mifflin Company, Boston MA, 1991.

Pipp, Frank J. "Management Commitment to Quality." *Quality Progress* 16, No. 8 (August 1983): 12-17.

Port, Otis and Geoffrey Smith. "Beg, Borrow—and Benchmark." *Business Week* (November 30, 1992): 74-75.

Prairie, Patti. "An American Express/IBM Consortium Benchmarks Information Technology." *Planning Preview* 21, No. 1 (January/February 1993): 22-27.

Press, Gil. "Benchmarking: Is Your Research Department Best?" *Marketing News* 25, No. 18 (September 2, 1991): 24.

Prokesch, Steven E. "Xerox Halts Japanese March." *New York Times*, November 7, 1985.

"Remaking the American CEO." *New York Times*, January 1987.

Pryor, Lawrence S. "Benchmarking: A Self-Improvement Strategy." *The Journal of Business Strategy* (November/December 1989): 28-32.

Pryor, Lawrence S. and Steven J. Katz. "How Benchmarking Goes Wrong (and How to Do It Right.) *Planning Preview* 21, No. 1 (January/February 1993): 6-11, 53.

"Pushing to Improve Quality." *Research-Technology Management* 33, No. 3 (May/June 1990): 19-22.

Quality & Productivity Management Association. "Benchmarking, The IBM-Rochester Way." *Commitment Plus* 66, No. 12 (August 1991): 2-4.

Ransley, Derek L. "Training Managers to Benchmark." *Planning Preview* 21, No. 1 (January/February 1993): 32-36.

Rigdon, Joan E. "Using New Kinds of Corporate Alchemy, Some Firms Turn Lesser Lights into Stars." *The Wall Street Journal* (May 3, 1993): B1, B13.

Saaty, Thomas L. *Decision-Making for Leaders: The Analytical Hierarchy Process for Decisions in a Complex World*, RWS Publications, 1988.

Sasenick, Susan M. "Benchmarking: Tales from the Front." *Healthcare Forum Journal* (January/February 1993): 37-52.

Schmid, Robert E. "Reverse Engineering a Service Product." *Planning Review* 15, No. 5 (September/October 1987): 33-35.

Scovel, Kathryn. "Learning from the Masters." *Human Resource Executive* (May 1991): 1, 28-29.

Sheridan, John H., "Where Benchmarkers Go Wrong," *Industry Week*, (March 15, 1993): 18-34.

Sillyman, Steve. "Guide to Benchmarking Resources: Benchmarking Is a Tool to Help Improve Processes and Thus Improve End Products and Services." *Quality* (March 1992): 17-18.

Spendolini, Michael J. *The Benchmarking Book*. AMACOM, May 1992.

Stertz, Bradley A. "Detroit's New Strategy to Beat Back Japanese Is to Copy Their Ideas. " *The Wall Street Journal* (October 1, 1992): A1, A12.

Stewart, Thomas A. "Brainpower." *Fortune* (June 3, 1991): 44-60.

—. "GE Keeps Those Ideas Coming." *Fortune* (August 12, 1991): 41-49.

Sullivan, Michael P. "Marketing/Management: Hidden Competitors." *United States Banker* 99, No. 8 (August 1988): 42-43.

Szulanski, Gabriel, "Intra-Firm Transfer of Best Practice, Appropriate Capabilities, and Organizational Barriers to Appropriation," Szulanski is a Ph.D. student at INSEAD, the European School of Business Administration in France. This paper was published by The Academy of Management Proceedings (1993).

Taylor, Bernard. "Corporate Planning for the 1990's: The New Frontiers." *Long Range Planning (UK)* 19, No. 6 (December 1986): 13-18.

Teitelbaum, Richard S. "The New Race for Intelligence." *Fortune* (November 2, 1992): 104-107.

Thomas, Philip R. *Competitiveness Through Total Cycle Time: An Overview for CEOs*. McGraw-Hill Publishing Company, 1990.

Thompson, Donald. "Xerox Puts Its Commitment on Display." *Industry Week* 231, No. 3 (November 10, 1986): 20-21.

TQM Magazine, special issue dedicated to benchmarking. Vol. 2, No. 3 (July/August 1992).

Tucker, Francis G., Seymour M. Zivan, and Robert C. Camp. "How to Measure Yourself Against the Best." *Harvard Business Review* 87, No. 1 (January-February 1987): 2-4.

Tyndall, Gene. "How You Apply Benchmarking Makes All the Difference." *Marketing News* (November 12, 1990): 10-19.

Vaziri, H. Kevin. "Questions to Answer Before Benchmarking." *Planning Preview* 21, No. 1 (January/February 1993): 37.

Walleck, A. Steven. "A Backstage View of World-Class Performers." *Wall Street Journal* (August 26, 1991): A10.

Walleck, A. Steven, J. David O'Halloran, and Charles A. Leader. "Benchmarking and World-Class Performance." *The McKinsey Quarterly*, No. 1(1991): 3-24.

Waston, Gregory H. "Benchmarking: Up-front Preparation and Strategic Perspective Lead to Benchmarking Success." *Productivity* 12, No. 9 (September 1991): 10-12.

"How Process Benchmarking Supports Corporate Strategy." *Planning Preview* 21, No. 1 (January/February 1993): 12-15.

——. *The Benchmarking Workbook*. Productivity Press, May 1992.

Xerox Corporation Booklet. "Leadership Through Quality: Implementing Competitive Benchmarking, Employee Involvement and Recognition—Part I," 1987.

——. "Competitive Benchmarking: The Path to a Leadership Position," 1988.

——. "Competitive Benchmarking: What It Is and What It Can Do For You," 1987.

Zivan, Seymour. "Benchmarking: The Effective Manager's Tool." *Boardroom Reports*, November 15, 1990.

Index of Companies

Subject Index

About the Authors

Christopher E. Bogan is CEO and founder of Best Practices Benchmarking and Consulting, Inc. of Lexington, Massachusetts, and Research Triangle Park, North Carolina. Coauthor of *The Baldrige: What It Is, How It's won, How to Use It to Improve Quality in Your Company,* Mr. Bogan is a leading authority on TQM and ways that companies can use best practice benchmarking to accelerate organizational learning and continuous performance improvement. A graduate of Amherst College and Harvard Business School, he speaks frequently before American and international business audiences, and he advises many leading service and manufacturing companies on their best practice strategies and benchmarking projects.

Michael J. English has served as director of quality positioning, director of quality measurement, and director of service program management during his 22-year career with GTE. His responsibilities have included developing and implementing companywide customer satisfaction measurement systems, benchmarking techniques, the GTE President's Quality Awards, customer service delivery methods, systematic continuous improvements, Baldrige Quality Award Criteria assessments and applications, and ISO 9000 quality management plans. He is a popular speaker, writer, and authority in the fields of service quality management, customer satisfaction, measurement, and benchmarking. Mr. English has served as an examiner for the Malcolm Baldrige National Quality Award since 1993. He is a 1970 M.A. (economics) graduate of California State University at Sacramento.